Smart Nanomaterials Technology

Nanotechnology is a rapidly growing scientific field and has attracted a great interest over the last few years because of its abundant applications in different fields like biology, physics and chemistry. This science deals with the production of minute particles called nanomaterials having dimensions between 1 and 100 nm which may serve as building blocks for various physical and biological systems. On the other hand, there is the class of smart materials where the material that can stimuli by external factors and results a new kind of functional properties. The combination of these two classes forms a new class of smart nanomaterials, which produces unique functional material properties and a great opportunity to larger span of application. Smart nanomaterials have been employed by researchers to use it effectively in agricultural production, soil improvement, disease management, energy and environment, medical science, pharmaceuticals, engineering, food, animal husbandry and forestry sectors.

This book series in Smart Nanomaterials Technology aims to comprehensively cover topics in the fabrication, synthesis and application of these materials for applications in the following fields:

- Energy Systems—Renewable energy, energy storage (supercapacitors and electrochemical cells), hydrogen storage, photocatalytic water splitting for hydrogen production
- Biomedical—controlled release of drugs, treatment of various diseases, biosensors,
- Agricultural—agricultural production, soil improvement, disease management, animal feed, egg, milk and meat production/processing,
- Forestry—wood preservation, protection, disease management
- Environment—wastewater treatment, separation of hazardous contaminants from wastewater, indoor air filters.

Amit Kumar Mandal · Hironmoy Sarkar ·
José Roberto Vega-Baudrit
Editors

Functionalized Nanoparticles Hydrogels for Wound Healing

Springer

Editors
Amit Kumar Mandal
Department of Sericulture
Raiganj University
Raiganj, West Bengal, India

José Roberto Vega-Baudrit
Laboratorio Nacional de
Nanotecnología LANOTEC
Centro Nacional de Alta
Tecnología CENAT, Pavas
San José, Costa Rica

Hironmoy Sarkar
Cell Biology and Bacteriology Laboratory
Department of Microbiology
Raiganj University
Raiganj, West Bengal, India

ISSN 3004-8273 ISSN 3004-8281 (electronic)
Smart Nanomaterials Technology
ISBN 978-981-95-0719-1 ISBN 978-981-95-0720-7 (eBook)
https://doi.org/10.1007/978-981-95-0720-7

This Springer imprint is published by the registered company Springer Nature Singapore Pte Ltd.
The registered company address is: 152 Beach Road, #21-01/04 Gateway East, Singapore 189721, Singapore

Preface

Wound healing, a vital yet intricate biological process, remains one of the most pressing challenges in modern medicine. Despite substantial advancements in healthcare, the burden of acute and chronic wounds-amplified by aging populations, diabetes, infections, and antibiotic resistance-continues to strain global healthcare systems. With an increasing need for efficient, cost-effective, and patient-friendly wound care solutions, this book presents a comprehensive overview of innovative strategies and emerging materials that are transforming the landscape of wound management and regenerative medicine.

This volume is the culmination of interdisciplinary research spanning biomaterials science, nanotechnology, biomedical engineering, and clinical medicine. Beginning with the foundational role of hydrogel polymeric matrices (Chapter "Introduction to Functionalized Nanocomposite Hydrogels and Their Applications"), the text delves into the evolution and design of nanocomposite hydrogels (NCHs), offering insights into their synthesis, properties, and biomedical applications. Chapters "Synthesis and Characterization of Nanocomposite Hydrogels"–"Metallic Nanoparticles for Wound Healing Management: A Promising Therapeutic Approach", explore the methods of hydrogel preparation, wound classification, biological mechanisms of healing, and the integration of metal nanoparticles to enhance therapeutic efficacy.

Subsequent chapters (Chapters "Ceramic Nanofibre-Based Materials for Chronic Wounds: Emphasis on Silica Based Material"–"Hydrogels as Combinational Therapy with Antimicrobial Peptides for Wound Healing") highlight the potential of advanced materials such as ceramic nanofibers, biopolymer-based nanotechnologies, and smart, stimuli-responsive hydrogels. These sections emphasize how cutting-edge materials science converges with biology to offer new solutions for chronic wounds, infection control, and tissue regeneration. From biosensor-enabled smart dressings to 3D-printed hydrogel scaffolds and combination therapies, the text charts the ongoing innovation that is shaping next-generation wound healing strategies.

The final chapters (Chapters "Hydrogels Against Wound Causing Opportunistic Pathogens"–"Recombinant Antimicrobial Peptides-Embedded Bacterial Cellulose Hydrogels for Advanced Wound Healing") explore emerging frontiers, including the application of antimicrobial hydrogels against drug-resistant pathogens, personalized wound care technologies, and the synergistic use of bacterial cellulose with recombinant antimicrobial peptides. These developments underscore a shift toward intelligent, multifunctional, and patient-specific solutions.

This book aims to serve as a valuable reference for researchers, clinicians, biomedical engineers, and students engaged in wound care, biomaterials development, and translational medicine. It provides not only a scientific foundation but also a forward-looking perspective on the materials and technologies that will shape the future of wound healing.

We hope that the insights presented here inspire continued research, foster cross-disciplinary collaboration, and ultimately contribute to improving clinical outcomes and patient quality of life.

Raiganj, India — Amit Kumar Mandal
Raiganj, India — Hironmoy Sarkar
San José, Costa Rica — José Roberto Vega-Baudrit

Contents

About the Editors

Dr. Amit Kumar Mandal is Leading Expert in chemical biology and nano-omics, is Head of the Centre for Nanotechnology Science at Raiganj University. With a Ph.D. from the University of North Bengal, he has authored 80+ SCI publications, holds three patents, and has an impact score over 400. Recognized among the top 0.5% of global researchers (ScholarGPS) and top 3% in India (AD Scientific Index), his research focuses on chemical biology and nano-omics. Recipient of multiple honors, including the RSC Top Cited Paper Award, he serves on editorial boards and was recently named to the RSC Applied Interfaces Early Career Advisory Board.

Dr. Hironmoy Sarkar is presently working as Assistant Professor, Department of Microbiology, Raiganj University, West Bengal. Dr. Sarkar is basically from Siliguri, West Bengal, and did his M.Sc. from Bangalore University and Ph.D. from Delhi University. Presently he is running an extramural research grant from Department of Biotechnology, Government of India. He was Awardee of community science activity grant 2023 from American Society of Microbiology. He has many impactful research papers in renowned journals like Journal of Antibiotics, ACS Omega, JBSD, etc. His present research includes development of different drugs combination including nano-molecules to counter antibiotic resistance.

Dr. José Roberto Vega-Baudrit is Leading Costa Rican Scientist in nanotechnology, polymer chemistry, and advanced materials. He holds a Ph.D. in nanotechnology from the University of Alicante, with specializations in biodegradable polymers and chemical engineering. As Acting Director of CeNAT-CONARE and Director of LANOTEC, he drives innovation, research, and public–private collaboration. Full Professor and Scientific Leader, he has published over 300 journal articles, holds five patents, and has mentored 50+ theses. Recognized as an "Innovation Champion," he has advanced nanotechnology standards, circular economy practices, and science education in Costa Rica and Latin America.

Introduction to Functionalized Nanocomposite Hydrogels and Their Applications

Debjyoti Paul and Kankana Das

Abstract In the realm of promising drug delivery systems and regenerative medicine, hydrogel polymeric matrices have emerged as a promising solution owing to their biocompatibility, high water retention, and ability to mimic natural tissues. However, conventional hydrogels face critical limitations, including poor mechanical strength, limited targetability, and lack of responsiveness. Nanocomposite hydrogels (NCHs), which integrate nanoscale materials into hydrogel matrices, have emerged as next-generation solutions with improved physicochemical and biomedical attributes. By incorporating diverse nanomaterials—ranging from carbon-based (e.g., graphene oxide, carbon nanotubes) to inorganic (e.g., clays, silica, hydroxyapatite) and metallic (e.g., silver, gold, Fe_3O_4) nanoparticles—these functionalized scaffolds achieve enhanced mechanical strength, drug-loading capacity, stimuli responsiveness, and targeted therapeutic delivery. Advances in synthesis strategies, including in situ nanoparticle formation, suspension incorporation, and surface-functionalization, have enabled the precise engineering of hydrogels with dynamic bio interfaces. Additionally, computational and theoretical models now provide a mechanistic understanding of nanoparticle-polymer interactions, guiding the rational design of structurally robust and functional hydrogels. Beyond structural reinforcement, NCHs are now tailored to meet specific biomedical challenges such as wound healing, regenerative medicine, biosensing, and neural tissue engineering, with emerging trends focusing on injectable, self-healing, and smart bio responsive systems. Despite challenges related to biocompatibility, scalability, and long-term stability, the interdisciplinary nature of NCH research promises new frontiers in precision medicine and soft tissue therapeutics. This chapter provides an integrated overview of the classification, molecular design, synthesis methods, and biomedical

D. Paul (✉)
Department of Dietetics and Applied Nutrition, Amity Institute of Applied Sciences Kolkata, Amity University Kolkata, Newtown, Kadampukur, West Bengal, India
e-mail: dpaul@kol.amity.edu; djpbiotech@gmail.com

K. Das
Department of Physiology, PES University Institute of Medical Sciences and Research, Konappana Agrahara, Electronic City, Bengaluru 560100, Karnataka, India
e-mail: daskank2013@gmail.com

A. K. Mandal et al. (eds.), *Functionalized Nanoparticles Hydrogels for Wound Healing*, Smart Nanomaterials Technology, https://doi.org/10.1007/978-981-95-0720-7_1

utilities of NCHs, outlining their transformative potential across future healthcare applications.

Keywords Nanocomposite hydrogels · Functional nanomaterials · Wound healing · Regenerative medicine · Nanomedicine · Hydrogels

Abbreviations

NCHs	Nanocomposite hydrogels
ECM	Extracellular matrix
CNTs	Carbon nanotubes
SWCNTs	Single-walled carbon nanotubes
MWCNTs	Multi-walled carbon nanotubes
GO	Graphene oxide
CQDs	Carbon quantum dots
nHA	Nano-hydroxyapatite
Fe_3O_4	Iron(III) oxide
PDMAA	Poly(N,N-dimethylacrylamide)
PEG	Polyethylene glycol
PAA	Poly(acrylic acid)
PVA	Poly(vinyl alcohol)
NC	Nanocellulose
CNCs	Cellulose nanocrystals
NFC	Nanofibrillated cellulose
MNs	Microneedles
ROS	Reactive oxygen species
MSD	Mean square displacement
AI	Artificial intelligence
0DM	Zero-dimensional nanomaterials
1DM	One-dimensional nanomaterials
2DM	Two-dimensional nanomaterials

1 Introduction

Designing novel therapeutic agents and drug-delivery systems are dependent on targeting either one or more than one of the physiological processes that include maintenance of homeostasis, managing inflammation, restricting cell proliferation, and remodeling of cells and tissues due to injury [1]. The evolution of drug delivery systems and scaffolds is determined by the efficacy of such systems in targeting these molecular axes in a more profound and efficient manner. Furthermore, an evolved drug-delivery system exhibits sustained drug-delivery, enhanced biocompatibility, bypassing mechanisms of rapid drug clearance without any side reactions or incurring adverse reactions.

In this regard, hydrogels as multidimensional polymeric scaffolds have been offering unique attributes for drug delivery. Hydrogels have unique properties that closely mimic the natural environment of living tissues. Their three-dimensional network structure allows them to retain significant amounts of water—up to 99% of their volume—making them soft, flexible, and highly biocompatible. This remarkable water retention capability not only enhances their mechanical properties but also facilitates a moist environment conducive to wound healing and tissue regeneration [2]. Consequently, hydrogels are increasingly utilized in various applications, including drug delivery systems, wound dressings, and tissue engineering scaffolds. Despite their numerous advantages, hydrogels face numerous shortcomings as well. Studies suggest that existing hydrogels work on the principles of reversible sol–gel transition causing immediate implantation of the cargo (drugs or bioactive peptides) and preventing rapid and uncontrolled release of the cargo by mechanism of immediate gelation. Therefore, the effectiveness of the hydrogel encapsulated systems is dependent on surgical precision in placing the hydrogel scaffolds at the direct site of infection or pathogenesis which might not be convenient or feasible at all times. This mechanism also explains the poor targetability, antitumor capability and poor responsiveness [3, 4]. Also, existing hydrogels have sub-par mechanical strength thereby rendering them incompatible for many biomedical applications, particularly for the purpose of engineering tissue scaffolds [27].

Nanocomposite hydrogels (NCHs) represent classes of organic–inorganic hybrid materials that provide a better alternative to the existing hydrogels scaffolds. The fabrication of NCHs involves incorporating nanoscale materials into polymer networks through chemical or physical crosslinking in aqueous environments. By combining these rigid nanoscale elements with flexible polymer frameworks, NCHs exhibit enhanced performance characteristics, including improved mechanical, physicochemical, and biological properties [5]. These features make NCHs particularly valuable as synthetic three-dimensional tissue scaffolds, effectively replicating the diverse mechanical, electrical, and biochemical characteristics of natural tissue environments. The prospect of low surface-to-volume ratio and successful integration of various types of nanomaterial bases that include carbon-based nanomaterials, inorganic/organic ceramic nanoparticles, metal/metal oxide in various combinations of functional nanomaterials with both natural and synthetic polymers lead to generation of wide spectrum of novel NCHs. In general, these nanocomposites exhibit enhanced mechanical strength, improved stability, and tailored release profiles for therapeutic agents. By incorporating nanoparticles—ranging from inorganic materials like clay and silica to organic polymers—researchers can fine-tune the physical and chemical properties of hydrogels, thereby expanding their applicability in complex biomedical scenarios. The importance of functionalized NCHs is underscored by their versatility in addressing longstanding challenges in medicine. For instance, they can be engineered to respond dynamically to environmental stimuli, such as pH or temperature changes, allowing for controlled drug release tailored to patient needs. Additionally, their inherent antimicrobial properties can significantly reduce infection risks associated with wound care. This innovative approach not only enhances the efficacy of existing treatments but also opens new avenues for developing advanced therapeutic modalities [6]. Therefore, the possibility of coming up with tailored and functional nanomaterials curated for specific biomedical applications continues to expand.

The incorporation of nanoparticles into hydrogel matrices results in providing enhanced functionalities. NCHs depending on their composition demonstrate responsive behaviour to various stimuli such as radiation, temperature, and electromagnetic radiation. The hydrogels can swell, expel or absorb water, and release embedded substances when exposed to these stimuli. The subsequent properties of the novel hydrogels thus generated are utilized in the targeted delivery of drug and gene delivery systems, as well as bio-sensing applications [7].

Furthermore, nanoparticles used in the composition of these hydrogels act as bridges between polymer chains thereby increasing their network entanglement. This modification therefore aptly endows the hydrogel matrix with enhanced durability and mechanical strength leading to crucial applications in wound healing, regenerative medicine and contact lens manufacturing.

NCHs also exhibit remarkable adhesion properties that provide critical biocompatibility with skin and soft tissues. This adhesiveness comes from the roughness offered by the nanoparticles that suitably interlock with the surface irregularities of bio-interfaces thus making them ideal candidates for wound dressing applications and easy applications as well as removal. Finally, these materials can be fabricated into precise micro-scale structures, which is particularly valuable for cell culture applications. Traditional two-dimensional cell culture surfaces often fail to replicate the natural three-dimensional tissue environment, potentially leading to abnormal cell behavior. The ability to create microstructured, cell-compatible 3D environments using these hydrogels helps better simulate natural tissue conditions [8]. Figure 1 represents a infographic description of different types of NCHs that are in use for biomedical applications.

As we delve deeper into this chapter, the design principles, synthesis methods, and diverse applications of functionalized NCHs will be explored. Insights on how advanced materials can revolutionize healthcare practices and improve patient outcomes across various medical fields can be achieved by understanding the intricate interplay between hydrogel matrices and incorporated nanoparticles.

2 Molecular and Structural Insights Into NCHs

Nanosized biomaterials, in general display range of diversities in terms of structural dimensionality. Thus, based on their size, length-to-diameter ratio, and diameter-to-thickness ratio nanoparticles can be clubbed as zero-dimensional (0DM), one-dimensional (1DM) or two-dimensional (2DM). 0DM nanomaterials are confined

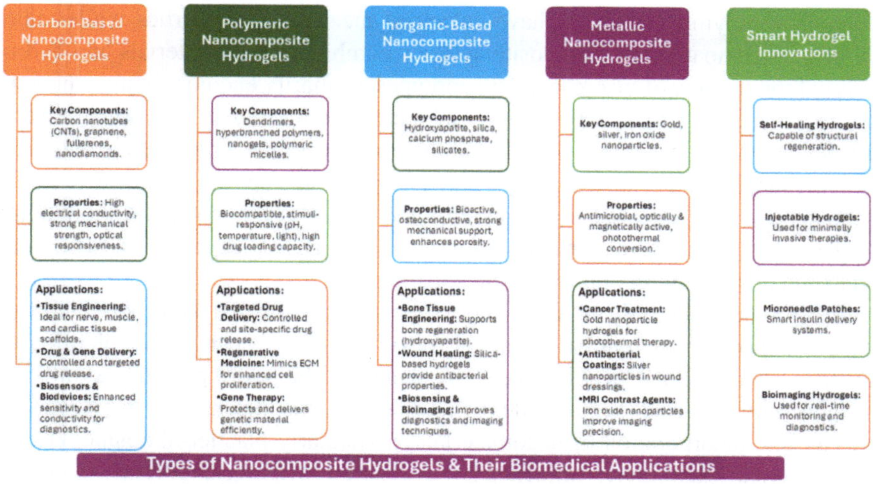

Fig. 1 An integrated perspective of types of NCHs and their biomedical applications. The image categorizes NCHs into four main types—carbon-based, polymeric, inorganic-based, and metallic—each characterized by distinct key components, properties, and applications. Additionally, it highlights smart hydrogel innovations, which are advancing biomedical research with self-healing, injectable, and bioimaging capabilities

below 100 nm in all three spatial dimensions and include quantum dots, nano crystals and all other types of nanoparticles. 0DM nanomaterials have high surface-to-volume ratio that enhances their interaction and reactivity with other materials. This also attributes them with high biocompatibility and photoluminescence and therefore are quite commonly used in biosensing, imaging and drug delivery [9]. Nanomaterials with relatively larger length-to-diameter ratio are 1DM and with relatively larger diameter-to-thickness ratio are classified as 2DM. 1DMs are aptly represented by carbon nanotubes (CNT) have hollow fibrous structures thus providing them with high tensile strength, high thermal and electrical conductivity and have inherent anisotropic nature. This allows them to have great applications in the synthesis of advanced biosensors. On the other hand, 2DM is represented by graphene and graphene-oxide based nanostructures finding applications as new-generation scaffolds in biomedicine [10].

Hydrogels, based on their polymeric composition are differentiated as natural, synthetic, and composite hydrogels. Natural hydrogels are derived from natural sources such as polysaccharides (e.g., chitosan, alginate, cellulose) and proteins (e.g., collagen, gelatin). Synthetic polymers include poly (ethylene glycol) (PEG), poly (acrylic acid) (PAA), and poly (vinyl alcohol) (PVA). While composite hydrogels entail blending natural and synthetic polymers to leverage the benefits of both. The differential polymeric composition makes natural polymer hydrogels biocompatible and biodegradable, synthetic polymer hydrogels offer tunable properties, and composite polymer hydrogels have enhanced mechanical properties while maintaining biocompatibility. Composite systems combining nanomaterials and hydrogels exhibit a range of macroscopic phenotypes, primarily encompassing nanocomposite hydrogel microneedles (MNs), which are used for transdermal drug delivery; injectable NCHs, designed for minimally invasive tissue engineering and drug delivery applications; self-healing NCHs, capable of repairing themselves after damage, offering potential in dynamic biomedical environments; and bioimaging NCHs, which integrate nanoparticles for enhanced imaging capabilities, facilitating real-time monitoring in biomedical diagnostics [10].

A computational study by Xu and Gersappe [11] investigated the structural formation and mechanical reinforcement of NCHs containing nanofillers of different shapes—spherical and platelet-like (disk-shaped). The study provided significant insights into the role of nanofillers in enhancing the mechanical properties of NCHs. The incorporation of nanomaterials, particularly platelet-like nanoplatelets and spherical nanofillers, influenced the gelation dynamics, polymer mobility, and stress response of these composite systems. The simulations revealed that platelet nanofillers facilitate early percolation (~0.8% volume fraction), leading to a well-structured organic–inorganic network at lower concentrations than spherical fillers (~1.5%). The anisotropic nature of nanoplatelets enabled intercalation structures, where layered platelet assemblies acted as effective crosslinking nodes, significantly enhancing the network connectivity and mechanical strength of the hydrogel. Comparative stress autocorrelation and step-strain tests indicated that platelet-filled hydrogels exhibited superior stress retention, slower stress relaxation decay, and higher viscosity, thereby demonstrated transition from a viscoelastic to a more solid-like behavior at concentrations above 1.5%. Additionally, polymer confinement effects, observed through mean square displacement (MSD) analysis, further confirmed that platelet-based hydrogels exhibited more restricted polymer mobility, reinforcing the hydrogel matrix. In contrast, spherical nanofillers aggregated randomly, forming heterogeneous, less effective crosslinked networks, resulting in comparatively weaker mechanical properties and delayed gelation behavior. These findings underscore the structural advantages of platelet nanofillers in tuning the molecular organization and mechanical robustness of NCHs, offering a fundamental basis for designing advanced hydrogel systems with tailored functional properties for biomedical and industrial applications.

Figure 2 illustrates a study conducted by Haraguchi et al. [12] on NCHs composed of poly(N,N-dimethylacrylamide) (PDMAA) and clay. The study highlighted the significant impact of clay and polymer contents on their mechanical properties. The hydrogels were synthesized via in-situ free-radical polymerization. Interestingly, the findings of this previous study agreed with the computational study conducted by Xu and Gersappe [11]. The NCHs exhibited superior mechanical strength compared to conventional chemically cross-linked hydrogels, with elongations at break exceeding 1500%. The study highlighted that mechanical properties are influenced by the composition of the hydrogels: increasing clay content enhances the modulus and ultimate tensile strength but slightly reduces elongation at break. Conversely, altering the polymer content reveals a critical threshold above which the hydrogels become

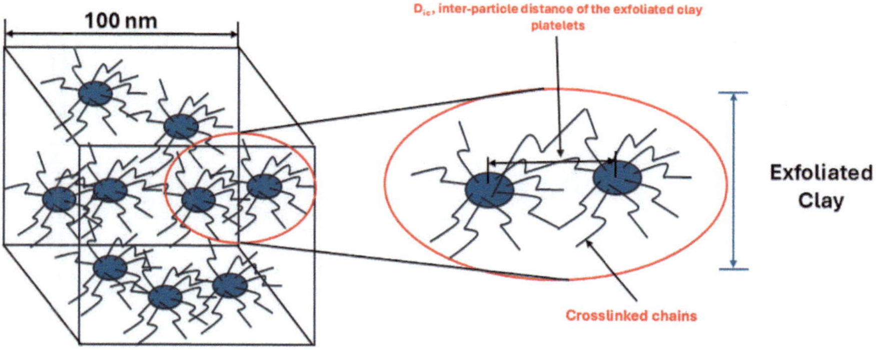

Fig. 2 Schematic representation of the organic (polymer)-inorganic (clay) network in NCHs. The image illustrates the dispersion of exfoliated clay platelets within a polymer matrix, where the clay particles serve as multifunctional cross-linking agents. The inter-particle distance (D_{ic}) of the exfoliated clay platelets influences the gelation process and mechanical reinforcement of the hydrogel. Polymer chains are physically crosslinked via non-covalent interactions with clay platelets, forming a highly structured, percolated network that enhances mechanical strength, elasticity, and resilience. The uniform dispersion of these clay platelets significantly improves the hydrogel's toughness and elongation at break compared to conventionally cross-linked hydrogels

remarkably tough, with strength and elongation at break-peaking at specific polymer concentrations. The network structure model explains these effects, emphasizing the role of clay as multifunctional cross-linking agents that enhance cross-link density and mechanical performance. Overall, these findings underscore the potential of NCHs for applications requiring high mechanical resilience and tunability.

3 Synthesis Strategies for Advanced Functionalization of NCHs

The synthesis of NCHs is usually done in either of four ways: (i) a suspension culture comprising of pre-assembly of nanoparticles and hydrogel crosslinking that curate a scaffold with defined distribution characteristics; (ii) in situ nanoparticle assembly followed by hydrogel cross-linking; (iii) a pre-mediated strategy that allows for inwards migration of nanoparticles within a pre-formed hydrogel scaffold; and (iv) nanoparticles acting as matrix crosslinkers between precursors of hydrogel assembly. Figure 3 represents an overarching scheme for developing a nanocomposite hydrogel for tailored applications.

Hydrogel crosslinking in the presence of nanomaterials typically results in uniform/homogenous distribution of nanoparticles. This strategy has resulted in the successful translational application of many nanomedicine formulations such as ONPATTRO (for delivery of siRNA drug for the treatment of polyneuropathy) [13],

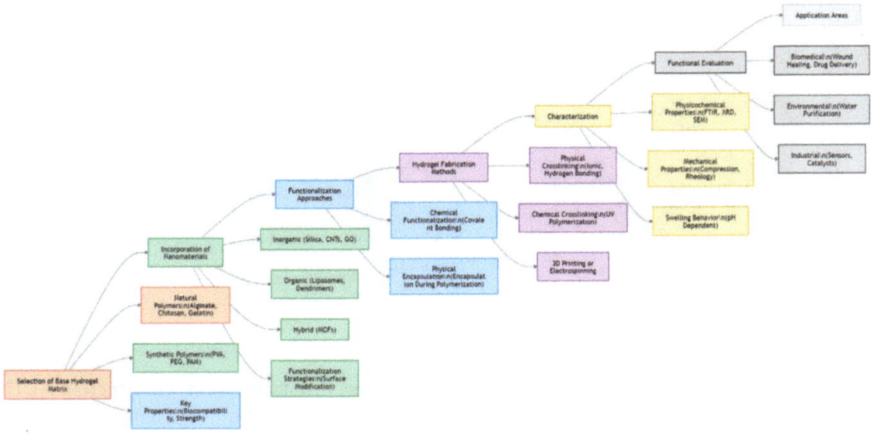

Fig. 3 Schematic representation of nanocomposite hydrogel development. This flowchart illustrates the comprehensive process of nanocomposite hydrogel development, beginning with the selection of a base hydrogel matrix (natural or synthetic polymers) and key properties (biocompatibility, strength). It highlights nanomaterial incorporation (inorganic, organic, hybrid), functionalization approaches (chemical, physical, and surface modifications), and hydrogel fabrication methods (crosslinking, encapsulation, and 3D printing). Additionally, it details characterization techniques (physicochemical, mechanical, and swelling behavior) and concludes with application areas, including biomedical, environmental, and industrial sectors

Doxil (the first PEGylated liposomal formulation of doxorubicin as a chemotherapeutic drug) [29] and Visudyne (a nanoliposomal formulation of verteporfin to treat abnormal blood vessels) [14]. However, this strategy can only be employed for nanomaterials that do not interfere with the hydrogel cross-linking and with further optimization measures if the chosen nanoparticles are prone to aggregation. Despite the challenges, this synthesis strategy remains a sought-after technique for researchers aiming for designing NCHs with dynamic pharmaco-kinetic attributes for targeted delivery of drugs and bioactives.

In situ nanoparticle formation within crosslinked hydrogel formation has been done in the synthesis of nanocomposites with high concentration of nanoparticles without undergoing aggregation, maintaining a uniform distribution with enhanced electrical conductivity. This approach involves crosslinking the hydrogel network first, followed by monomer diffusion and nanoparticle synthesis within the matrix, as demonstrated in electro-responsive hydrogels. For example, pyrrole nanorods were synthesized in situ within polyacrylamide-chitosan hydrogels via oxidation-mediated polymerization using $FeCl_3$, significantly improving electrical conductivity and enabling on-demand dexamethasone release upon electrostimulation, promoting murine myoblast activity [15]. Similarly, cardiac regenerative hydrogels were developed by integrating porcine gelatin, dopamine, and pyrrole-pendant polymers, where Fe^{3+} facilitated polypyrrole nanoparticle formation and polymerization, enhancing electrophysiological signal transmission and myocardial revascularization [16]. While effective, this method is limited by material compatibility, cell viability concerns, and diffusion constraints in thick hydrogels, necessitating further optimization for broader biomedical applications.

Another approach to generating NCHs involves the use of porous hydrogel matrix that are loaded with prefabricated nanomaterials with dimensions less than the pore size. Consequently, such a matrix allows for inward diffusion of nanoparticles inside its core leading to swelling when submerged in concentrated nanoparticle solutions. Importantly, such a set-up inhibits interference between formation of either of the hydrogel formation or of the aggregation between the nanoparticles, a viable option for biomedical applications engineered to function under electrical stimulus. Despite the many advantages of the inward diffusion technique to generate nanocomposite hydrogel, the technique has presented certain practical challenges as well. For instance, strategies utilizing cyclic deswelling and swelling of hydrophilic networks under organic solvents (e.g., acetone) and aqueous nanoparticle dispersions require careful modulation; failure to perform a deswelling cycle typically results in negligible nanoparticle uptake due to limited accessibility within the swollen matrix [17, 18]. Moreover, while this approach has proven effective in thin hydrogel films, its application to bulkier hydrogels—such as injectable alginate cryogels—often leads to substantial nanoparticle accumulation at the surface with minimal penetration into the interior. This results in non-uniform distribution and concentration gradients from the surface inward. Efficient incorporation is further complicated by dependencies on the physicochemical properties of both the nanoparticles (e.g., size, charge, solubility) and the hydrogel matrix (e.g., pore size, interconnectivity, charge density, and hydrophilicity). These limitations must be carefully addressed to ensure homogenous loading and consistent performance in NCHs, especially when intended for therapeutic or stimuli-responsive biomedical applications.

4 Materials Used in Biomedicine for Fabricating NCHs

The nanomaterial bases selected in the generation of NCHs determine the functional attributes of such scaffolds. These key functionalities include mechanical reinforcement, biocompatibility, drug delivery capability, self-healing, and stimuli responsiveness. The selection of nanomaterials directly influences how well the NCH mimics the extracellular matrix (ECM), supports cellular behavior, and responds to physiological cues in biomedical settings.

4.1 Carbon-Based NCHs

Carbon based nanocomposites such as graphene, single-walled (SWCNTs) or multi-walled (MWCNTs) carbon nano tubes (CNTs), nanodiamonds, and buckminsterfullerene (C_{60}) have been developed for a range of biomedical applications due to their inherent physico-chemical properties. These nanomaterials improve electrical conductivity, mechanical integrity, and drug encapsulation efficiency. Specifically, graphene oxide (GO), carbon nanotubes (CNTs), and carbon quantum dots (CQDs), have emerged as powerful reinforcements in NCH fabrication [19, 20].

Graphene oxide (GO), due to its high surface area and abundance of oxygen-containing functional groups, enhances the crosslinking density of hydrogels and imparts photothermal responsiveness. GO-functionalized hydrogels exhibit superior tensile strength, cytocompatibility, and antimicrobial activity, which are essential for wound healing and neural interface applications [28, 31].

Carbon nanotubes (CNTs), classified into single-walled (SWCNTs) and multi-walled (MWCNTs) structures, are tubular carbon nanostructures with exceptional tensile strength, electrical conductivity, and nanoscale diameter-to-length ratios. SWCNTs consist of a single graphene sheet rolled into a cylinder, whereas MWCNTs comprise multiple concentric graphene layers. Their unique mechanical and electronic characteristics make them highly suitable for biomedical applications when properly functionalized.

SWCNTs exhibit superior electrical properties, making them ideal for fabricating electrically conductive hydrogels for neural stimulation and signal transduction. Their nanoscale diameter and surface properties facilitate cellular interaction and intracellular delivery of therapeutics. Studies have shown that SWCNTs promote neurite outgrowth and support the maturation of neural networks when incorporated into bioelectronic interfaces and hydrogel matrices (Lavrador et al. 2020).

MWCNTs, on the other hand, provide enhanced mechanical reinforcement due to their thicker structure and higher Young's modulus. They have been widely used in tissue scaffolds for muscle, bone, and cardiac regeneration. Functionalized MWCNTs integrated into gelatin, chitosan, or PEG hydrogels improve not only mechanical resilience but also bioactivity. For instance, Sun et al. [30] demonstrated that carbon nanotube-incorporated collagen hydrogels could significantly enhance cardiomyocyte alignment and contraction force in engineered cardiac patches.

Despite their advantages, pristine CNTs are hydrophobic and can induce cytotoxic effects. Surface functionalization with carboxyl, hydroxyl, or PEG moieties is essential to improve dispersion, reduce toxicity, and facilitate biocompatibility. Tailored surface chemistry enables CNTs to be used safely in applications such as biosensors, drug delivery carriers, and 3D cell culture platforms.

Carbon quantum dots (CQDs), known for their tunable fluorescence and antioxidant properties, have recently been integrated into hydrogels to enhance imaging, ROS scavenging, and drug delivery capabilities, particularly for diabetic wound healing and cancer therapy [20]. Table 1 provides a list of successful endeavors with carbon-based NCHs employed for biomedical applications in the last 5 years.

4.2 Inorganic Particle-Based NCHs

Inorganic particles (e.g. silicate clays, silica nanoparticles, bioactive glass, hydroxyapatite (nHA), metal oxides) are the other commonly employed nanofillers used in the

Table 1 Functional attributes of carbon-based nanomaterials in NCHs

Nanomaterial	Key functionalities	Biomedical applications	References
Graphene oxide	Mechanical strength, conductivity, photothermal	Wound healing, neural scaffolds	Tarashi et al. [31], Lavrador et al. [28]
Carbon nanotubes	Anisotropic conductivity, mechanical reinforcement	Nerve repair, cardiac scaffolds	Sun et al. [30], Lavrador et al. [28]
Carbon quantum dots	Fluorescence, antioxidant activity	Cancer therapy, diabetic wound healing	Karchoubi et al. [20]

fabrication of NCHs. Nanosilicates such as *montmorillonite* and *Laponite* (a synthetic hectorite clay) are widely used to reinforce hydrogels. They carry a high density of surface hydroxyl groups that form hydrogen bonds with polymer chains, acting as physical crosslinkers. These interactions can create a secondary network within the hydrogel, dramatically improving mechanical properties and imparting self-healing via reversible bonding. For example, a fully physically crosslinked polyacrylamide/cellulose nanofiber hydrogel was achieved by strengthening a chemically polymerized network with ferric ion–nanocellulose coordination (ionic bonding) and H-bonding—conceptually like adding clays. Consequently, silicate nanoclays enhance hydrogel toughness, water retention, and injectability with rapid recovery and self-healing capacity. Furthermore, they form reversible physical bonds with polymer chains, offering energy dissipation and shape memory properties [21, 22]. In addition, silica nanoparticles, especially mesoporous variants, serve as reinforcement and drug reservoirs. Their surfaces can be functionalized with pH- or redox-sensitive linkers to enable smart drug release [28].

Reportedly, inorganic particles such as nHA provide calcium and phosphate ions for bone matrix formation. When incorporated into biopolymers, nHA improves compressive modulus and supports osteogenic differentiation (Kumar et al. 2019) [23].

Metals and Metal oxide nanofillers endow formulated NCHs with variety of desirable Metallic nanoparticles like Ag and Au provide antimicrobial and angiogenic benefits, while Fe_3O_4 nanoparticles support magnetic targeting and hyperthermia therapy [3, 28]. Table 2 provides an account of few studies carried out in the past few years that underscore the importance of inorganic material based nanocomposites that have shown promise for use in biomedicine.

Table 2 Inorganic nanoparticles and their functions in NCHs

Nanoparticle	Functional contribution	Application area	References
Silicate nanoclays	Mechanical reinforcement, shape memory	Tissue scaffolds, drug carriers	Anadão [21], Chang et al. [22]
Silica NPs	Smart drug delivery, reinforcement	Post-surgical therapy	Lavrador et al. [28]
Hydroxyapatite	Osteoconductivity, stiffness	Bone tissue engineering	Kumar et al. (2019), Mozafari et al. [23]
Silver NPs	Antimicrobial, anti-inflammatory effects	Wound healing, infection control	Cui et al. [3]
Gold NPs	Angiogenesis, biofunctionalization	Regenerative medicine	Lavrador et al. [28]
Fe_3O_4 NPs	Magnetic responsiveness, imaging	Tumor therapy, spinal scaffolding	Lavrador et al. [28]

4.3 Next-Generation Functional NCHs

Next-generation nanocomposite hydrogels are engineered to exhibit intelligent, multifunctional behaviors by integrating a diverse range of nanomaterials into polymeric networks. These advanced systems are increasingly tailored to meet the evolving demands of biomedical applications—offering attributes such as self-healing, shape-memory, stimuli-responsiveness, and hierarchical bioactivity.

Nanocellulose (NC), encompassing cellulose nanocrystals (CNCs) and nano fibrillated cellulose (NFC), contribute significantly to the mechanical strength and flexibility of NCHs. With abundant surface hydroxyl groups, NC facilitates chemical functionalization and supports the fabrication of 3D printable, stimuli-responsive hydrogels, making it an ideal component in tissue scaffolds and bioinks [20, 26].

Chitosan, a cationic polysaccharide, stands out due to its ability to form dynamic covalent bonds and respond to pH variations. When combined with redox-sensitive moieties or magnetic nanoparticles, chitosan-based NCHs demonstrate site-specific drug release, thereby enhancing therapeutic precision [24]. A remarkable example is presented by Wang et al. [25], who formulated a chitosan–graphene oxide (GO) hydrogel as a biomimetic wound dressing. Incorporation of GO almost doubled the hydrogel's tensile strength (from ~141 to 300 kPa) and improved elasticity, attributed to strong interfacial interactions between the polymer chains and GO sheets. Additionally, the hydrogel's porous microstructure (with reduced pore size due to GO integration) facilitated cell infiltration and proliferation. In vitro, the assays confirmed high biocompatibility and fibroblast adhesion, while in vivo trials in rats demonstrated 92% wound closure by day 21, significantly surpassing control treatments.

These findings illustrate how chitosan-based NCHs synergize mechanical integrity with regenerative capacity, positioning them as promising candidates for skin and soft tissue repair.

Chitosan's versatility extends to neural tissue engineering. Owing to its positive charge, chitosan forms polyelectrolyte complexes with anionic polymers and nanoparticles, producing soft, ECM-mimetic matrices favorable for neural cell growth. When integrated with conductive nanomaterials like graphene or CNTs, these hydrogels facilitate electrical signaling, essential for neuronal regeneration. For example, chitosan/cellulose hybrid hydrogels with self-healing capability have supported neural stem cell differentiation, while oxidized chitosan scaffolds embedded with graphene or polypyrrole have promoted neurite extension by providing a conductive framework for cell signaling.

Polyethylene glycol (PEG) serves as a foundation for injectable, inert hydrogel systems. Its hydrophilic and non-immunogenic nature makes it highly suitable for clinical use, especially when coupled with functional nanomaterials like nanoclays or GO, which impart improved mechanical durability, bioadhesion, and responsiveness. PEG-based NCHs have shown promise in orthopedic, ophthalmic, and tissue replacement therapies [22, 28].

Beyond individual polymer-nanoparticle pairings, hybrid NCHs that combine both organic (e.g., dendrimers, micelles) and inorganic (e.g., clay platelets, metal oxides) nanofillers are emerging as next-generation biomaterials. These hierarchically structured, self-adaptive matrices offer enhanced drug encapsulation, dual-release kinetics, cell-guided architecture, and biosensing capabilities. Their applications span smart wound dressings, responsive biosensors, and controlled delivery systems that respond to physiological stimuli such as pH, temperature, or oxidative stress [20]. Table 3 represents tabulated details of endeavors that closely match with the desirable attributes of next-gen nanocomposite hydrogels for a wide range of biomedical applications.

In essence, next-gen NCHs represent a paradigm shift—from passive, single-function gels to dynamic, multifunctional bio interfaces that can sense, respond, and adapt to the biological environment, thereby aligning closely with the functional demands of precision medicine and regenerative therapies.

Table 3 Functional performance of next-gen NCH platforms

Base material	Integrated features	Biomedical focus	References
Nanocellulose	Mechanical reinforcement, self-healing	Neural regeneration, 3D printing	Cheng et al. [26], Karchoubi et al. [20]
Chitosan	Self-healing, redox/pH responsiveness	Drug delivery, wound management	Hu et al. [24]
PEG	Injectable, photothermal/magnetic responsiveness	Orthopedic fillers, implants	Chang et al. [22], Lavrador et al. [28]

5 Conclusion and Perspectives

5.1 Unifying Theoretical Frameworks

The recent convergence of nanotechnology, biomaterials science, and soft matter physics has fostered the development of unifying theoretical frameworks to explain the structural behavior and functionality of nanocomposite hydrogels (NCHs). Classical polymer physics is being extended to account for nanoscale interactions, such as polymer–nanoparticle interface dynamics, interfacial adhesion, and mechanical percolation thresholds [19]. The incorporation of quantum mechanics, molecular dynamics, and multiscale modeling has enabled the rational design of NCHs with tailor-made functionalities [21, 23].

Theoretical models, such as the Vaia-Giannelis free energy framework, predict clay layer separation in polymer matrices based on enthalpic and entropic contributions [21]. These frameworks are now being augmented with computational fluid dynamics and machine learning tools to anticipate material behavior under physiological stress or degradation conditions.

5.2 Emerging Interdisciplinary Collaborations

NCH research is no longer siloed within materials science; it is being vigorously shaped by interdisciplinary collaborations across optoelectronics, bioengineering, artificial intelligence, and clinical medicine. For instance, researchers at the Guangdong Laboratory of Artificial Intelligence have fused optoelectronic nanomaterials with bioactive polymers to enable real-time biosensing and theranostics [6]. Similarly, the integration of quantum dots, carbon nanomaterials, and DNA-functionalized surfaces into hydrogels has been championed by teams in biomedical engineering and nano photonics [19].

Collaborations between clinicians and materials scientists are driving innovation in organ-on-chip, stimuli-responsive implants, and microneedle systems for vaccine delivery, emphasizing clinical translation [23]. Biophysicists and pharmacologists are also leveraging these platforms to deliver anticancer, antimicrobial, and anti-inflammatory agents with high precision.

5.3 Grand Challenges and Frontier Areas of Research

Despite the promising outlook, the field of NCHs faces significant scientific and translational challenges:

- *Biosafety and Long-Term Biocompatibility:* While in vitro biocompatibility is well documented, the long-term fate of nanofillers in vivo, particularly those based on CNTs, graphene, and metallic NPs, remains a concern due to potential cytotoxicity and immune responses [6, 19].
- *Controlled Multi-Functionality:* Engineering NCHs that can simultaneously achieve targeted drug delivery, electrical conductivity, tissue mimicry, and degradation is an unresolved challenge. The interactions between embedded nanostructures and the host biological system must be better understood [23].
- *Scale-Up and Reproducibility:* The reproducible fabrication of NCHs with consistent nanoscale architecture remains a bottleneck. Techniques like electrospinning, 3D bioprinting, and microfluidics need optimization for industrial scale-up [21].
- *Smart Integration with Digital Health:* Future NCHs are anticipated to act as bioresponsive, AI-integrated platforms, capable of dynamic sensing, reporting, and modulation of local microenvironments. Research in this area requires collaboration with data scientists and biomedical device developers [6].

In biomedical applications, NCHs play a pivotal role in a wide range of applications. With the discovery of several attributes that improve the functionalization of such scaffolds and mimic natural systems. Specifically, these novel attributes prove to be very useful in the domain of regenerative medicine, wound healing and imaging. Ergo, the scope of NCHs continues to expand and offer exciting opportunities for further research.

References

1. Singh AP, Biswas A, Shukla A, Maiti P (2019) Targeted therapy in chronic diseases using nanomaterial-based drug delivery vehicles. Signal Transd Targ Therapy 4(1):33
2. Guo Y, Bae J, Fang Z, Li P, Zhao F, Yu G (2020) Hydrogels and hydrogel-derived materials for energy and water sustainability. Chem Rev 120(15):7642–7707
3. Cui X, Wei TC, Guo LM, Xu GY, Zhang K, Zhang QS et al (2024) Vancomycin-loaded sol–gel system for in situ coating of artificial bone to prevent surgical site infections. Macromol Biosci 24(9):2400078
4. Malizos K, Blauth M, Danita A, Capuano N, Mezzoprete R, Logoluso N et al (2017) Fast-resorbable antibiotic-loaded hydrogel coating to reduce post-surgical infection after internal osteosynthesis: a multicenter randomized controlled trial. J Orthop Traumatol 18:159–169
5. Motealleh A, Kehr NS (2017) Nanocomposite hydrogels and their applications in tissue engineering. Adv Healthcare Mater 6(1):1600938
6. Huang S, Hong X, Zhao M, Liu N, Liu H, Zhao J et al (2022) Nanocomposite hydrogels for biomedical applications. Bioeng Transl Med 7(3):e10315. https://doi.org/10.1002/btm2.10315
7. Rafieian S, Mirzadeh H, Mahdavi H, Masoumi ME (2019) A review on nanocomposite hydrogels and their biomedical applications. Sci Eng Comp Mater 26(1):154–174
8. Wahid F, Zhao XJ, Jia SR, Bai H, Zhong C (2020) Nanocomposite hydrogels as multifunctional systems for biomedical applications: current state and perspectives. Comp B Eng 200:108208
9. Wang Z, Hu T, Liang R, Wei M (2020) Application of zero-dimensional nanomaterials in biosensing. Front Chem 8:320
10. Golshirazi A, Mohammadzadeh M, Labbaf S (2025) The synergistic potential of hydrogel microneedles and nanomaterials: breaking barriers in transdermal therapy. Macromol Biosci 25(1):2400228

11. Xu D, Gersappe D (2017) Structure formation in nanocomposite hydrogels. Soft Matter 13(9):1853–1861
12. Haraguchi K, Farnworth R, Ohbayashi A, Takehisa T (2003) Compositional effects on mechanical properties of nanocomposite hydrogels composed of poly (N, N-dimethylacrylamide) and clay. Macromolecules 36(15):5732–5741
13. Lim SA, Cox A, Tung M, Chung EJ (2022) Clinical progress of nanomedicine-based RNA therapies. Bioact Mater 12:203–213
14. Abdel Gaber SA, Fadel M (2023) Nanotechnology and photodynamic therapy from a clinical perspective. Transl Biophot 5(1):e202200016
15. Cui Z, Ni NC, Wu J, Du GQ, He S, Yau TM et al (2018) Polypyrrole-chitosan conductive biomaterial synchronizes cardiomyocyte contraction and improves myocardial electrical impulse propagation. Theranostics 8(10):2752
16. Liang S, Zhang Y, Wang H, Xu Z, Chen J, Bao R et al (2018) Paintable and rapidly bondable conductive hydrogels as therapeutic cardiac patches. Adv Mater 30(23):1704235
17. Guo JS, Liang HP, Wan LJ, Bai CL (2003) Highly dispersed metal nanoparticles in porous anodic alumina films prepared by a breathing process of polyacrylamide hydrogel. Chem Mater 15(22):4332–4336
18. Pardo-Yissar V, Gabai R, Shipway AN, Bourenko T, Willner I (2001) Gold nanoparticle/hydrogel composites with solvent-switchable electronic properties. Adv Mater 13(17):1320–1323
19. Gaharwar AK, Peppas NA, Khademhosseini A (2014) Nanocomposite hydrogels for biomedical applications. Biotechnol Bioeng 111(3):441–453. https://doi.org/10.1002/bit.25160
20. Karchoubi F, Ghotli RA, Pahlevani H, Salehi MB (2024) New insights into nanocomposite hydrogels; a review on recent advances in characteristics and applications. Adv Ind Eng Polym Res 7(1):54–78
21. Anadão P (2016) Polymer/clay nanocomposites: concepts, researches, applications and trends for the future. In: Ebrahimi F (ed) Nanocomposites: new trends and developments. InTech
22. Chang CW, Van Spreeuwel A, Zhang C, Varghese S (2010) PEG/clay nanocomposite hydrogel: a mechanically robust tissue engineering scaffold. Soft Matter 6(20):5157–5164
23. Mozafari M, Mehraien M, Vashaee D, Tayebi L (2016) Electroconductive nanocomposite scaffolds: a new strategy into tissue engineering and regenerative medicine. In: Ebrahimi F (ed) Nanocomposites: new trends and developments. InTech
24. Hu JL, Zhang YX, Song CH, Wei HG, Zhong C, Chu LQ (2023) Preparation and characterization of carboxymethyl chitosan/polyvinyl alcohol composite hydrogel with improved mechanical and antibacterial properties. React Funct Polym 187:105592
25. Wang Y, Liu S, Yu W (2021) Functionalized graphene oxide-reinforced chitosan hydrogel as biomimetic dressing for wound healing. Macromol Biosci 21(4):2000432
26. Cheng Y, Zang J, Zhao X, Wang H, Hu Y (2022) Nanocellulose-enhanced organohydrogel with high-strength, conductivity, and anti-freezing properties for wearable strain sensors. Carbohydr Polym 277:118872
27. Bao Z, Xian C, Yuan Q, Liu G, Wu J (2019) Natural polymer-based hydrogels with enhanced mechanical performances: preparation, structure, and property. Adv Healthc Mater 8(17):1900670
28. Lavrador P, Esteves MR, Gaspar VM, Mano JF (2021) Stimuli-responsive nanocomposite hydrogels for biomedical applications. Adv Funct Mater 31(8):2005941
29. Rivankar S (2014) An overview of doxorubicin formulations in cancer therapy. J Cancer Res Ther 10(4):853–858
30. Sun H, Zhou J, Huang Z, Qu L, Lin N, Liang C et al (2017) Carbon nanotube-incorporated collagen hydrogels improve cell alignment and the performance of cardiac constructs. Int J Nanomedicine 12:3109–3120
31. Tarashi S, Nazockdast H, Sodeifian G (2019) Reinforcing effect of graphene oxide on mechanical properties, selfhealing performance and recoverability of double network hydrogel based on κ-carrageenan and polyacrylamide. Polymer 183:121837

Synthesis and Characterization of Nanocomposite Hydrogels

Enish Kujur and Anoop Kumar

Abstract Hydrogels are materials that can hold significant volumes of water, which makes them of great value for many purposes, among them drug delivery and wound healing. The particles added to these hydrogels enable improvement of mechanical strength and sensitivity to environmental change and increase in a number of functions, including the creation of antibacterial or healing functionalities. This chapter explains the methods which can be used to prepare these nanocomposite hydrogels and the characterizations techniques to inspect their structural properties, mechanical properties, and other key properties. Thus, the aim is to achieve a very definite understanding of what exactly they are, how they are prepared and why and how they are of importance in the medical field.

Keywords Nanocomposite hydrogels · Nanoparticles · Drug delivery · Wound healing · Biomedical materials · Tissue engineering · Material synthesis

1 Introduction

Hydrogels are 3D network structures made of hydrophilic polymer chains capable of absorbing and retaining significant amounts of water, primarily characterized by their swelling behavior when submerged in water, allowing them to hold large volumes of liquid within their structure due to the interactions between the polymer's functional groups and water molecules [1]. As these materials contain high levels of water and are biocompatible, they can be applied in various industries, including drug delivery, agriculture and wound healing [1]. Synthesis of hydrogels can be done from natural polymers like alginate, chitosan, and gelatin and synthetic polymers such as PVA, PEG, and PAM selecting based on swelling behavior, mechanical strength, rate of degradation, and biocompatibility [2–4].

E. Kujur · A. Kumar (✉)
Department of Biotechnology, University of North Bengal, Raja Rammohanpur, Darjeeling, West Bengal, India
e-mail: anoopnbu@gmail.com

A. K. Mandal et al. (eds.), *Functionalized Nanoparticles Hydrogels for Wound Healing*, Smart Nanomaterials Technology, https://doi.org/10.1007/978-981-95-0720-7_2

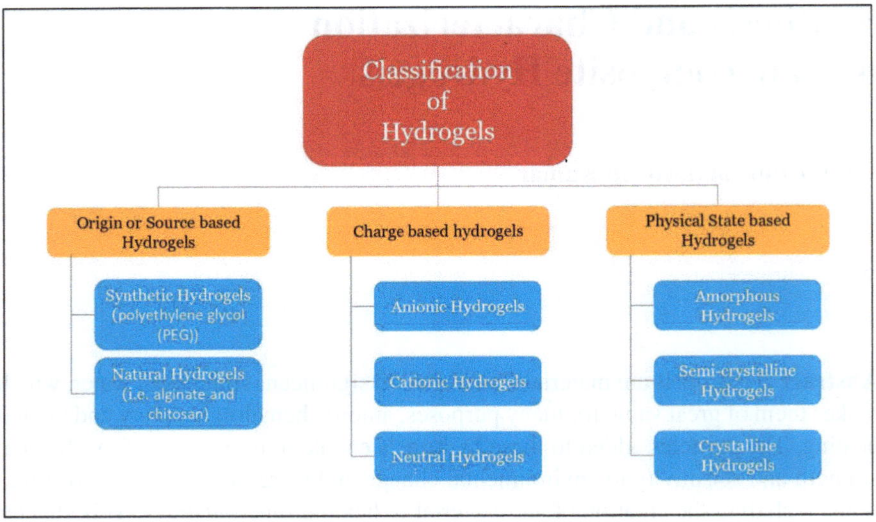

Fig. 1 Types of hydrogels based on origin, charge and physical state

Hydrogels are classified into three groups: origin (synthetic and natural), charge (anion, cation, neutral), and physical state (amorphous, semi-crystalline, and crystalline), where each type of hydrogel provides diverse properties for a specific application of the category [2, 5–7]. Figure 1 categorizes hydrogels according their source (natural or synthetic), charge (cationic, anionic, neutral), and physical state (amorphous, semi-crystalline, crystalline).

Hydrogel formation occurs through crosslinking, either physical; hydrogen bonds, ionic interactions, or chemical; covalent bonds, with ionic crosslinking often utilized in materials such as alginate [1, 8–10].

Nanoparticles in a diametric range from 1 to 100 nm have excellent surface area-to-volume ratio and quantum effects, for which the nanoparticles are very good for modifying hydrogels to support biomedical applications. Metallic, polymeric, and inorganic nanoparticles are probably the most frequently used for incorporation into hydrogels to support applications of this kind of hydrogels in drug delivery, imaging, diagnostics [11–14]. Hydrogels infused with nanoparticles, for example silica or carbon nanotubes or graphene oxide, transform into a mechanically strengthened, stimuli-responsive, antimicrobial, and significant drug delivery agent toward wound healing, tissue engineering, and biosensing studies [15–19].

Combinations of hydrogels and nanoparticles that form nanocomposite hydrogels can utilize the former's best features, such as structural strength, tunability, and biology-enhanced performance. Such characteristics are the needs of modern biomedicine applications, for example tissue engineering scaffolds, targeted drug delivery, and biosensors [19, 20]. These exhibits mechanical strength and bioactivity, stimuli-responsive behavior, and multifunctionality; all very important for advanced biomedical interventions [16, 21–25].

2 Synthesis of Nanocomposite Hydrogels

The synthesis of nanocomposite hydrogels involves carefully selecting polymeric materials and nanoparticles, optimizing functionalization methods, and choosing the appropriate synthesis techniques. Each step is essential for achieving the desired mechanical, chemical, and biological characteristics essential for certain biomedical applications. Figure 2 provides an overview of the steps for synthesizing nanocomposite hydrogels, including polymer and nanoparticle selection, their properties, and their biomedical applications.

2.1 *Selection of Polymer and Nanoparticle Materials*

The proper screening of polymers and nanoparticles is essential for determining the performance of nanocomposite hydrogels [2]. The interaction between the polymer matrix and nanoparticles, as well as their individual properties, must be carefully considered to design hydrogels with enhanced bioactivity, mechanical strength, and stimuli-responsiveness [26].

2.1.1 Polymers for Hydrogel Matrix

The hydrogel matrix's swelling capacity, mechanical strength, and biocompatibility of is mainly influenced by the kind of polymer used [2]. Natural polymers such as alginate, chitosan, and collagen are highly used for their biocompatibility and unique properties [27]. Alginate is a brown seaweed-based derivative that forms hydrogels

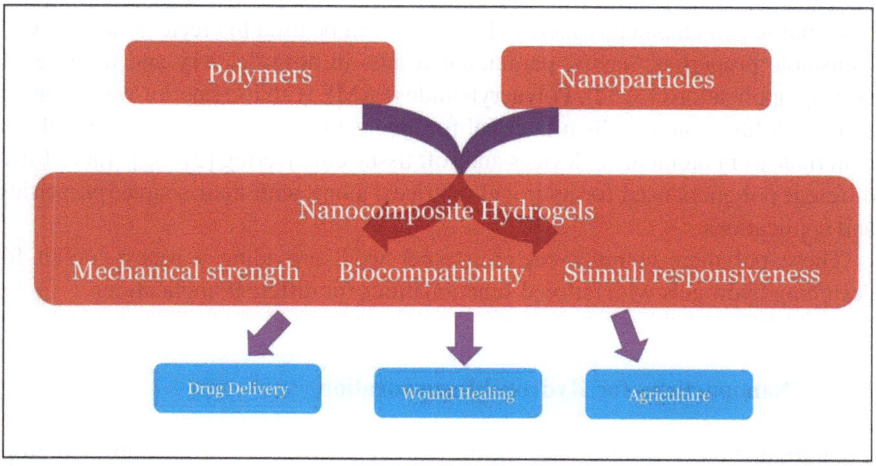

Fig. 2 General overview of synthesis of nanocomposite hydrogels, their properties and applications

Table 1 Polymers for hydrogel matrix

Polymer type	Source/category	Key properties	Applications	References
Alginate	Natural (seaweed-derived)	Biocompatibility, gels with divalent cations	Wound healing, drug delivery	Ahmad Raus et al. [28]
Chitosan	Natural (chitin derivative)	Antimicrobial, gels in acidic environments	Wound healing	Pramanik et al. [29]
Collagen	Natural (ECM protein)	Tissue resemblance, biocompatibility	Tissue engineering	Dubey and Deng [30]
PVA	Synthetic	Non-toxic, stable, transparent	Contact lenses, wound dressings	Zhong et al. [31]
PEG	Synthetic	Hydrophilic, chemically inert	Targeted therapy, tissue scaffolding	Raina et al. [3], Sun et al. [32]
PAM	Synthetic	High water content, flexibility	Biomedical devices, soft tissues	Hong et al. [33]

when exposed to divalent cations, such as calcium, making it ideal for applications in wound healing and drug delivery [3, 28]. Chitosan, which is a chitin derivative, gels in acidic environments and is an antimicrobial agent, making it even more useful for application in wound healing [29]. Collagen represents the most ample extracellular matrix protein and closely resembles tissues, therefore, collagen-based hydrogels are the best-perceived scaffolds used in tissue engineering applications [3, 30]. Synthetic polymers consists of polyethylene glycol or PEG, polyvinyl alcohol or PVA and polyacrylamide or PAM [5]. These can provide tunable properties and stability. Polyvinyl alcohol or PVA is also non-toxic, stable chemically, and transparent, hence widely used in contact lenses and wound dressings [3, 31]. Polyethylene glycol or PEG, a hydrophilic and chemically inert polymer that can be used to create hydrogels with adjustable properties, having paramount results in drug delivery and tissue engineering applications [3, 32]. Polyacrylamide (PAM), with its superior water content and flexibility, is an excellent material for applications that require soft and elastic materials, as in biomedical devices and soft tissue engineering [27, 33]. Table 1 lists different polymers used for hydrogel matrices, along with their source, properties, and applications.

These polymers complement each other well, providing a robust toolkit for designing hydrogels with tailored functionalities for different applications.

2.1.2 Nanoparticles for Hydrogel Incorporation

The introduction of nanoparticles within hydrogels enhances considerably their properties. They can introduce improved mechanical strength, antimicrobial activity, and stimuli-responsiveness. Among the metallic particles often used within hydrogel systems, silver and gold nano-metallic particles are among the widely used [27]. Silver nanoparticles are especially valued for their well-known potent antimicrobial properties, which makes them frequent choices for the design of wound dressings intended to prevent infections [3, 34]. Gold nanoparticles (AuNPs) are biocompatible and versatile, equipped to be functionalized with drugs, proteins, or nucleic acids, making them useful in targeted drug delivery and diagnostic applications [2, 34]. Polymeric nanoparticles also offer significant improvements in hydrogel functionalities. PLGA nanoparticles are biodegradable and biocompatible, thus suitable for sustained and controlled drug delivery systems [35]. Chitosan nanoparticles possess mucoadhesive properties making them ideal for drug delivery through mucosal surfaces-which may include the nasal or oral route [36]. Another important category of inorganic nanoparticles used in hydrogels is inorganic nanoparticles. Among these, the silicate nanoparticles have a considerable surface area and are thus utilized to carry drugs or as carriers for enzymes in biosensing applications [37]. Hydroxyapatite is a natural component found in bone and has been the focus of extensive study for bone tissue regeneration and regenerative medicine; it contributes extensively to the developing scaffold that closely mimics the bone matrix [3, 38]. Table 2 lists nanoparticles used for hydrogel incorporation, along with their categories, properties, and applications.

These diverse nanoparticles, incorporated into hydrogels, expand their potential applications, especially in biomedical and therapeutic fields, by incorporating functionalities tailored to specific needs.

Table 2 Nanoparticles for hydrogel incorporation

Nanoparticle type	Category	Key properties	Applications	References
Silver nanoparticles	Metallic	Potent antimicrobial	Wound dressings	Hammami et al. [34]
Gold nanoparticles	Metallic	Biocompatibility, functionalization ability	Targeted drug delivery, diagnostics	Hammami et al. [34]
PLGA nanoparticles	Polymeric	Biodegradable, controlled release	Sustained drug delivery	Roberts et al. [35]
Chitosan nanoparticles	Polymeric	Mucoadhesive	Mucosal drug delivery	Omidian et al. [36]
Silicate nanoparticles	Inorganic	High surface area	Drug delivery, biosensing	Ways et al. [37]
Hydroxyapatite	Inorganic	Bone matrix mimicry	Bone tissue engineering	Shi et al. [38]

2.2 *Functionalization of Nanoparticles*

The functionalization of nanoparticles is a critical step in enhancing their integration into hydrogel matrices, stability, and optimized performance of nanocomposite hydrogels. Functionalization is mainly through surface modification, which enhances the chemical reactivity of a nanoparticle or the possibility of interaction within biological environments or hydrogel matrices [39]. Among other methods, surface coating is the most commonly known, whereby nanoparticles are coated by biocompatible polymers like polyethylene glycol (PEG). This method increases their stability in biological fluids, decreases their immune detection, and increases their circulation time in the bloodstream, making them very effective for biomedical uses [40, 41]. Another strategy for functionalizing nanoparticles is through ligand conjugation, targeting ligands such as antibodies, peptides, or aptamers are attached to the surface of the nanoparticle. This allows for targeted delivery toward specific cells or tissues, significantly enhancing the accuracy and effectiveness for controlled drug delivery [42, 43]. Nanoparticles possesses reactive groups such as carboxyl and amine groups, which enable them to covalently bind to the matrix hydrogel [44]. It then ensures a stable and homogeneous distribution of nanoparticles dispersed within the hydrogel, thereby enhancing its mechanical and functional properties for diverse biomedical applications [20, 45]. It not only enhances the capability of nanoparticles to be integrated in hydrogel systems but also their potential for targeted, efficient, and stable applications in drug targeting, tissue repair, and other advanced biomedical interventions [3].

Figure 3 illustrates different nanoparticle functionalization techniques, like surface coatings and ligand attachment, and their advantages for integrating into hydrogels.

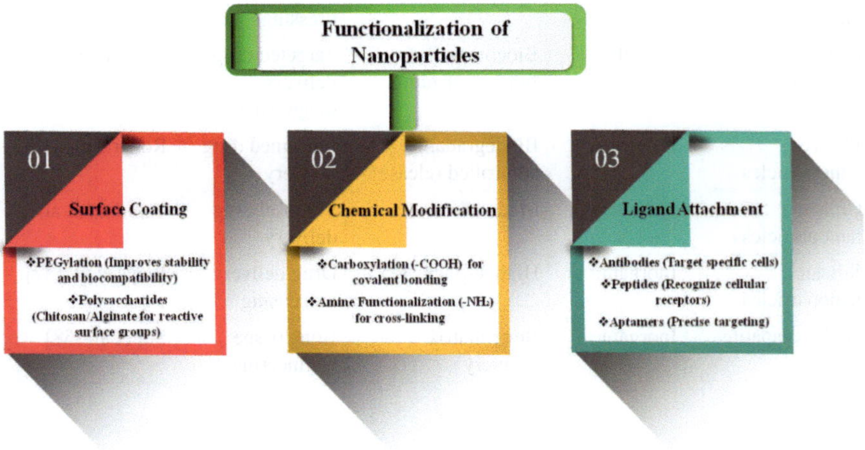

Fig. 3 Functionalization of nanoparticles

Table 3 Surface coating techniques for nanoparticles

Technique	Coating material	Benefits	Applications	References
PEGylation	Polyethylene glycol (PEG)	Enhances biocompatibility, reduces immune detection, extends circulation time	Drug delivery systems	Suk et al. [47]
Polysaccharide coating	Chitosan, alginate	Improves biocompatibility, provides functional groups for further modifications	Medical and environmental uses	Visan and Cristescu [50]

2.2.1 Surface Coating Techniques

Surface coatings play a crucial role in enhancing the bonding of nanoparticles within the hydrogel environment thereby improving the stability and biocompatibility of the synthesized nanocomposite hydrogels [46]. One of the popular methods is PEGylation which involves attachment of materials encapsulated in polyethylene glycol (PEG) onto the surface nanoparticles [47]. This technique enhances biocompatibility while minimizing the chances of immune system detection and extending blood circulation time, making it suitable for drug release mechanisms [47, 48]. Polysaccharide coatings, which include natural polysaccharides specifcally chitosan or alginate, are other very effective materials that enhance the biocompatibility of nanoparticles [2]. Such protective layers also provide surface functional groups available for further chemical changes, broadening the functions of nanoparticles in medicine and the environment [49, 50].

Table 3 lists surface coating techniques for nanoparticles, including the coating materials used and their applications.

2.2.2 Chemical Modification of Nanoparticles

Chemical surface modification of nanoparticles helps introduce reactive functional groups on their surface, for better integration into hydrogel matrices [49]. Alteration in this nature facilitates better covalent bonding, allowing the functions of nanoparticles in specific roles embedded in the hydrogel [2]. The most applied methods involve the introduction of carboxyl (–COOH) groups into the nanoparticle surface through carboxylation. This functionalization enables covalent interaction with amine groups ($-NH_2$) integrated into the hydrogel, whereby stable and robust nanocomposite hydrogels with improved structural integrity are produced [3, 51]. Similarly, amine functionalization, by adding amine groups ($-NH_2$) onto the nanoparticle's surface, allows for interaction with carboxyl (–COOH) or aldehyde (–CHO) groups of the hydrogel [44]. This kind of functionalization may promote

Table 4 Chemical modification techniques for nanoparticles

Modification method	Functional groups introduced	Purpose	Advantages	References
Carboxylation	Carboxyl (–COOH)	Enhances covalent bonding with amine groups (–NH_2) in hydrogels	Produces hydrogels with stable structural integrity and improved performance	Thoniyot et al. [51]
Amine functionalization	Amine (–NH_2)	Promotes interaction with carboxyl (–COOH) or aldehyde (–CHO) groups	Improves cross-linking, blending, and mechanical strength	Mahmoudi et al. [52]

cross-linking and blending of nanoparticles, improving the tensile properties and functionality of the final composite material [2, 52]. Table 4 lists chemical modification techniques for nanoparticles, including the functional groups used and their applications.

Chemical modifications enable specially designed interactions between nanoparticles and hydrogels; their preparation can open paths for creating advanced multifunctional materials destined for a variety of biomedical and technological applications.

2.2.3 Ligand Attachment

Ligand attachment to nanoparticles: It's one of the advanced approaches to enhance specificity in therapeutic delivery systems [48]. With ligands like peptides, antibodies and aptamers attached to the outer surface of the nanoparticles, a targeted interaction is established. Antibody conjugation is an integration of antibodies into nanoparticles, which facilitates drug's targeted delivery to specific cell types, such as tumor cells, increasing therapeutic efficacy by reducing off-target effects [53, 54]. Similarly, peptide functionalization involves attachments of peptides to recognize specific cellular receptors, thus enhancing the accuracy of the controlled release of drugs in affected tissues and optimizing the outcome of drugs [3, 55]. Table 5 lists ligand attachment techniques for nanoparticles, including the types of ligands and their applications.

This strategic use of ligands also opens potential for the development of more advanced and targeted therapeutic applications besides furthering the optimization of drug delivery mechanisms.

Table 5 Ligand attachment techniques for nanoparticles

Technique	Ligands used	Purpose	Advantages	References
Antibody conjugation	Antibodies	Targeted delivery to specific cell types (e.g., tumor cells)	Reduces off-target effects, enhances therapeutic efficacy	Kumari et al. [53]
Peptide functionalization	Peptides	Recognition of specific cellular receptors	Enhances delivery accuracy, improves outcomes in targeted tissues	Kim and Park [55]

2.3 Synthesis Methods for Nanocomposite Hydrogels

The synthesis of nanocomposite hydrogels involves several methods, each with its own set of advantages and limitations. The choice of method depends on the desired properties of the final material, such as uniformity, mechanical strength, and functionality.

2.3.1 In Situ Polymerization

In in-situ polymerization technique nanoparticles are diffused in a monomer or polymer solution with crosslinking agents which starts polymerization [2]. This method ensures the even distribution of nanoparticles within the hydrogel matrix and promotes strong interactions between nanoparticles and polymer chains. The process starts by dissolving the monomer or polymer in a suitable solvent to form a homogeneous solution. Functionalized nanoparticles is added, evenly dispersed through ultrasonication or mechanical stirring. Polymerization is initiated by a cross-linking agent and also by the application of external stimuli: heat, UV light, chemical initiators. The cross-linking agent react with monomer and/or polymer and the functionalized moieties on the nanoparticles to form very stable gel matrices. The technique has several advantages, including nanoparticle homogeneity distribution and increased mechanical strength due to strong interaction between the nanoparticles and the hydrogel matrix [20]. However, the complex process requires careful control of reaction conditions in order to develop the desired properties. Moreover, an inappropriate functionalization may cause nanoparticle agglomeration during polymerization, which leads to heterogenous distribution and properties [56]. Figure 4 provides in-situ polymerization process.

Table 6 outlines the procedures followed in in situ polymerization for synthesizing nanocomposite hydrogels.

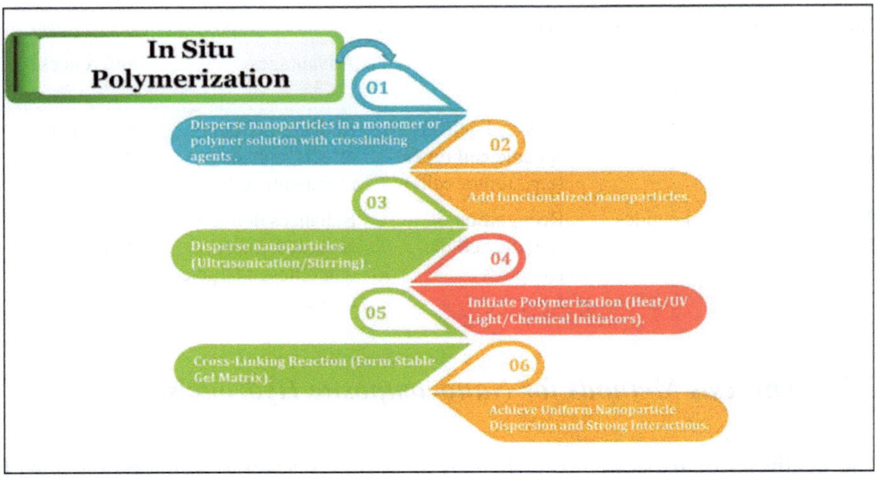

Fig. 4 In situ polymerization

Table 6 In situ polymerization for nanocomposite hydrogel synthesis

Procedure	Description	Advantages	Challenges
Monomer/polymer dissolution	Dissolve monomer/ polymer in a solvent to form a homogeneous solution	Ensures uniform dispersion of nanoparticles	Requires careful solvent choice and preparation
Nanoparticle dispersion	Functionalized nanoparticles added and dispersed via ultrasonication or mechanical stirring	Achieves homogeneity and strong interaction with polymer chains	Nanoparticle agglomeration if functionalization is improper
Polymerization initiation	Initiated by cross-linking agents or external stimuli (heat, UV light, chemical initiators)	Produces stable, well-crosslinked gel matrices	Control of reaction conditions is critical to avoid issues
Cross-linking	Cross-linking agent reacts with nanoparticles and monomers/polymer	Increases mechanical strength and stability of hydrogel	Improper functionalization or reaction conditions can cause heterogeneity

2.3.2 Physical Blending

In physical blending, pre-synthesized nanoparticles are simply mixed with a hydrogel precursor solution either before or after the gelation process. Due to its ease of usage and flexibility, this method has an advantage because it allows for the incorporation of

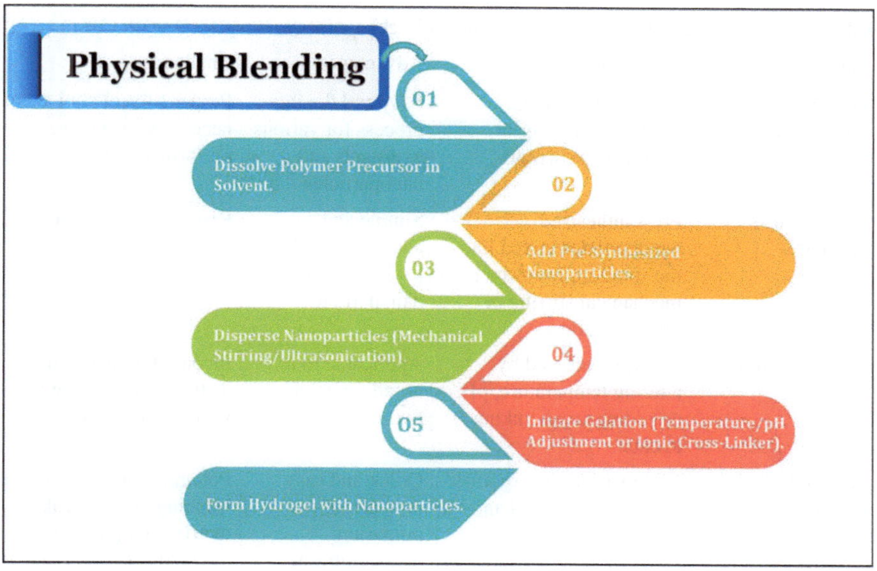

Fig. 5 Physical blending

nanoparticles into different hydrogels. It starts with the polymer precursor dissolved in water or an appropriate solvent to form a clear and uniform solution. These are then mixed with the solution and mechanically stirred or ultrasonicated till they are homogenously mixed, thereby uniformly dispersing the pre-synthesized nanoparticles. On basis of some characteristics of the precursor of hydrogel, the entire process can be initiated by altering the temperature, readjusting the pH, and adding ionic cross linker for initiating gelation. This method gives ease and flexibility and therefore is generally applicable to a broad range of hydrogels and nanoparticles. However, challenges include achieving uniform nanoparticle distribution, particularly if the nanoparticles tend to aggregate, and weaker interactions between the nanoparticles and the hydrogel matrix, which may result in the leaching of nanoparticles [20, 57]. Figure 5 presents a diagram of the physical blending method, illustrating how pre-made nanoparticles are mixed with a hydrogel precursor solution.

Table 7 explains the physical blending method for synthesizing nanocomposite hydrogels, highlighting the process and associated challenges.

2.3.3 Covalent Cross-Linking

Covalent cross-linking is the generation of covalent bonds between nanoparticles and the polymers chains during the gelation process, thereby making a nanocomposite hydrogel with enhanced stability and superior mechanical strength [20]. The process starts from the functionalization of nanoparticles through attaching reactive groups like carboxyl, amine, or thiol groups, which can engage in covalent bonding with the

Table 7 Physical blending for nanocomposite hydrogel synthesis

Step	Description	Advantages	Challenges
Polymer precursor dissolution	Polymer precursor dissolved in water or appropriate solvent	Easy and flexible process for various hydrogels and nanoparticles	Proper solvent choice is necessary to achieve a uniform solution
Nanoparticle mixing	Pre-synthesized nanoparticles mixed into the solution, followed by mechanical stirring or ultrasonication	Simple and adaptable method for incorporating nanoparticles	Risk of nanoparticle aggregation if not well-dispersed
Gelation initiation	Gelation initiated by altering temperature, pH, or adding cross-linking agents	Applicable to a broad range of hydrogels	Inconsistent nanoparticle distribution, leading to weak interactions
Final gel formation	Nanoparticle-incorporated hydrogel is formed and solidified	Quick and easy method for preparing nanocomposite hydrogels	Potential for nanoparticle leaching due to weak matrix-nanoparticle interactions

polymers chains. Functionalized nanoparticles are then dispersed into the polymer solution, ensuring the uniform dispersal of these particles. Then the cross-linking agent is added to the solution with nanoparticles which reacts with polymer chains in the solution. The stimuli heat and light are then applied to induce gelation hence forming covalent bonds between nanoparticles and the polymer matrix. This method offers enhanced mechanical characteristics and long term stability, but the nanocomposite hydrogel acquired is fit for application in prolonged conditions. It, however, requires a more complicated step of synthesis and sometimes even specific functionalization of the nanoparticles by that means. The method therefore must not be practiced for any nanoparticle-polymer combination, hence being not very universal [57].

Figure 6 shows the covalent cross-linking method, where nanoparticles bond with polymer chains to form highly stable nanocomposite hydrogels.

Table 8 outlines the procedures associated with covalent cross-linking for synthesizing nanocomposite hydrogels and challenges.

2.4 *Post-Synthesis Processing*

The post-synthesis processing of nanocomposite hydrogels encompasses various stages to improve the functionality or tailor them for specific applications [58]. The drying process is important when hydrogels are applied in a dried form, such as in wound dressings and tissue engineering scaffolds. Lyophilization, or freeze-drying,

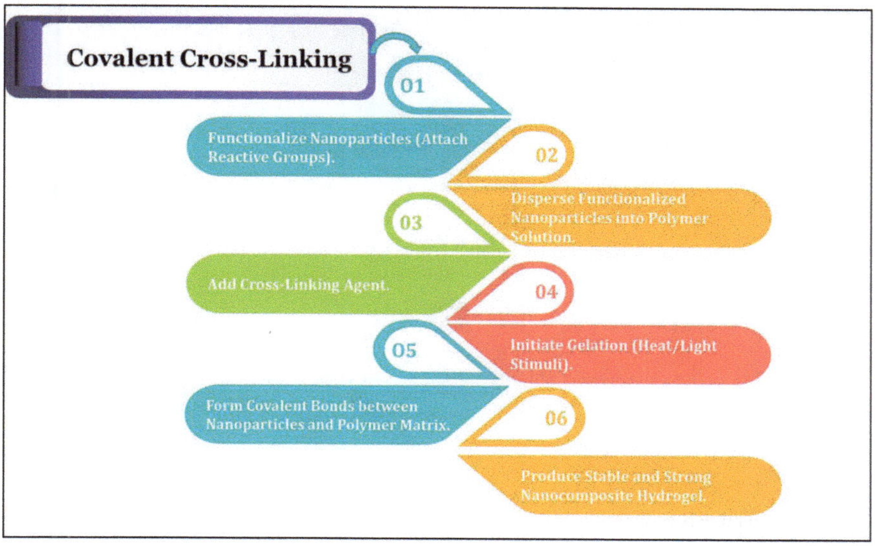

Fig. 6 Covalent cross-linking

Table 8 Covalent cross-linking for nanocomposite hydrogel synthesis

Step	Description	Advantages	Challenges
Nanoparticle functionalization	Attach reactive groups (e.g., carboxyl, amine, thiol) to nanoparticles	Ensures strong interaction between nanoparticles and polymers	Requires specific functionalization for different nanoparticle-polymer combinations
Nanoparticle dispersion	Disperse functionalized nanoparticles into polymer solution	Achieves uniform nanoparticle distribution	Potential issues with uniformity if dispersion is not optimal
Cross-linking	Add cross-linking agent to polymer solution, apply stimuli (heat/light)	Stronger mechanical properties and improved stability	Complex synthesis process, requiring careful control of reaction conditions
Final gel formation	Formation of a stable nanocomposite hydrogel with covalent bonds	Suitable for long-term stability and use in challenging conditions	Limited applicability due to specific requirements for nanoparticle and polymer compatibility

is often used, wherein the hydrogel is frozen and then sublimed at low pressure, which leaves a porous, dried hydrogel that retains much of its structural identity [59]. Air drying is more of a simple method that allows the hydrogel to dry from ambient temperature or under controlled conditions, though it is likely to shrink and

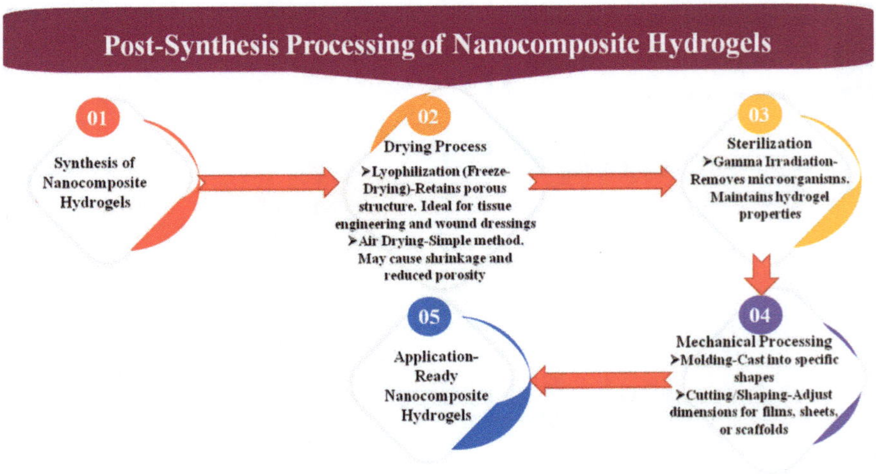

Fig. 7 Post-synthesis processing

reduce porosity [25]. Sterilization is another important step in biomedical applications to ensure the hydrogel does not contain microbial contamination. Gamma irradiation, which is a commonly employed sterilization method, is known to effectively remove microorganisms with minimal degradation of the functional characteristics of the hydrogel [60]. Finally, mechanical processing needs to be applied to produce desired forms of hydrogels like films, sheets, or scaffolds. Through such methods as molding, which entail the casting of hydrogels into predetermined shapes during or prior to gelation, and cutting and shaping, wherein dried hydrogels are mechanically adjusted to meet specified dimensions, such nanocomposite hydrogels may be modified to meet specific application requirements [61, 62]. These post-synthesis processing techniques are majorly imperative for improved nanocomposite hydrogels in applications across different avenues, though predominantly within the biomedical sector.

Figure 7 illustrates the post-synthesis processing steps of nanocomposite hydrogels.

3 Characterization of Nanocomposite Hydrogels

Characterization of nanocomposite hydrogels are essential to understand their structural, mechanical, thermal, and functional properties. The choice of characterization technique depends on the specific properties being evaluated [63].

3.1 Structural Characterization

Structural characterization provides insights into the distribution and interaction of nanoparticles embedded within the hydrogel matrix, in addition to the overall morphology and porosity of the hydrogel [64]. Figure 8 shows the structural characterization techniques used for analyzing nanocomposite hydrogels.

3.1.1 Scanning Electron Microscopy (SEM)

SEM is a very powerful tool to acquire surface morphology images of nanocomposite hydrogels at the nanoscale resolution, thereby detailed information related to size, shape, and distribution of nanoparticles within the hydrogel matrix. Freeze-drying of hydrogel sample is done first by removal of water to preserve its internal structure. Over the dried sample a thin layer of conductive material, such as gold or platinum, is spread for preventing charging in the imaging process. The sample is mounted inside the SEM chamber. The scanning of the sample's surface is done by an electron beam focused on the sample. Signals generated due to the interactions between the electrons and the sample are detected to generate a high-resolution image of the surface morphology and the nanoparticle distribution. Simultaneously, SEM presents advantages of high resolution, which allows detailed observation of surface features, and versatility, since SEM may be utilized to image several materials. Sample preparation techniques, freeze-drying and coating, however, tend to alter morphology and affect the accuracy of the images taken. Last, SEM primarily delivers surface information and offers no detailed insights into the inner morphology of the hydrogel [65].

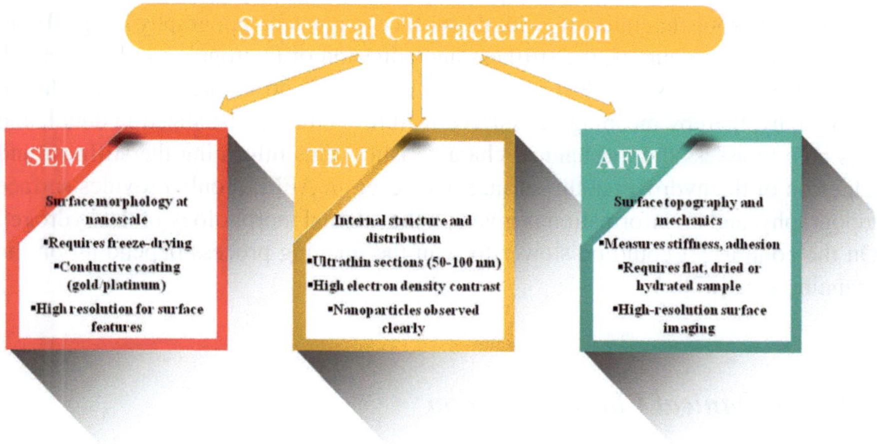

Fig. 8 Structural characterization

3.1.2 Transmission Electron Microscopy (TEM)

TEM is a high-resolution technique that can provide clear images of the internal structure of the nanocomposite hydrogels, hence clearly observable nanoparticles at the nanolevel. It is very effective in proving the existence, size, and distribution nature of the nanoparticles within the hydrogel matrix. The process usually involves sectioning the hydrogels into ultrathin sections, generally 50–100 nm in thickness using a microtome. These sections are mounted on a TEM grid, which usually has a carbon film covering it for supporting purposes. The sample is then put inside the TEM and subjected to a beam of high-energy electrons, which pass through it. The electrons interact with the sample and the image produced will be captured and magnified to be seen on a screen or camera. The advantage of TEM is that resolution is in the nanoscale, and hence lots of detail can be obtained inside the structure and nanoparticle distribution with high contrast, especially when the nanoparticles show a higher electron density than the hydrogel matrix. The preparation of ultrathin sections is equipment intensive and relatively challenging, particularly with soft materials. Additionally, TEM only enables the working with small sample sizes, and it cannot give a worldwide view of the whole hydrogel [66].

3.1.3 Atomic Force Microscopy (AFM)

This technique offers highly resolved topographic images of the hydrogel surface and evaluates its mechanical properties at the nanoscale, which includes surface stiffness and adhesion. The preparation of a flat substrate for the hydrogel sample that dries according to requirements for analysis or remains kept as a hydrated sample is regarded as the first step in the procedure. The surface is then scanned using a sharp AFM probe that interacts with the surface to create a topography image. From the force-distance curves, the stiffness and adhesion of hydrogels can be assessed. The advantage of using AFM is that it is capable of providing a high-resolution topography, thereby meaning the surface features could be visualized in detail, and it is able to assess the mechanical characteristics, thus informing the stiffness and adhesion of the hydrogel at the surface. However, in AFM, it only provides surface topography, and no information is given on the internal morphology of the hydrogel. On the contrary, it could be slower to scan the imaging process depending on the scanning area [67].

3.2 Mechanical Characterization

Mechanical characterization is crucial for evaluating the strength, elasticity, and durability of nanocomposite hydrogels. These properties are important for applications

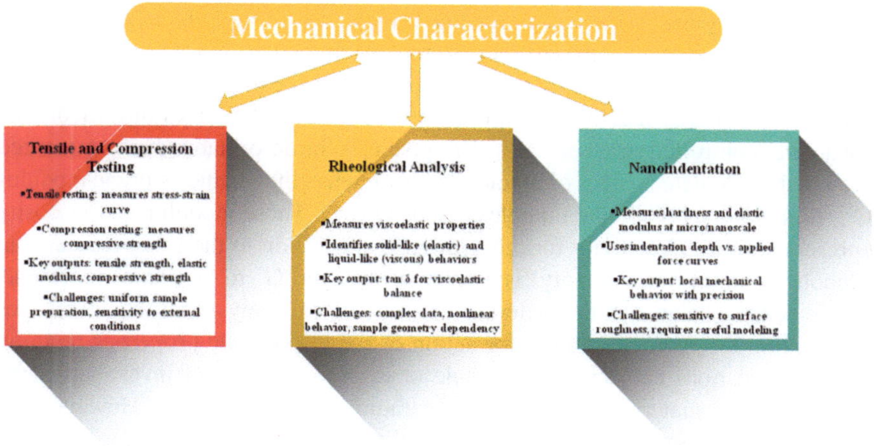

Fig. 9 Mechanical characterization

where the hydrogel will be subjected to mechanical stress, such as in tissue engineering or wound dressings [47]. Figure 9 depicts the mechanical characterization techniques used to characterize nanocomposite hydrogels.

3.2.1 Tensile and Compression Testing

The tensile and compression properties measure mechanical strength, elasticity, and compressibility of nanocomposite hydrogels. Standardized hydrogels samples are prepared to obtain crucial measurements. Tensile testing prepares dog-bone-shape samples, whereas compression samples are provided in cylindrical shapes. Tensile testing clamps the sample to elongate it up to breaking point; thus, it indicates a stress–strain curve. Compression testing compresses a test sample between two plates with recording of force and displacement. Key mechanical properties are determined by calculations based on stress–strain curves through the tensile, elastic modulus, and compressive strength determinations. The method gives quantitative data on comparison hydrogel formulations for simulating mechanical conditions that exist in the real world. This provides valuable insight into performance. Preparing uniform samples is a challenge; the viscoelastic nature of hydrogels sensitizes their mechanical properties to loading rates, temperature, and hydration levels-problems for consistent results [68].

3.2.2 Rheological Analysis

Rheological analysis tests the flow and deformation behavior of nanocomposite hydrogels under various conditions of stress and strain in terms of unveiling viscoelastic properties. These are significant for understanding the performance of this hydrogel in dynamic biological environments. In preparing for this process, the hydrogel sample is prepared in a shape that fits the rheometer geometry; disc or cylindrical is appropriate. The sample is introduced into the plates of a rheometer that applies controlled strain or stress. This results in the deformation being measured through the calculated values of rheological parameters that include storage modulus (G'), loss modulus (G''), and complex viscosity (η^*). These modulus and viscosity represent solid-like elasticity for the hydrogel, in addition to the loss modulus that symbolizes liquid-like viscosity. The value of the ratio G'' to G' ($\tan \delta$) represents the viscoelastic balance of the material. The rheological analysis has the advantages of providing the more accurate information of the viscoelastic properties required by the applications subjected to dynamic or cyclic loads and of investigating the behavior of the material for various temperatures and frequencies. However, the rheological data are undeniably complex and must be treated with a great deal of care where the viscoelastic material behavior is nonlinear. Additionally, the sample geometry will affect the measurements. Therefore, proper sample preparation and testing conditions are crucial to obtain accurate results [69].

3.2.3 Nanoindentation

Macroscopic mechanical characteristics of nanocomposite hydrogels are measured using nanoindentation from which the hardness and elastic modulus are derived. The experiment starts with preparation of a hydrogel specimen on an optically flat substrate so that an appropriately smooth hydrogel surface is provided for indentation by the indenter. A sharp indenter was forced into the hydrogel surface under controlled force, allowing the depth of indentation to be measured based on applied force. The deduced mechanical properties from these parameters- indentation depth, including force through suitable fitting of the loading curve according to a specified model and obtained through the unloading curve-on the material's elastic recovery, allow measurement at the microscale of mechanical properties. Among the strengths of nanoindentation, it allows for local mechanical behavior in hydrogels with high precision and sensitivity, particularly suited for soft materials like hydrogels. However, the method is very sensitive to surface roughness and imperfections affecting the quality of measurement being obtained. Moreover, interpretation of nanoindentation curves requires an intensive consideration of viscoelastic characteristics of the investigated material as well as indenter geometry [70, 71].

3.3 Swelling Behavior

Swelling behavior is a critical characteristic of hydrogels, reflecting their capability to absorb biological fluids and water. The swelling ratio and kinetics provide insights into the hydrogel's structure, porosity, and potential applications in targeted delivery and tissue regeneration [72].

3.4 Thermal Characterization

The thermal characterization for nanocomposite hydrogels is essential to obtain a deep understanding of thermal stability, phase transitions, and degradation behavior for assessing performances within different environments and applications. The following are some examples of measured thermal transitions by DSC: the glass transition temperature (Tg), melting temperature (Tm), and crystallization behavior may assess the thermal stability of a hydrogel and processing conditions [73]. Table 9 explains the use of Differential Scanning Calorimetry (DSC) for thermal characterization of nanocomposite hydrogels.

Thermogravimetric analysis is used to assess the thermal stability and the decomposition pattern based on the variation of weight loss with temperature, giving an idea about the onset of decomposition as well as the temperature at which maximum weight loss occurred [74].

Table 10 explains the use of Thermogravimetric Analysis (TGA) to assess the thermal resistance of nanocomposite hydrogels.

DMA determines the viscoelastic behavior of hydrogels through measurements of storage modulus, loss modulus, and damping factor or tan δ, in terms of dependence on temperature or frequency, which is useful in probing the behavior of the material under dynamic stress [75].

Table 9 Differential scanning calorimetry (DSC)

Aspect	Details
Purpose	Measures thermal transitions (Tg, Tm, crystallization)
Key outputs	Tg, Tm, and crystallization behavior
Applications	Assessing thermal stability and processing conditions
Advantages	Detects precise phase transitions
Limitations	Sensitive to moisture content and sample quality

Table 10 Thermogravimetric analysis (TGA)

Aspect	Details
Purpose	Measures weight loss with temperature
Key outputs	Onset of decomposition, maximum weight loss temperature
Applications	Predicting thermal resistance and disintegration behavior
Advantages	Provides insights into thermal breakdown patterns
Limitations	Affected by sample preparation and residual moisture

Table 11 Dynamic mechanical analysis (DMA)

Aspect	Details
Purpose	Evaluates viscoelastic properties
Key outputs	Storage modulus (G′), loss modulus (G″), tan δ
Applications	Probing behavior under dynamic stress and temperature changes
Advantages	Provides detailed viscoelastic behavior
Limitations	Requires precise geometry and careful data interpretation

Table 11 outlines the application of Dynamic Mechanical Analysis (DMA) for viscoelastic characterization of nanocomposite hydrogels.

Although these methods are deemed to be valuable tools in both thermal and mechanical probing, their applicability is bound by such factors as sample size, moisture level, and geometry, among others, in promoting consistency in accuracy and interpretation.

3.5 Functional Characterization

Functional characterization evaluates the specific properties imparted by nanoparticles to the hydrogel matrix, such as drug release profiles, antimicrobial activity, and biocompatibility. These properties are crucial for assessing the hydrogel's potential in biomedical applications [21]. It is defined through the study of diffusion, degradation, and interaction of the drug with the hydrogel matrix in drug releasing studies. Such studies would involve loading the hydrogel with drugs or immersing it in a release medium and then analyze the kinetics by means of models like Higuchi or Korsmeyer-Peppas [76, 77]. Table 12 details the approaches and uses of drug release studies for nanocomposite hydrogels.

Antimicrobial activity is tested by exposure to bacterial cultures like *E. coli* or *S. aureus*, to the hydrogel with evaluation of growth inhibition either by measuring zones of inhibition on agar plates or by assessing bacterial growth in broth cultures [78].

Table 12 Drug release studies

Aspect	Details
Purpose	Evaluates drug release profiles from the hydrogel
Methods	Drug loading or immersion in release medium
Analysis	Kinetics modeled by Higuchi or Korsmeyer-Peppas models
Applications	Drug delivery systems
Advantages	Provides insights into diffusion and interaction behavior
Limitations	Requires precise conditions for reproducibility

Table 13 Antimicrobial activity

Aspect	Details
Purpose	Assesses hydrogel's ability to inhibit microbial growth
Methods	Zone of inhibition (agar plates), bacterial growth (broth cultures)
Applications	Infection control in wound healing and medical devices
Advantages	Simple and effective for initial evaluations
Limitations	Dependent on microbial strains and experimental conditions

Table 14 Cytotoxicity and biocompatibility

Aspect	Details
Purpose	Confirms safety and compatibility with biological systems
In vitro methods	MTT assay to assess cell viability
In vivo methods	Tissue response studies (inflammation, fibrosis)
Applications	Tissue engineering and regenerative medicine
Advantages	Validates safety for biomedical applications
Limitations	In vivo methods are time-intensive and require animal models

Table 13 outlines the methods for assessing the antimicrobial activity of nanocomposite hydrogels.

Several cytotoxicity and biocompatibility tests affirm the safety of hydrogel for use and further confirmed by in-vitro assays via which the viability of cells is assessed either by MTT technique, or in vivo studies of tissue responses such as inflammation or fibrosis in the animal models [79, 80]. Table 14 outlines the methods for evaluating the cytotoxicity and biocompatibility of nanocomposite hydrogels.

These evaluations are not only useful to show that hydrogels are safe and effective for use but they help in the optimization of hydrogels for applications in drug delivery, infection control, or tissue regeneration.

4 Applications of Nanocomposite Hydrogels

Nanocomposite hydrogels retain high water content, are biocompatible, and especially multifunctional, thus possessing diverse biomedical applications. In addition, in case of wound healing, their application would be as an advanced dressing for retaining moisture, managing exudate, using nanoparticles like silver or zinc oxide to prevent bacterial infections, and delivering therapeutic agents such as antibiotics or growth factors in controlled release for which proper healing could be incorporated for faster recovery [81–83]. In tissue engineering, these hydrogels serve as a matrix extracellular in bone, skin, cartilage, and soft tissues for their regeneration. Frequently, such hydrogels have nanoparticles that are bioactive to catalyze the

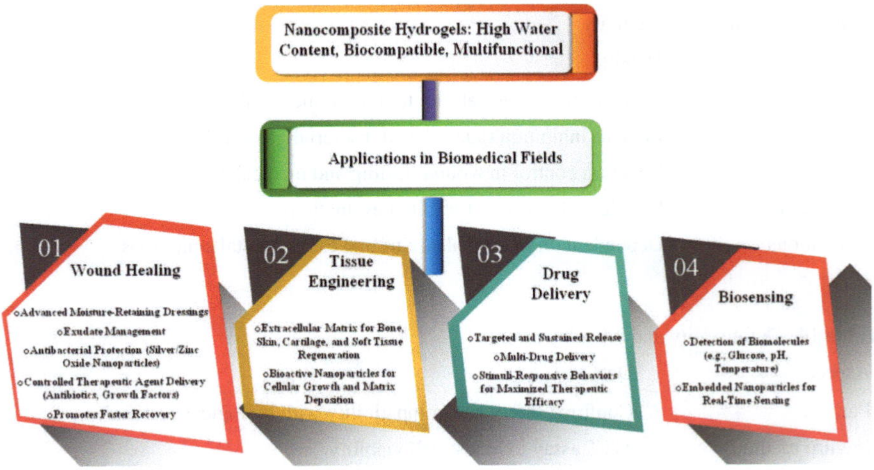

Fig. 10 Applications of nanocomposite hydrogels

cellular growth and additional deposition of the matrix [84–86]. They also demonstrate brilliance in drug delivery, providing targeted, sustained, and multi-drug release capabilities via utilization of nanoparticle functionalization and stimuli-responsive behaviors that can mitigate side effects and maximize therapeutic efficacy [87–89]. Nanocomposite hydrogels also have biosensing applications to detect biomolecules or environmental transformations, like glucose levels, pH variations, or temperature shifts, through embedded sensing nanoparticles [90–92]. These are multifunctional hydrogels of tremendous potential that may transform the healthcare through versatility and targeted functionality.

Figure 10 illustrates the main uses of nano-composite hydrogels, such as wound healing, targeted delivery, tissue regeneration and bio-sensing.

5 Challenges and Future Perspectives

Despite the significant advances in the development and application of nanocomposite hydrogels, several challenges remain that must be addressed to fully realize their potential in biomedical applications [93].

Figure 11 shows the main challenges in making and scaling up nanocomposite hydrogels, along with new trends and future directions in this field.

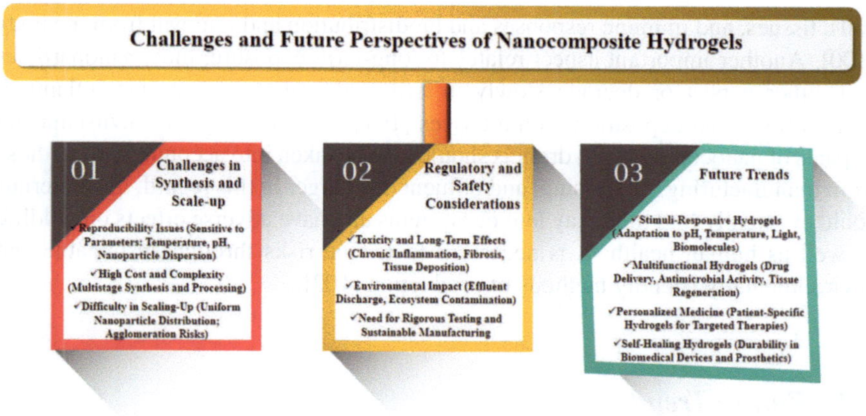

Fig. 11 Challenges and future perspectives of nanocomposite hydrogels

5.1 Challenges in Synthesis and Scale-Up

The large scale manufacturing of nanocomposite hydrogels is associated with a few major challenges, among them, reproducibility, cost, and complexity. This material synthesis is highly challenging to reproduce due to its sensitivity towards several parameters concerning temperature, pH, and nanoparticle dispersion, which could result in variations within the properties of the final product and its performance [94]. In addition, synthesis may involve several stages, such as nanoparticle functionalization, formation of hydrogel, and post-synthesis processing, all which contribute to the cost and complexity. Scaling up such processes for commercial production remains pricey and potentially requires innovative ways of manufacturing [95, 96]. Actually, uniformity in nanoparticle distribution within the hydrogel matrix is critical for homogeneous properties, including mechanical strength, drug release behavior, and bioactivity. However, scale-up uniformity is tough to sustain since nanoparticles tend to agglomerate during synthesis and processing, as monitored in the works of Thoniyot et al. [51] and Gao et al. [97].

5.2 Regulatory and Safety Considerations

Regulatory and safety considerations are critical for the approval of nanocomposite hydrogels in medical applications. There needs to be adequate testing to clear doubts surrounding the toxicity and long-term effects of nanoparticles along with their environmental impact [98, 99]. These nanoparticles, apart from showing a different profile than their counterpart in bulk conditions, because of their small size and distinctive physicochemical properties, require tremendous evaluations of their effects on cells, tissues, and immune responses and biodistribution and removal from the body [100]. Another important aspect relates to long-term exposure, thus, nanoparticles that either persist or degrade slowly in hydrogels can cause chronic inflammation, fibrosis, and deposition within tissues [101]. Alternatively, the environmental impacts of nanocomposite hydrogels should also be taken into account, such as those from manufacturing, processing, and effluent discharge. In this regard, the materials could eventually find their way into ecosystems and have adverse effects on wildlife as well as human health. It is necessary to reduce risks through sustainable and environmentally friendly methods of production [102].

5.3 Future Trends

The future of nanocomposite hydrogels is in sophisticated smart materials, potentially responsive to physiological changes or specific environmental cues. New trends which have recently emerged are the stimuli-responsive hydrogels, engineered for responses to some pH, temperature, light, or biomolecules changes, such as for instance on-demand drug delivery and adaptable tissue engineering scaffolds [103]. Another exciting area includes multifunctional hydrogels, where several therapeutic properties are packed into a single device, such as the treatment of tough clinical conditions associated with chronic wound situations, along with antimicrobial activity, drug release, and tissue regeneration [104]. Personalized medicine is also gaining grounds wherein hydrogels can be engineered to individual patient profiles, such as genetic data or molecular detail, for delivering targeted therapies while minimizing side effects [105]. Further, the self-healing hydrogels can be engineered with much potential to render biomedical devices to become hardy and durable by allowing the material to self-heal after any damage it has suffered, it is an important characteristic in prosthetic implants and wound healing applications [106].

6 Conclusion

Nanocomposite hydrogels are hybrid materials that aim to bring together the best characteristics of both hydrogels and nanoparticles. They can absorb large volumes of water, they are biocompatible for in vivo applications, and engineered to achieve performance in a variety of medical applications including wound healing, drug delivery, or tissue regeneration. However, so far, bulk preparation of these materials remains challenging because of the complexity of the process involved and the high cost. Additionally, researchers must validate that such materials are safe for long-term use. And amidst such challenges, research in this area is on the rise, and maybe in the near future, there will be smarter materials that can sense changes in the body, thus making them very useful in medicine.

References

1. Zainal SH, Mohd NH, Suhaili N et al (2021) Preparation of cellulose-based hydrogel: a review. J Mater Res Technol 10:935–952
2. Bashir S, Hina M, Iqbal J et al (2020) Fundamental concepts of hydrogels: synthesis, properties, and their applications. Polymers 12:2702
3. Raina N, Pahwa R, Bhattacharya J et al (2022) Drug delivery strategies and biomedical significance of hydrogels: translational considerations. Pharmaceutics 14:574
4. Thang NH, Chien TB, Cuong DX (2023) Polymer-based hydrogels applied in drug delivery: an overview. Gels 9:523
5. Dodda JM, Deshmukh K, Bezuidenhout D, Yeh Y-C (2023) Hydrogels: definition, history, classifications, formation, constitutive characteristics, and applications. Booksrscorg 13:1144–1154
6. Gomez-Florit M, Pardo A, Domingues RMA et al (2020) Natural-based hydrogels for tissue engineering applications. Molecules 25:5858
7. Revete A, Aparicio A, Cisterna BA et al (2022) Advancements in the use of hydrogels for regenerative medicine: properties and biomedical applications. Int J Biomater 2022:1–16
8. Bustamante-Torres M, Romero-Fierro D, Arcentales-Vera B et al (2021) Hydrogels classification according to the physical or chemical interactions and as stimuli-sensitive materials. Gels 7:182
9. Malektaj H, Drozdov AD, de Claville-Christiansen J (2023) Mechanical properties of alginate hydrogels cross-linked with multivalent cations. Polymers 15:3012
10. Parhi R (2017) Cross-linked hydrogel for pharmaceutical applications: a review. Adv Pharm Bullet 7:515–530
11. Begines B, Ortiz T, Pérez-Aranda M et al (2020) Polymeric nanoparticles for drug delivery: recent developments and future prospects. Nanomaterials 10:1403
12. Burlec AF, Corciova A, Boev M et al (2023) Current overview of metal nanoparticles' synthesis, characterization, and biomedical applications, with a focus on silver and gold nanoparticles. Pharmaceuticals 16:1410
13. Khan I, Saeed K, Khan I (2019) Nanoparticles: properties, applications and toxicities. Arab J Chem 12:908–931
14. Unnikrishnan G, Joy A, Megha M et al (2023) Exploration of inorganic nanoparticles for revolutionary drug delivery applications: a critical review. Discov Nano 18:157
15. Afshar M, Rezaei A, Eghbali S et al (2024) Nanomaterial strategies in wound healing: a comprehensive review of nanoparticles, nanofibres and nanosheets. Int Wound J 21:e14953
16. El-Husseiny HM, Mady EA, Hamabe L et al (2022) Smart/stimuli-responsive hydrogels: Cutting-edge platforms for tissue engineering and other biomedical applications. Mater Today Bio 13:100186
17. Patra JK, Das G, Fraceto LF et al (2018) Nano based drug delivery systems: recent developments and future prospects. Journal of Nanobiotechnology 16:71
18. Wahid F, Zhao X-J, Jia S-R et al (2020) Nanocomposite hydrogels as multifunctional systems for biomedical applications: current state and perspectives. Comp B Eng 200:108208
19. Xing W, Tang Y (2022) On mechanical properties of nanocomposite hydrogels: searching for superior properties. Nano Mater Sci 4:83–96
20. Gaharwar AK, Peppas NA, Khademhosseini A (2013) Nanocomposite hydrogels for biomedical applications. Biotechnol Bioeng 111:441–453
21. Ahmad N, Bukhari SNA, Hussain MA et al (2024) Nanoparticles incorporated hydrogels for delivery of antimicrobial agents: developments and trends. RSC Adv 14:13535–13564
22. Asadi N, Alizadeh E, Salehi R et al (2017) Nanocomposite hydrogels for cartilage tissue engineering: a review. Artif Cells Nanomed Biotechnol 46:465–471
23. Gounden V, Singh M (2024) Hydrogels and wound healing: current and future prospects. Gels 10:43
24. Jiang Y, Krishnan N, Heo J et al (2020) Nanoparticle–hydrogel superstructures for biomedical applications. J Control Release 324:505–521

25. Li J, Huang Y, Gao M et al (2024) Shrinkage properties of porous materials during drying: a review. Front Mater 11:686
26. Dannert C, Stokke BT, Dias RS (2019) Nanoparticle-hydrogel composites: from molecular interactions to macroscopic behavior. Polymers 11:275
27. Bahram M, Mohseni N, Moghtader M (2016) An introduction to hydrogels and some recent applications
28. Ahmad Raus R, Wan Nawawi WMF, Nasaruddin RR (2021) Alginate and alginate composites for biomedical applications. Asian J Pharm Sci 16:280–306
29. Pramanik S, Aggarwal A, Kadi A et al (2024) Chitosan alchemy: transforming tissue engineering and wound healing. RSC Adv 14:19219–19256
30. Dubey NK, Deng W-P (2018) Polymeric gels for cartilage tissue engineering. Polym Gels 41:505–525
31. Zhong Y, Lin Q, Yu H et al (2024) Construction methods and biomedical applications of PVA-based hydrogels. Front Chem 12:1376799
32. Sun S, Cui Y, Yuan B et al (2023) Drug delivery systems based on polyethylene glycol hydrogels for enhanced bone regeneration. Front Bioeng Biotechnol 11:1117647
33. Hong Y, Lin Z, Yang Y et al (2022) Flexible actuator based on conductive PAM hydrogel electrodes with enhanced water retention capacity and conductivity. Micromachines 13:1951
34. Hammami I, Alabdallah NM, Jomaa AA, Kamoun M (2021) Gold nanoparticles: synthesis properties and applications. J King Saud Univ Sci 33:101560
35. Roberts R, Smyth JW, Will J et al (2020) Development of PLGA nanoparticles for sustained release of a connexin43 mimetic peptide to target glioblastoma cells. Mater Sci Eng C 108:110191
36. Omidian H, Gill EJ, Dey Chowdhury S, Cubeddu LX (2024) Chitosan nanoparticles for intranasal drug delivery. Pharmaceutics 16:746
37. Ways TM, Ng KW, Lau WM, Khutoryanskiy VV (2020) Silica nanoparticles in transmucosal drug delivery. Pharmaceutics 12:751
38. Shi H, Zhou Z, Li W et al (2021) Hydroxyapatite based materials for bone tissue engineering: a brief and comprehensive introduction. Crystals 11:149
39. Elkhoury K, Russell CS, Sanchez-Gonzalez L et al (2019) Soft-nanoparticle functionalization of natural hydrogels for tissue engineering applications. Adv Healthcare Mater 8:1125
40. Ahmad F, Salem-Bekhit MM, Khan F et al (2022) Unique properties of surface-functionalized nanoparticles for bio-application: functionalization mechanisms and importance in application. Nanomaterials 12:1333
41. Ahmad Z, Salman S, Khan SA et al (2022) Versatility of hydrogels: from synthetic strategies, classification, and properties to biomedical applications. Gels 8:167
42. Fu Z, Xiang J (2020) Aptamer-functionalized nanoparticles in targeted delivery and cancer therapy. Int J Mol Sci 21:9123
43. Wang L-W, Gao Y, Feng Z et al (2024) Designing drug delivery systems for cell therapy. Nat Rev Bioeng 2:944–959
44. Oliveira JP, Prado AR, Keijok WJ et al (2019) Impact of conjugation strategies for targeting of antibodies in gold nanoparticles for ultrasensitive detection of 17β-estradiol. Sci Rep 9:125
45. Wang Q, Zhang Y, Ma Y et al (2023) Nano-crosslinked dynamic hydrogels for biomedical applications. Mater Today Bio 20:100640
46. Cho H, Jeon S, Yang J et al (2020) Hydrogel nanoparticle as a functional coating layer in biosensing, tissue engineering, and drug delivery. Coatings 10:663
47. Suk JS, Xu Q, Kim N et al (2016) PEGylation as a strategy for improving nanoparticle-based drug and gene delivery. Adv Drug Deliv Rev 99:28–51
48. Pessoa T, Kishima JOF, Silva JBMD et al (2023) Crosslinking methods in polysaccharide-based hydrogels for drug delivery systems. Biomed Mater Dev 2:288–306
49. Tian L, Li X, Ji H et al (2022) Melanin-like nanoparticles: advances in surface modification and tumour photothermal therapy. J Nanobiotechnol 20:485
50. Visan AI, Cristescu R (2023) Polysaccharide-based coatings as drug delivery systems. Pharmaceutics 15:2227

51. Thoniyot P, Tan MJ, Karim AA et al (2015) Nanoparticle–hydrogel composites: concept, design, and applications of these promising, multi-functional materials. Advanced Science 2:1400010
52. Mahmoudi C, Tahraoui-Douma N, Mahmoudi H et al (2024) Hydrogels based on proteins cross-linked with carbonyl derivatives of polysaccharides, with biomedical applications. Int J Mol Sci 25:7839
53. Kumari M, Acharya A, Krishnamurthy PT (2023) Antibody-conjugated nanoparticles for target-specific drug delivery of chemotherapeutics. Beilstein J Nanotechnol 14:912–926
54. Van Zundert I, Bravo M, Deschaume O et al (2021) Versatile and robust method for antibody conjugation to nanoparticles with high targeting efficiency. Pharmaceutics 13:2153
55. Kim K, Park M-H (2024) Role of functionalized peptides in nanomedicine for effective cancer therapy. Biomedicines 12:202
56. Kelarakis A (2024) In situ generation of nanoparticles on and within polymeric materials. Polymers 16:1611–1611
57. Alshangiti DM, El-damhougy TK, Zaher A et al (2023) Revolutionizing biomedicine: advancements, applications, and prospects of nanocomposite macromolecular carbohydrate-based hydrogel biomaterials: a review. RSC Adv 13:35251–35291
58. Jacob S, Nair AB, Shah J et al (2021) Emerging role of hydrogels in drug delivery systems, tissue engineering and wound management. Pharmaceutics 13:357
59. Sornkamnerd S, Okajima MK, Kaneko T (2017) Tough and porous hydrogels prepared by simple lyophilization of LC gels. ACS Omega 2:5304–5314
60. Tapia-Guerrero YS, Del Prado-Audelo ML, Borbolla-Jiménez FV et al (2020) Effect of UV and gamma irradiation sterilization processes in the properties of different polymeric nanoparticles for biomedical applications. Materials 13:1090
61. Bian W, Liau B, Badie N, Bursac N (2009) Mesoscopic hydrogel molding to control the 3D geometry of bioartificial muscle tissues. Nat Protocols 4:1522–1534
62. Wu J, Yun Z, Song W et al (2024) Highly oriented hydrogels for tissue regeneration: design strategies, cellular mechanisms, and biomedical applications. Theranostics 14:1982–2035
63. Raghuwanshi VS, Garnier G (2019) Characterisation of hydrogels: linking the nano to the microscale. Adv Colloid Interf Sci 274:102044
64. Ahmed EM (2015) Hydrogel: preparation, characterization, and applications: a review. J Adv Res 6:105–121
65. Huang S, Hong X, Zhao M et al (2022) Nanocomposite hydrogels for biomedical applications. Bioeng Transl Med 7:548
66. Zengin A, Castro JPO, Habibovic P, van Rijt SH (2021) Injectable, self-healing mesoporous silica nanocomposite hydrogels with improved mechanical properties. Nanoscale 13:1144–1154
67. Joshi J, Homburg SV, Ehrmann A (2022) Atomic force microscopy (AFM) on biopolymers and hydrogels for biotechnological applications—possibilities and limits. Polymers 14:1267
68. Hu X, Liang R, Li J et al (2019) Mechanically strong hydrogels achieved by designing homogeneous network structure. Mater Des 163:107547
69. Stojkov G, Niyazov Z, Picchioni F, Bose RK (2021) Relationship between structure and rheology of hydrogels for various applications. Gels 7:255
70. Cuomo F, Cofelice M, Lopez F (2019) Rheological characterization of hydrogels from alginate-based nanodispersion. Polymers 11:259
71. Hu C, Li Z (2015) A review on the mechanical properties of cement-based materials measured by nanoindentation. Constr Build Mater 90:80–90
72. Chai Q, Jiao Y, Yu X (2017) Hydrogels for biomedical applications: their characteristics and the mechanisms behind them. Gels 3:6
73. Jadhav N, Gaikwad V, Nair K, Kadam H (2009) Glass transition temperature: basics and application in pharmaceutical sector. Asian J Pharm 3:82
74. Ng HM, Saidi NM, Omar FS et al (2018) Thermogravimetric analysis of polymers. Encycl Polym Sci Technol 34:1–29

75. Patra S, Ajayan PM, Narayanan TN (2020) Dynamic mechanical analysis in materials science: the Novice's Tale. Oxford Open Mater Sci 1:itaa001
76. Denzer BR, Kulchar RJ, Huang RB, Patterson J (2021) Advanced methods for the characterization of supramolecular hydrogels. Gels 7:158
77. Li J, Mooney DJ (2016) Designing hydrogels for controlled drug delivery. Nat Rev Mater 1:785
78. Cadinoiu AN, Rata DM, Daraba OM et al (2022) Silver nanoparticles biocomposite films with antimicrobial activity: in vitro and in vivo tests. Int J Mol Sci 23:10671
79. Kyriakides TR, Raj A, Tseng TH et al (2021) Biocompatibility of nanomaterials and their immunological properties. Biomed Mater 16:042005
80. Liao C, Li Y, Tjong S (2018) Graphene nanomaterials: synthesis, biocompatibility, and cytotoxicity. Int J Mol Sci 19:3564
81. Firlar I, Altunbek M, McCarthy C et al (2022) Functional hydrogels for treatment of chronic wounds. Gels 8:127
82. Nunes D, Andrade S, Ramalho MJ et al (2022) Polymeric nanoparticles-loaded hydrogels for biomedical applications: a systematic review on in vivo findings. Polymers 14:1010
83. Pino P, Bosco F, Mollea C, Onida B (2023) Antimicrobial nano-zinc oxide biocomposites for wound healing applications: a review. Pharmaceutics 15:970
84. Aldhaher A, Shahabipour F, Shaito A et al (2023) 3D hydrogel/ bioactive glass scaffolds in bone tissue engineering: status and future opportunities. Heliyon 9:e17050
85. Ao Y, Zhang E, Liu Y et al (2022) Advanced hydrogels with nanoparticle inclusion for cartilage tissue engineering. Front Bioeng Biotechnol 10:951513
86. Arabpour Z, Abedi F, Salehi M et al (2024) Hydrogel-based skin regeneration. Int J Mol Sci 25:1982
87. Dreiss CA (2020) Hydrogel design strategies for drug delivery. Curr Opin Colloid Interf Sci 48:1–17
88. Mitchell MJ, Billingsley MM, Haley RM et al (2020) Engineering precision nanoparticles for drug delivery. Nat Rev Drug Discov 20:101–124
89. Shirakura T, Kelson TJ, Ray A et al (2014) Hydrogel nanoparticles with thermally controlled drug release. ACS Macro Lett 3:602–606
90. Haidari H, Kopecki Z, Sutton AT et al (2021) PH-responsive "smart" hydrogel for controlled delivery of silver nanoparticles to infected wounds. Antibiotics 10:49
91. Volpatti LR, Matranga MA, Cortinas AB et al (2019) Glucose-responsive nanoparticles for rapid and extended self-regulated insulin delivery. ACS Nano 14:488–497
92. Zhang K, Xue K, Loh XJ (2021) Thermo-responsive hydrogels: from recent progress to biomedical applications. Gels 7:77
93. Ansar R, Saqib S, Mukhtar A et al (2022) Challenges and recent trends with the development of hydrogel fiber for biomedical applications. Chemosphere 287:131956
94. Al-Rashid A, Khan SA, Al-Ghamdi S, Koç M (2021) Additive manufacturing of polymer nanocomposites: needs and challenges in materials, processes, and applications. J Mater Res Technol 14:910–941
95. Newham G, Evans SD, Ong ZY (2022) Mechanically tuneable physical nanocomposite hydrogels from polyelectrolyte complex templated silica nanoparticles for anionic therapeutic delivery. J Colloid Interf Sci 617:224–235
96. Zhang J, Wang Z (2022) Nanoparticle–hydrogel based sensors: synthesis and applications. Catalysts 12:1096
97. Gao W, Zhang Y, Zhang Q, Zhang L (2016) Nanoparticle-hydrogel: a hybrid biomaterial system for localized drug delivery. Ann Biomed Eng 44:2049–2061
98. Choi J, Kim BH (2024) Ligands of nanoparticles and their influence on the morphologies of nanoparticle-based films. Nanomaterials 14:1685–1685
99. De Jong WH, Geertsma RE, Borchard G (2022) Regulatory safety evaluation of nanomedical products: key issues to refine. Drug Deliv Transl Res 12:2042–2047
100. Abbasi R, Shineh G, Mobaraki M et al (2023) Structural parameters of nanoparticles affecting their toxicity for biomedical applications: a review. J Nanopart Res 25:43

101. Jooken S, Deschaume O, Bartic C (2023) Nanocomposite hydrogels as functional extracellular matrices. Gels 9:153
102. Martínez G, Merinero M, Pérez-Aranda M et al (2020) Environmental impact of nanoparticles' application as an emerging technology: a review. Materials 14:166
103. Zhang Y, Wu BM (2023) Current advances in stimuli-responsive hydrogels as smart drug delivery carriers. Gels 9:838
104. Mao Y, Zhang Y, Wang Y et al (2023) A multifunctional nanocomposite hydrogel with controllable release behavior enhances bone regeneration. Regener Biomater 10:685
105. Alghamdi MA, Fallica AN, Virzì N et al (2022) The promise of nanotechnology in personalized medicine. J Person Med 12:673
106. Rumon MdMH, Akib AA, Sultana F et al (2022) Self-healing hydrogels: development, biomedical applications, and challenges. Polymers 14:4539

Classification of the Wound and Their Nature

Sandipa Tripathi, Tuhin Manna, Subhamoy Dey, Suparna Majumder, Mousumi Manna, Ankita Das, Amiya Kumar Panda, and Chandradipa Ghosh

Abstract A wound arises from the damage to biological tissues, such as the skin, mucous membranes, and organs. Wounds can result from a variety of injuries; therefore, to prevent infections and further harm, the wounds should be cleaned and dressed properly. Wound classification is an essential step for wound diagnosis that is exceptionally vital for clinicians for planning the appropriate treatment strategies. Considering the importance of types of wounds, these wounds are categorized into different classes. Due to the unique nature, structure, and pathophysiology of each type of wound, the treatment plan designed for one wound cannot be used for another. This present chapter is composed to classify them into multiple groups, including pressure, diabetic, venous and surgical ulcers. Wounds can primarily be considered as acute and chronic types. Acute and chronic wounds are challenges in healthcare frameworks all over the world and affect many people's lives every year. Compared to acute wounds, which heal in a balanced measure and brief amount of time, chronic wounds require a long time to restore. This write-up would be a significant part of information for tissue engineers to design specific substitute for individual type of wound which might help for innovating proper treatment strategies.

Keywords Wound classification · Acute wounds · Chronic wounds

S. Tripathi · S. Dey
Centre for Life Sciences, Vidyasagar University, Midnapore, West Bengal, India

S. Tripathi · T. Manna · S. Dey · S. Majumder · M. Manna · C. Ghosh (✉)
Department of Human Physiology, Vidyasagar University, Midnapore, West Bengal, India
e-mail: ch_ghosh@mail.vidyasagar.ac.in

A. K. Panda
Department of Chemistry, Vidyasagar University, Midnapore, West Bengal, India

A. Das
Department of Physiology, Vidyasagar College for Women, Kolkata, West Bengal, India

A. K. Panda
Rani Rashmoni Green University, Singur, West Bengal, India

A. K. Mandal et al. (eds.), *Functionalized Nanoparticles Hydrogels for Wound Healing*, Smart Nanomaterials Technology, https://doi.org/10.1007/978-981-95-0720-7_3

1 Introduction

From ancient times to the present day evaluating wounds has remained a difficult task for the medical community. Wounds commonly lead to higher morbidity and significant mortality consequences in many instances [1, 2]. Any damage or injury to structure and function of the anatomy caused by harsh damage in biological tissues like skin, mucous membranes, and organs are referred to as wounds. This fracture can also involve different tissues and structures like subcutaneous tissue, tendons, muscles, vessels, nerves and bones. It may be stated that the skin, which is the biggest organ in humans, is the most commonly vulnerable for any harm as it can easily get burnt or injured from surgery or trauma. Generally, wounds are classified as either acute or chronic.

Dealing with acute and chronic wounds proves to be a difficult and heavy load on healthcare systems globally. Characterizing a wound is crucial in diagnosing wounds as it assists clinicians in determining the best treatment plan and evaluating treatment effectiveness [3]. Classifying a wound into various groups is known as wound classification (such as diabetic, venous, pressure) or conditions (like infection vs. non-infection, ischemia vs. non-ischemia,) that are crucial for assessing and characterizing wounds [4].

Acute wounds adhere to the normal course of wound healing and show distinct healing indicators within 4 weeks, while chronic wounds do not follow the normal course of healing stages and do not exhibit clear healing indicators within that time frame. Burns, bites, minor cuts, abrasions, surgical wounds, and severe traumatic lacerations from gunshot or crush injuries are examples of acute wounds, which are produced by damage to unbroken skin. It has been well documented that acute wounds have a consistent healing time regardless of the type of skin injury. Although there have been notable improvements in wound care in recent years, researchers and medical professionals are still working on developing new treatments to prevent and manage acute and chronic wounds infection [5]. In general medical practice, antibiotics can be applied through intravenous injection or oral administration for acute wounds [6], while for chronic wounds born infection, antimicrobial ointments, creams, or gels are recommended to remove deep bacterial or fungal infections migrating to subcutaneous tissues [7].

In case of acute wounds where wounds occur in roughly 5.6–26.0% of wounds due to microbial colonization and the breakdown of the skin barrier (moisture, temperature, and nutrient conditions) [8]. Burn, traumatic and surgical site wounds are the three most susceptible wounds to infection [9]. Different types of pathogens have been discovered in acute wounds after burn injury, such as *Staphylococcus aureus, Escherichia coli, Pseudomonas aeruginosa*, coagulase-negative *Staphylococci* [10], and various aerobic and anaerobic microorganisms. Acute surgical site wound infections (SSWIs) can arise in a variety of surgical wounds, from planned surgeries to emergency surgeries arisen from trauma. *Staphylococcus epidermidis* and *S. aureus* are the most common pathogenic microbes for acute SSWIs. Additional microorganisms have been cultured from severe SSWIs, such as *Clostridium difficile, E.*

coli, coagulase-negative *Staphylococci* [11], *Mycobacterium chelonae, Legionella pneumophila, Clostridium perfringens, P. aeruginosa, Mycobacterium fortuitum* and *Acinetobacter baumanii* [12]. Traumatic wounds include abrasions or lacerations that cause significant harm to tissues, bones, and internal organs [13]. Gram-positive *S. aureus* and Gram-negative *P. aeruginosa* are the distinctive pathogenic microorganisms found in infections from traumatic wounds [14]. Invasive fungal infection, which has a severe traumatic wounds complication, is commonly caused by agricultural accidents, natural disasters and war bombings leading to deep injuries to the tissues [15]. Acidic and iron-rich conditions are suitable for fungi's survival, which may lead to serious pelvic wounds or limb amputations [16, 17]. Fungal infections cause amputation and death at rates of 31 and 38%, respectively, despite being less commonly reported than bacterial infections [17].

Conversely, chronic wounds are frequently brought on by internal causes because of an underlying illness that eventually compromises the integrity of the dermal and epidermal tissues. Chronic wounds such as pressure ulcers, leg ulcers, and foot ulcers can develop due to metabolic diseases like diabetes mellitus, poor arterial supply (peripheral vascular disease), or poor venous drainage (venous hypertension). An increase in age, smoking, obesity, inadequate nutrition, and weakened immune system due to conditions like AIDS or treatment like chemotherapy or radiation therapy may worsen chronic ulcers. Conversely, pressure or decubitus ulcers develop due to continual external pressure on areas such as the buttocks, sacrum, and heels. However, the root cause of the issue is frequently to blame for the long-lasting nature of pressure sores, which similar to other types of chronic wounds, tend to heal slowly and in an unpredictable fashion [18]. The primary problem with chronic wounds is that they are susceptible to bacterial contamination. Nevertheless, it appears that the impact of bacteria on wound healing is dependent on the interaction between their presence and the patient. Clinicians must have the ability to distinguish between colonization, contamination, and infection of wounds. Every chronic wound contains bacteria. These contaminants are sourced from the native microbiota and/or the environment. The term colonization describes the existence of rapidly spreading bacteria with no response from the host and is a frequently occurring event. The majority of these organisms are typical skin bacteria such as *S. epidermidis,* various coagulase-negative *Staphylococcus sp.* , *Corynebacterium sp., Brevibacterium sp., Propionibacterium acnes,* and *Pityrosporum sp.* [2, 18]. In wound infections, proliferating bacteria not only damage the wound's surface but also invade the surrounding, healthy tissue, triggering a response from the host. Mainly, the pathogens discovered in infected wounds include *S. aureus,* Beta-hemolytic *Streptococcus sp., E. coli, Proteus sp., Klebsiella sp., Pseudomonas sp., Acinetobacter sp.,* and *Stenotrophomonas sp.* (Xanthomonas) [18, 19]. It has been reported that in the initial stages of acute wounds, the predominant bacteria present are from the normal skin microbiota. *S. aureus* will be followed shortly by beta-hemolytic *Streptococcus sp.* These organisms are frequently seen in diabetic foot ulcers. The wound will become infected with facultative anaerobic gram-negative rods such as *Proteus sp., E. coli,* and *Klebsiella sp.* ~4 weeks later. As the injury worsens, it begins to impact underlying structures. Frequently, infections involve multiple types of microorganisms [2]. Chronic wounds that last a long time usually have a higher concentration of anaerobic bacteria compared to aerobic bacteria. For example, *Acinetobacter sp.*, *Pseudomonas sp.*, and *Stenotrophomonas sp.* (Xanthomonas) are often present in chronic wounds [18, 20].

In this chapter, the challenges faced by clinicians when treating both acute and chronic wounds have been addressed. Considering the importance of types of wounds different types of acute and chronic wound are outlined depending of their different pathobiology. This write-up would be a significant piece of information for tissue engineers to design specific substitute for individual type of wound which might help for innovating proper treatment strategies.

2 Classification of Wounds and Their Nature

Wounds can be categorized based on their etiology, nature of occurrence, depth and the process of healing (Table1). The etiological classification includes mainly the traumatic wounds like abrasions, crush or cut wounds, diabetic wounds etc. [21]. In contrast to these surgical wounds are intentionally created during medical procedures [21]. Wounds can also be categorized by the duration of healing as acute wounds which heal promptly and chronic wounds that persist due to conditions like diabetes, long term pressure etc. (Fig. 1) [22]. Abrasions are the wounds caused by mechanical injuries like frictions against any hard or rough surfaces. In case of abrasions a quick healing mechanism can be found in <2 weeks without any scar formation [23]. But a resultant scarring can be formed in cases of deep abrasions. By daily cleaning and dressing this can be prevented. Additionally, to avoid excessive cicatrisation, debridement must be performed.

2.1 Acute Wounds

Acute wounds are injuries that occur suddenly and require prompt medical attention. They can arise from various causes, including accidents, surgical procedures, or thermal injuries like burns. In contrast to chronic wounds, which can continue for months or even years, acute wounds usually heal in a more predictable timescale. The acute wound healing process encompasses three main phases: inflammation, proliferation, and maturation [21]. During the inflammation phase, the body responds to injury with swelling, redness, and pain, signaling the immune system to start healing the damaged tissue. This is an important phase for initiating the healing process. During the proliferation stage new tissue is formed and the wound begins to close. This is when granulation tissue, a mixture of freshly created blood vessels and connective tissue, serves as a framework for healing by filling the wound area [28]. Patients often feel encouraged as they notice visible improvements in their wounds.

Table 1 Different types of acute and chronic wounds and the associated microorganisms

Wounds	Types of wounds	Associated microorganisms	References
Acute wounds	Abrasions or scrapes Avulsions or contusions Crush or cut wounds Missile or gunshot wounds Radiation wounds Surgical or traumatic wounds	*S. aureus, P. aeruginosa, S. epidermidis, E. coli, C. difficile,* Coagulase-negative *Staphylococci, L. pneumophila, M. chelonae, C. perfringens, M. fortuitum*, and *A. baumanii*	Ding et al. [24]
Chronic wounds	Venous or vascular ulcer Diabetic ulcer Pressure ulcer Ischemic ulcer	*S. aureus, P. aeruginosa, Streptococcus haemolyticus, Streptococci sp., Enterococcus* sp., *P. aeruginosa,* Gram-negative *bacilli, Proteus mirabilis, E. coli, K. pneumoniae,* Coagulase-negative *Staphylococcus, Enterococcus sp., Corynebacterium sp., Enterobacteriaceae*	Kalan and Brennan [25], Pugliese [26], Trivedi et al. [27]

Finally, during the maturation phase, the newly developed tissue strengthens and reorganizes over time. This stage may continue for months or even years, leading to a scar that may fade but often remains as a reminder of the injury [23]. Initial care typically includes cleaning the wound to prevent infection, which is crucial in the early hours post-injury. Depending on the wound type and severity, dressings are applied to protect the area and maintain a moist environment, which promotes faster healing. In recent years, innovative treatments have emerged to enhance healing outcomes. Suction pressure is used in negative pressure wound treatment to promote blood flow and tissue regeneration as well as bioengineered skin substitutes are now helping to treat complex wounds [29].

2.1.1 Abrasions or Scrapes

Abrasions, commonly known as scrapes, are among the most prevalent types of acute wounds encountered in both everyday life and clinical settings. These superficial injuries occur when the skin is rubbed or scraped off, often resulting from falls, sports injuries, or accidents. While they might seem minor at first glance, abrasions

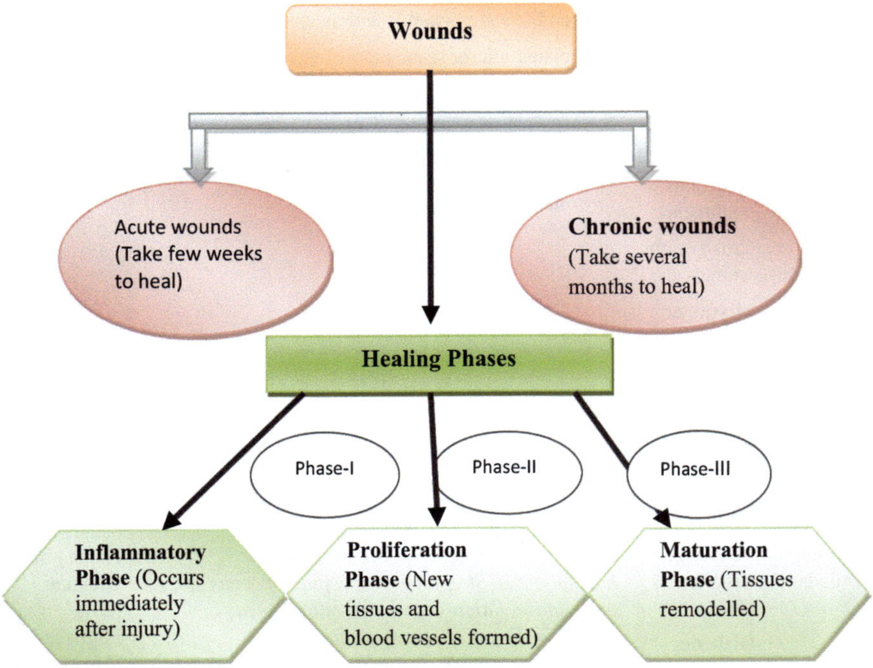

Fig. 1 Types of wounds and the healing phases

can lead to discomfort, increased risk of infection, and in some cases, scarring. Abrasions primarily affect the outermost layer of skin, the epidermis and may extend into the upper part of the dermis. This makes them different from deeper wounds like lacerations or punctures, which can involve more extensive damage to underlying tissues. The appearance of an abrasion typically includes red or pink skin, often accompanied by minor bleeding or oozing. The affected area can also be painful and sensitive, particularly when exposed to air or friction. Common scenarios for abrasions include athletic activities like falling while running, cycling, or playing contact sports. Children are particularly prone to these injuries, as their adventurous nature often leads them to explore their surroundings with little regard for potential hazards. The emotional aspect of experiencing an abrasion, especially for children can be significant. Fear of pain, the anxiety of receiving treatment, and concerns about appearance all contribute to the experiences. The healing process of abrasive wounds often follows the same stages as other acute wounds (i) hemostasis, (ii) inflammation, (iii) proliferation, and (iv) maturation [30]. Immediately after an abrasion occurs, the body starts a process to stop bleeding. Blood vessels constrict, and Platelets gather at the site of the injury, forming a clot that helps protect the underlying tissues. Following hemostasis, the inflammation phase begins, characterized by redness, warmth, and swelling. This response is crucial for fighting off potential infections and initiating the healing process. Immune cells migrate to the area to remove pathogens and debris, which is why you might notice increased sensitivity around the area [30]. During this phase, the body works to form new tissue which helps to close the wound. Skin cells multiply and migrate throughout the surface of the wound, while blood vessels expand to provide the area with necessary nutrients. For superficial abrasions, this process is relatively quick, often leading to the formation of a new layer of skin within a week or so. The final phase can last for months to years. Though the new skin appears healed, it may still undergo changes in texture and colour. Scarring may occur particularly if the abrasion was deep or if the area experiences friction during healing. Proper abrasion management can have a substantial impact on healing results and lower the problematic consequences like infection or scarring [31]. The initial step following abrasion is to gently clean the wound area [31]. Considering the size and location of the abrasion a sterile bandage or dressing may be used for protection from further injury. For larger abrasions, using a non-stick dressing can prevent the bandage from adhering to the wound, minimizing discomfort during changes. Regular checking should be done for signs of infection, such as swelling, increased redness, or pus [32]. Educating patients and caregivers about what to look for can empower them to manage the wound effectively. Although abrasions are usually not severely painful, discomfort can arise, particularly during the initial healing stages. Acetaminophen and ibuprofen are examples of over-the-counter pain relievers that can help with pain relief and inflammation reduction [23]. To promote optimal healing, it's important to avoid activities that may cause friction on the affected area. For example, wearing loose clothing over a knee abrasion or avoiding certain heavy works until the skin has healed can be beneficial.

2.1.2 Avulsion

An avulsion occurs when skin or another body part, such as nail or teeth is torn off by force. It occurs when skin layers are pulled off to reveal subcutaneous tissue, bone, tendon, and muscle. An avulsion also refers to an amputation, especially when a limb is trapped in a large piece of equipment [33]. An automobile accident-related skin avulsion is a prevalent kind of avulsion that can cause a "torn" flap of skin. A finger that has been partly cut with a knife is an additional illustration of an avulsion. Additional varieties comprise the ear, eyelid, teeth, and nail avulsion [34].

One of the risks associated with contemporary industrial growth and vehicle travel is skin avulsion wound to the appendages. Skin avulsions may vary in intensity from minor skin flaps to moderate degloving and severe amputation of a hand or leg [35]. Different terms for this injury include "pneumatic tyre torsion avulsion injury" or "roller wringer injury". The terminology utilized here pertains to the mechanism of damage, yet the fundamental patho-physiology and therapeutic philosophy remain the same [36]. Skin avulsion injuries will occur from any tangent shearing force applied to a confined limb. Sometime the appendages of our body get compressed between rollers, heavy wheel, and the road surface. The sub-cutaneous tissues of those compressed appendages become ruptured at a point distal to the compression site. A "flapper" in the context of rock climbing is an injury where a portion of the skin is torn off. This often happens when a climber slips off a grip and the friction forces build up between their fingers and the holds [37].

Ears are more susceptible to severe injuries because of their conspicuous positions. The disfiguring distortion caused by a traumatic ear avulsion (TEA) is a relatively rare condition that can have severe psychological effects over time. The intricate form, distinct anatomical arrangement, and tiny size of the arteries supplying its perfusion make reconstructing an ear that has been traumatized as a significant operational task even today [38]. In an effort to maximize the chances of the replanted and reattached auricle surviving, a variety of approaches have been reported. The selection of the suitable reconstructive approach depends critically on the structure of ear and the underlying process of causing the injury. When juxtaposed to other procedures, the success rate for micro vascular repair is greater. However, it is more difficult, and venous congestion after surgery might happen. This is because the veins have tiny branches that are difficult to identify following damage [39].

Severe eyelid wounds are rather uncommon among face tissue complications, and they are typically the consequence of assaults or car crashes. If eyelid deformities are not properly treated, they might cause serious morbidity [40]. Every time an injury to the eyelid is seen, it's critical to determine whether the globe has been injured as well. Any damage to the canalicular network should be assessed during eyelid avulsion. When there is periocular trauma, canalicular lacerations are frequently seen and are linked to eyelid wound. The main goal of treating these instances is to get the lacerated canaliculus repaired precisely and stabilized temporarily in order to re-establish a patent canalicular network. Making sure the eyelid returns to its natural alignment is the secondary goal [41]. Besides that a tooth that has been partly or fully torn out of its orbit is called an avulsion. Dental professionals may adjust and implant secondary tooth. Since primary teeth often get damaged and impede the formation of secondary teeth, they are not restored. Permanent retention is possible for a fully avulsed tooth that is restored within an hour of the incidence taking place [42]. The protective coating on the finger may tear away from the root as a result of wound to the nail. Nail loss is not usually repaired, in contrast to other forms of avulsion. The nail bed creates a germinal layer when the nail is lost, and further the cells grow keratin and produce a new nail. The exposed part is extremely sensitive until this covering is formed [43].

2.1.3 Crush or Cut Wound

Wound types and their effects vary on the human body. When the area of the body experiences severe compression especially from large objects or heavy machine, collapsed structure could get a crush injury. Continuous pressure that disrupts alongside normal tissue function and blood circulation is the primary mechanism causing ischemia and cellular necrosis [44].

Crush wound is frequent in bomb blast, earthquake and car crashes among other incidences. The majority of laceration, severe bruises, broken bones and compartment syndrome happen by incidental crush injuries [45]. When compared to other kinds of wounds, it results in greater acute tissue damage. It is probable that the skin may be broken and sometimes tissues may appear bruised and swollen [45]. Compartment syndrome a dangerous condition involving elevated pressure in a muscle compartment which is frequently happened in case of crush injuries. Elevation of pressure in the area that causes bleeding exudates and raised interstitial pressure, which causes more ischemia damage and swelling leading to a permanent injury in muscle and nerve [45]. The severity of the injury enhances the possibility of infection, especially if any particle has contaminated the site [46].

Cut wound also referred to as lacerations, are inflicted by sharp things like metal, glass or knives. Depending on the object and the force used, these wounds might differ in depth and length [43]. In the medical terms, a laceration is typically used to indicate an irregular rupture in the skin, whereas a cut is equivalent to an incision, which is a purposeful division created during surgical treatment [47]. The nature of cut wound can vary significantly based on the mechanism of injury, the sharpness of the object and the force applied during the incidence. It is notable that while the majority of stab wound are inflicted by weapons with sharp edges, it is recognized that blunt weapon can also lead to deadly stabbings [48]. Cut wound typically involves the phases of hemostasis, inflammation, proliferation and remodeling with each phase playing a critical part in the healing process [49].

2.1.4 Gunshot Wound

A gunshot wound is a deep wound from a projectile, such as bullet fired from a gun usually air gun or rifle [50]. This wound is significant concern in both forensic practice and clinical management due to their complexity and unusual appearances. Differentiating between entry and exit of wounds is critical in forensic situations, although it can be challenging due to variable factors including maintained injuries, disintegration and unusual wound patterns [51].

Gunshot wound can lead to a range of complications and long-term effects including spinal cord injuries that are the second most important cause of such injuries in the United States, often resulting in significant morbidity [52]. When it concerns to the management of gunshot wound, decision about bullet removal has not been confirmed to reduce pain in case of spinal cord injuries [53].

Depending on the bullets, its velocity, mass, entry site, path, damaged anatomy and exit site, a gunshot wound can cause wide ranges of trauma. Because of unpredictable nature of bullet fragmentation and path after entry, gunshot wound can be more destructive than other types of penetrating injuries [54]. Usually, a gunshot wound causes substantial bleeding immediately, which increases the risk of hypovolemic shock, a condition in which the body fails to supply enough oxygen to crucial organ such as heart, lungs, liver and the brain and spinal cord, which are parts of the central nervous system [55].

2.1.5 Radiation Wound

Radiation wound, often resulting from exposure to ionizing radiation, can appear as cutaneous radiation injuries with different levels of severity, potentially leading to skin necrosis and ulceration [56]. These injuries can arise from medical treatments, industrial incidents or radiological incidents and may involve not only the skin but also deeper tissues and organs [57]. Such wound requires significant clinical therapy, including quick, precise diagnosis, dose calculation and right medication to reduce the damage [58].

Acute radiation dermatitis often occurs after 90 days of exposure, however chronic dermatitis includes both intrinsic variables as genetics and external factors such the amount of radiation received [59]. Aplastic anemia, a reduction in the number of blood cells, is the characteristic of this medical condition. This could lead to bleeding with insufficient number of platelets, anemia for a shortage of red blood cells in the blood stream, and infections because of low number of white blood cells (WBCs). Through blood testing acute dosage as low as 0.25 grays (25 rad) can show these changes [60]. The clinical indication of the radiation-induced damage of GI tracts includes symptoms such as nausea, vomiting, anorexia and discomfort in the stomach [61]. After absorbed dosage of 6–30 grays (600–3000 rad), this type of condition frequently arises [60]. This syndrome usually happens at absorbed doses >30 grays (3000 rad), but it can happen at doses as low as 10 grays (1000 rad). This type of effect starts within minutes to a few hours, without vomiting, and appears with neurological symptoms including headaches, dizziness, reduced consciousness. Even with through urgent care, the outcome is often life-threatening [60]. Radiation exposure at high levels damages DNA, which when left unrepaired can result in severe and potentially fatal chromosomal defects. Ionizing radiation may result reactive oxygen species production and direct cell damage. The initial event causes significant damage to DNA, while subsequent occurrences cause cluster DNA damage [60].

2.1.6 Surgical or Traumatic Wounds

Surgical wounds are intentional cuts made during medical procedures or any type of operation that involves making a cut into the skin, including procedures carried out by doctors or surgeons. The surgical wounds can be categorized as clean, contaminated, clean-contaminated and dirty wounds, based on their level of contamination and the method of closure [62]. Surgical wounds typically follow a predictable healing trajectory involving haemostasis, inflammation, proliferation, and remodeling. Factors such as blood supply, nutrition, and overall health significantly influence recovery process.

Traumatic wounds are generally caused due to damage of skin and underlying tissues. If left unattended or neglected, they may cause severe damage to the underlying soft tissues. Traumatic wounds are a type of sudden injury that can cause trauma to the patient. They may be the result of any form of violence or accidents. These injuries include burns, bites, skin tears, abrasions, lacerations, and penetrating wounds *etc.* The acute cut and penetrating wounds are the primary types of traumatic wounds. These kind of cut wounds occur when some sharp object penetrates the skin and the underlying tissues. Among these traumatic wounds, penetrating wounds are considered as the most severe and deep, as in case of a stab wound. They can vary widely in size and depth and can broadly be classified as open (lacerations, abrasions *etc.*) or closed (bruises) wounds.

Both surgical and traumatic wounds carry risks of complications such as infection, delayed healing, and scarring. Factors influencing these outcomes include patient health, wound care practices, and the presence of foreign bodies [63]. And thus, the treatment of such wounds focuses on preventing infections, controlling bleeding, and promoting healing process. Understanding the nuances of surgical and traumatic wounds is crucial for healthcare providers. Effective assessment and management can significantly enhance healing outcomes and minimize complications. Emphasizing proper wound care practices is essential in promoting recovery and ensuring patient safety.

2.2 *Chronic Wound*

A chronic wound forms when an acute wound does not heal properly within an expected timeframe, typically 4–6 weeks. In these situations, the healing process may be delayed, halted, or worsened over time. These are also heterogeneous in pathogenesis, size, body part location, susceptibility to infection, and risk of amputation [64]. Such wounds require specialized treatment to promote healing. Chronic wounds are recognized as a significant public health issue worldwide, consuming considerable time, resources, and care. They adversely affect patients' quality of life, often leading to ongoing care needs, pain, and reduced mobility for many individuals. Research by studies indicates that ~1–2% of individuals in developed nations experience chronic wounds at any given moment. In developing countries like India, the issue of chronic wounds comes with other challenges such as low rates of literacy, restricted access to high-quality medical care and dependence on imported medical devices, economical class, and the lack of health insurance policies [65]. A study by [66], identified systemic conditions like diabetes and atherosclerosis as primary contributors to chronic wounds, along with other significant factors such as pressure ulcers, vasculitis, and trauma.

2.2.1 Venous or Vascular Ulcer

Venous or vascular ulcers are open sores on the skin that typically occur in venous hypertension-affected area of the body. Venous hypertension is the increase in venous pressure caused by venous blockage, which might be the underlying phenomenon for venous ulcer formation [67]. Cardiac valve dysfunction along with obstructive outflow and arteriovenous malformation further aid to the venous hypertension pathogenesis [68]. In addition, inflammation in the venous circulation and venous hypertension, along with the extravasation of inflammatory cells and molecules into the interstitial space, cause a complex process that leads to venous ulcers. [69]. The development of venous leg ulcers (VLU) is among the most serious side effects of chronic venous insufficiency, which accounts for about 70–90% of all other chronic leg ulcers (CLUs). Some of the primary risk factors that lead to venous ulcers include age above 55 years, Chronic ulcers in the family history, high Body Mass Index, superficial or deep venous thrombosis, lower extremity joint disease, parental history of leg ulcers, less physical activity, venous obstruction in deep veins, etc. [70].

According to [71], venous ulcers usually have a distinct border and an irregular shape. The most commonly reported symptoms include pain, edema, pruritus, and limb heaviness which gets worse with time. The majority of venous ulcers develop on bony areas, including the gaiter area (over the medial malleolus), telangiectasias, atrophie blanche (atrophic, white scarring), corona phlebectatica (abnormally dilated veins around the ankle and foot) and lipodermatosclerosis and the lower leg's inverted champagne bottle deformity. Venous ulcer complications include skin cancer and infections which may also cause morbidity [71]. This also has financial and personal effects, such as frequent visits to medical facilities, decreased productivity, more discomfort, the need for frequent dressing changes, and repeated hospital visits [72]. Age, an elevated body mass index, persistent lower limb arterial disease, ulcer length of 10 cm or more, and wound duration exceeding three months are some poor prognostic indicators for ulcer healing [73].

Venous ulcer recurrence has been reported as around 70% worldwide. Compression stockings and venous intervention have been shown to be important in preventing recurrence, and leg elevation combined with compression stockings can be highly advantageous. Exercise or regular physical activity is advised to improve overall leg muscle or specifically calf muscle pump function. Timely healing of ulcers and preventing other complications related to ulcers depend on early detection and appropriate treatment of these underlying conditions. For this, it is helpful to think about treating each patient individually, taking into account their underlying condition, co-morbidities, and general health. A multidisciplinary approach is frequently the most effective way to address the particular difficulties faced by geriatric patients [74]. The course of treatment consists of compression, leg elevation, local therapy, maximum bed rest and surgical procedures such as deep vein reconstruction and superficial and perforating vein ablation.

2.2.2 Diabetic Ulcer

Diabetic wound ulcer also profoundly known as diabetic foot ulcers (DFUs) present a significant challenge in medical care because of their complex pathophysiology and the elevated chance of complications, including infections and impaired healing. These wounds are characterized by prolonged healing times, often exacerbated by factors like oxidative stress, bacterial colonization, and dysregulated molecular pathways [75]. Pathogenesis of diabetic foot ulcers (DFUs) is a multifaceted process involving several interrelated mechanisms primarily driven by the chronic complications of diabetes mellitus. Central to the development of DFUs are neuropathy, vascular insufficiency, and infection, which collectively contribute to the ulcerative process. Neuropathy, resulting from prolonged hyperglycemia, leads to sensory, motor, and autonomic nerve damage, causing loss of protective sensation, foot deformities, and skin breakdown due to autonomic dysfunction [76]. Vascular complications, including both macrovascular and microvascular changes, result in ischemia and impaired wound healing due to endothelial dysfunction and reduced nitric oxide production, which are exacerbated by oxidative stress and inflammation [76]. Inflammatory response in DFUs is characterized by an imbalance in cytokine production with elevated levels of pro-inflammatory cytokines, such as, TNF-α and IL-6, and a dysregulated immune response that impairs the transition from the neutrophilic to macrophage phase of healing [77]. Additionally, the over expression of matrix metalloproteinases (MMPs), particularly MMP-9, lead to excessive extracellular matrix degradation, further hindering the healing process [78]. The presence of pathogenic microorganisms, such as *Staphylococcus sp.* and *Streptococcussp.*, exacerbate the inflammatory milieu and contribute to chronic infection complicating the healing of DFUs [79]. The interplay of these factors results in a chronic wound environment that is difficult to heal, often necessitating advanced therapeutic interventions such as growth factor therapy, stem cell applications and novel drug formulations to promote healing and prevent amputations [80]. Understanding these complex pathophysiological mechanisms is crucial for developing effective management strategies and innovative treatments for DFUs [78].

Recent advances in treatment strategies have focused on innovative materials and technologies to address these challenges. For instance, pH/glucose dual-responsive hydrogels have been developed to clear reactive oxygen species (ROS) and deliver antibacterial agents in response to the wound's microenvironment, thereby promoting healing and reducing inflammation. Similarly, self-healing hydrogels offer the ability to maintain prolonged contact with wound sites, mimicking the regenerative capabilities of native tissues and providing a stable healing environment [81]. Additionally, glucose-responsive smart gels and nanocomposite-integrated microneedles have been explored for their abilities to respond to high glucose levels in diabetic wounds offering targeted drug delivery and antibacterial effects [82, 83].

Despite these advancements, challenges remain in optimizing these therapies for clinical use, necessitating further research to enhance their efficacies, stabilities and regulatory compliances [84]. Overall, the integration of these innovative strategies holds promise for enhancing the treatment and results of diabetic wounds, possibly revolutionizing wound care procedures, and improving diabetic patients' quality of life [85].

2.2.3 Pressure Ulcer

Pressure ulcers are localized wounds to the skin and underlying tissue that are also referred to as bedsores or decubitus ulcers that are mostly brought on by prolonged pressure on bony prominences like the sacrum, heels, and hips, frequently combined with shear and friction. The complexity of PUs is underscored by their multifactorial etiology, which includes risk factors such as old age, sedentary lifestyles, and comorbidities like diabetes and hypertension [86]. Prevalence of pressure ulcers remains significant, affecting ~5–15% of patients in healthcare settings with a notable burden on individuals over 70 years of age [87]. The pathophysiology of pressure ulcers involves ischemia and subsequent tissue necrosis due to sustained pressure that occludes blood flow, leading to anoxia and irreversible tissue damage if not relieved [88]. Muscle tissue is particularly susceptible to ischemic damage compared to skin [88]. Risk factors include immobility, spinal cord injuries, and conditions such as diabetes and neurological disorders, which exacerbate the risk by impairing sensation and mobility [89]. Additionally, prolonged pressure on biological tissues can cause capillary bed blockage, which lowers the area's oxygen concentrations. Although tissues can endure abnormally high external pressure, the main contributing factor is constant pressure applied for a long period of time. Blood flow obstruction requires external pressure to be greater than the arterial capillary pressure of 32 mm Hg. For the pressure to prevent venous blood return, it must be higher than the venous capillary occlusion pressure, which is 8–12 mm Hg. Maintaining pressure above these limits causes necrosis and tissue ischemia [90]. With the passage of time, the ischemic tissue commences to amass deleterious metabolites. Consequently, tissue ulceration and necrosis transpire. A lack of mobility for merely two hours in a patient confined to bed or an individual undergoing surgical procedures is adequate to establish the groundwork for a pressure ulcer [91]. The pathology of pressure ulcers is very complex involving various proteins that play roles in tissue regeneration, as identified through proteomic analyses, which may offer new therapeutic targets [92].

Risk assessment, the use of pressure-relieving equipment, and staff training are important prevention and management techniques [93]. The NODITE strategy, which emphasizes nutrition, offloading, debridement, infection control, tissue management, and education, has been shown to be effective in managing severe cases in palliative care settings [89]. Despite advancements in treatment and prevention, the prevalence of pressure ulcers has remained stable over the years with an increase in severity and a decrease in surgical interventions. Overall, a multidisciplinary approach involving comprehensive care strategies including psychological support is essential to address the complex challenges posed by pressure ulcers and improve patient outcomes [94].

2.2.4 Arterial or Ischemic Wound

Other kind of chronic wound is an ischaemic wound, which is brought on by a reduction in the blood circulation to the specific organ, which leaves the tissues lacking in the glucose and oxygen needed for the metabolism [95]. Because there is less oxygen and nutrition getting to the tissues as a result of the blood loss, the death of cells follows. This can thus result in the development of uncomfortable ulcers that either never heal at all or heal very slowly. The lower limbs especially the feet and toes are the most frequently affected by ischaemic ulcers [96]. Since these regions are the furthest from the heart, circulation issues may cause a decrease in blood flow in these places. Additionally, susceptible to inflammation are the lower limbs, which might worsen circulation and hasten the onset of ischaemic ulcers. Although they mostly affect the skin, ischaemic ulcers can sometimes harm interior organs like the colon. Restricted blood supply to the colon causes ischaemic colitis, a disorder that results in intestinal ulcer [97].

However, a number of complex illnesses, such as arterial entrapment, adventitial cyst, dissection, dysplasia, embolism, fibromuscular, trauma, vasospasm, and vasculitis may additionally trigger skin hypo-perfusion, resulting in wounds that fail to heal [98]. The majority of ischaemic wounds are triggered by mild to serious peripheral arterial disease (PAD). Management of the fundamental illnesses, whether it is PAD or another condition like autoimmune illnesses, chronic venous insufficiency, diabetes, hypertension, and trauma is the first step towards the successful management of any ischaemic lesion. A complete vascular assessment is a vital part of the investigation of any wound that appears below the knee in order to rule out PAD as a contributory cause [99].

3 Conclusion

Practically every person has experienced an open wound throughout their life. The majority of these wounds could be easily treated while others required medical intervention. Wound management is crucial in the medical field because a wound that becomes infected may result in severe complications. Precise identification of wound types can assist clinicians in promptly diagnosing wound issues and determining appropriate treatment strategies. In this chapter, our goal has been to present the existing understandings in wound care and evaluations to aid researchers, clinicians, and anyone interested in the field of wound healing in comprehending the intricate processes of wound healing. The group of authors have sketched here the basic principles of treating wounds and focused on the most recent recommendations.

To recommend the best course of treatment for wound infections, a precise diagnosis is essential. However, current methods have variable specificity and sensitivity, are time-consuming, and are speculative. The techniques of PCR and autofluorescent imaging have been used to create new diagnostic techniques. Monotherapies, such as the use of antibiotics, for the management of acute or chronic wound infections still carry a significant risk of developing antimicrobial resistance. Additionally, the bacterial community behaviors are dynamic, the spatial arrangements of the microorganisms in the wound are complex, and the relationships between human immunity and polymicrobials are undefined. Dual therapies are anticipated to enhance the creation of efficacious treatments for wound infections in response to these. For example, it has been demonstrated that topical negative pressure and silver-impregnated foam work in concert to eliminate bacterial biofilms in the wound. In clinical trials, certain combination therapies, such as vacuum pump therapy and ultrasound-assisted debridement, have also been investigated to treat infections in deep sternal wounds. These pioneering studies all showed great promise for treating wounds and preventing infections. Additional combinations consist of immune-based antimicrobial molecules (antimicrobial peptides), antimicrobial agents (antibiotics, herbal medicines, and synthetics), therapeutic microorganisms (bacteriophages and probiotics), and cell therapy. Future advancements in wound infection treatment are anticipated as a result of these combinations. External stimuli like laser therapy, light emitting diodes, high frequency ultrasound, micro current electrical stimulation and antimicrobial phototherapy (NIR- based therapies) could be applied to wound infection treatment as well.

Nevertheless, clinicians have a strong belief that factors like diet and the presence of other illnesses also impact the process of recovery. Therefore, it is important to treat patients for all of their disorders, not just the wound. Prescription of nutritional components should be applied in conjunction with additional factors. One of the prime reasons for the growing interests in medicinal plants is their minimal side effects and cost effective in comparison to usage of commercial products. Nevertheless, employing medicinal herbs for wound healing is a comprehensive method that is still in its early stages. Understanding wound classification and assessment mechanisms, wound microbiology, and current wound dressings could assist the medical and pharmaceutical communities in developing more effective dressings for individual patients.

References

1. Gonzalez ACDO, Costa TF, Andrade ZDA, Medrado ARAP (2016) Wound healing: a literature review. An Bras Dermatol 91(5):614–620
2. Siafaka PI, Zisi AP, Exindari MK, Karantas ID, Bikiaris DN (2016) Porous dressings of modified chitosan with poly (2-hydroxyethyl acrylate) for topical wound delivery of levofloxacin. Carbohydr Polym 143:90–99
3. Shenoy VN, Foster E, Aalami L, Majeed B, Aalami O (2018) Deepwound: automated postoperative wound assessment and surgical site surveillance through convolutional neural networks. In: Proceedings of the 2018 IEEE international conference on bioinformatics and biomedicine (BIBM). IEEE, pp 1017–1021
4. Russell L (1999) The importance of wound documentation and classification. Br J Nurs 8(20):1342–1354
5. Fan Z, Yao B, Ding Y, Xu D, Zhao J, Zhang K (2022) Rational engineering the DNA tetrahedrons of dual wavelength ratiometric electrochemiluminescence biosensor for high efficient detection of SARS-CoV-2 RdRp gene by using entropy-driven and bipedal DNA walker amplification strategy. Chem Eng J 427:131686
6. Bassetti M, Peghin M, Castaldo N, Giacobbe DR (2019) The safety of treatment options for acute bacterial skin and skin structure infections. Exp Opin Drug Saf 18(8):635–650
7. Jamaledin R, Yiu CK, Zare EN, Niu LN, Vecchione R, Chen G et al (2020) Advances in antimicrobial microneedle patches for combating infections. Adv Mater 32(33):2002129

8. Rahim K, Saleha S, Zhu X, Huo L, Basit A, Franco OL (2017) Bacterial contribution in chronicity of wounds. Microb Ecol 73:710–721
9. PLOS Genetics Editors (2015) Expression of concern: RNAi-dependent and independent control of LINE1 accumulation and mobility in mouse embryonic stem cells. PLoS Genet 11(6):e1005377
10. Reid G, Younes JA, Van der Mei HC, Gloor GB, Knight R, Busscher HJ (2011) Microbiota restoration: natural and supplemented recovery of human microbial communities. Nat Rev Microbiol 9(1):27–38
11. Branch-Elliman W, Ripollone JE, O'Brien WJ, Itani KM, Schweizer ML, Perencevich E et al (2017) Risk of surgical site infection, acute kidney injury, and *C. difficile* infection following antibiotic prophylaxis with vancomycin plus a beta-lactam versus either drug alone: a national propensity-score-adjusted retrospective cohort study. PLoS Med 14(7):e1002340
12. Percival SL (2017) Importance of biofilm formation in surgical infection. J Br Surg 104(2):e85–e94
13. Iheozor-Ejiofor Z, Newton K, Dumville JC, Costa ML, Norman G, Bruce J (2018) Negative pressure wound therapy for open traumatic wounds. Cochrane Database Syst Rev 7:1154
14. Dai T, Kharkwal GB, Tanaka M, Huang YY, de Arce VJ, Hamblin MR (2011) Animal models of external traumatic wound infections. Virulence 2(4):296–315
15. Ganesan A, Shaikh F, Bradley W, Blyth DM, Bennett D, Petfield JL et al (2019) Classification of trauma-associated invasive fungal infections to support wound treatment decisions. Emerg Infect Dis 25(9):1639
16. Kronen R, Liang SY, Bochicchio G, Bochicchio K, Powderly WG, Spec A (2017) Invasive fungal infections secondary to traumatic injury. Int J Infect Dis 62:102–111
17. Warkentien T, Rodriguez C, Lloyd B, Wells J, Weintrob A, Dunne JR et al (2012) Invasive mold infections following combat-related injuries. Clin Infect Dis 55(11):1441–1449
18. Bowler PG, Duerden BI, Armstrong DG (2001) Wound microbiology and associated approaches to wound management. Clin Microbiol Rev 14(2):244–269
19. Ovington L (2003) Bacterial toxins and wound healing. Ostomy Wound Manag 49:8–12
20. Edwards R, Harding KG (2004) Bacteria and wound healing. Curr Opin Infect Dis 17(2):91–96
21. Milne K, Penn-Barwell JG (2023) Classification and management of acute wounds and open fractures. Surgery 41(2):93–99. https://doi.org/10.1016/J.MPSUR.2022.12.002
22. Kumar S, Leaper DJ (2005) Classification and management of acute wounds. Surgery 23(2):47–51. https://doi.org/10.1383/SURG.23.2.47.60348
23. Slowik V, Müller S, Wagner C, Mechtcherine V (2023) Abrasion. RILEM State Art Rep 22:119–124. https://doi.org/10.1007/978-94-024-1013-6_7
24. Ding X, Tang Q, Xu Z, Xu Y, Zhang H, Zheng D et al (2022) Challenges and innovations in treating chronic and acute wound infections: from basic science to clinical practice. Burns Trauma 10:tkac014
25. Kalan LR, Brennan MB (2019) The role of the microbiome in non-healing diabetic wounds. Ann N Y Acad Sci 1435(1):79–92
26. Pugliese DJ (2016) Infection in venous leg ulcers: considerations for optimal management in the elderly. Drugs Aging 33:87–96
27. Trivedi U, Madsen JS, Rumbaugh KP, Wolcott RD, Burmølle M, Sørensen SJ (2017) A post-planktonic era of in vitro infectious models: issues and changes addressed by a clinically relevant wound like media. Crit Rev Microbiol 43(4):453–465
28. Schultz GS, Chin GA, Moldawer L, Diegelmann RF (2011) Principles of wound healing. Diab Foot Probl 14:395–402. https://doi.org/10.1142/9789812791535_0028
29. Novak A, Khan WS, Palmer J (2014) Suppl 1: the evidence-based principles of negative pressure wound therapy in trauma and orthopedics. Open Orthopaed J 8:168
30. Velnar T, Bailey T, Smrkolj V (2009) The wound healing process: an overview of the cellular and molecular mechanisms. J Int Med Res 37(5):1528
31. Bechert K, Abraham SE (2009) Pain management and wound care. J Am College Cert Wound Spec 1(2):65–71. https://doi.org/10.1016/J.JCWS.2008.12.001

32. Eriksson E, Liu PY, Schultz GS, Martins-Green MM, Tanaka R, Weir D, Gould LJ, Armstrong DG, Gibbons GW, Wolcott R, Olutoye OO, Kirsner RS, Gurtner GC (2022) Chronic wounds: treatment consensus. Wound Rep Regener 30(2):156–171. https://doi.org/10.1111/WRR.12994
33. McCoy JS, Nelson R (2024) Avulsion fractures. In: StatPearls. StatPearls Publishing. http://www.ncbi.nlm.nih.gov/books/NBK559168/
34. Sun H, Kim M, Kim J (2023) Salvage of a nasal chondrocutaneous flap after a traumatic avulsion injury using a combination of treatments polydeoxyribonucleotide injection, continuous non-rebreather mask oxygen therapy, and chemical leeching: a case report. Arch Aesth Plast Surg 29:50–54
35. Yang L, Guo J, He J, Shao J (2022) Skin grafting treatment of adolescent lower limb avulsion injury. Front Surg 9:38. https://doi.org/10.3389/fsurg.2022.953038
36. Innis CO (1957) Treatment of skin avulsion injuries of the extremities. Br J Plast Surg 10:122–140. https://doi.org/10.1016/S0007-1226(57)80020-4
37. Beal P (2011) Bouldering: movement, tactics, and problem solving. The Mountaineers Books, New York
38. Al-Ali MA, Abu-Zidan FM (2022) Auricular avulsion injuries: literature review and management algorithm. Turk J Emerg Med 22(2):59. https://doi.org/10.4103/2452-2473.342811
39. Allam KA, Abdelmegeed AG (2021) Retrospective study of management of near total avulsion of auricles and review of the literature. Egypt J Plast Reconstruct Surg 45(4):259–264
40. Herford AS, Cicciu M, Clark A (2009) Traumatic eyelid defects: a review of reconstructive options. J Oral Maxill Surg 67(1):3–9. https://doi.org/10.1016/j.joms.2008.09.009
41. Tint NL, Alexander P, Cook AE, Leatherbarrow B (2011) Eyelid avulsion repair with bi-canalicular silicone stenting without medial canthal tendon reconstruction. Br J Ophthalmol 95(10):1389–1392. https://doi.org/10.1136/bjo.2010.196246
42. Petrovic B, Marković D, Peric T, Blagojevic D (2010) Factors related to treatment and outcomes of avulsed teeth. Dental Traumatol 26(1):52–59. https://doi.org/10.1111/j.1600-9657.2009.00836.x
43. Pandhi D, Verma P (2012) Nail avulsion: indications and methods (surgical nail avulsion). Indian J Dermatol Venereol Leprol 78:299. https://doi.org/10.4103/0378-6323.95444
44. Long B, Liang SY, Gottlieb M (2023) Crush injury and syndrome: a review for emergency clinicians. Am J Emerg Med 69:180–187
45. Mohil RS (2012) Classification of wounds. Princip Pract Wound Care 1:42–52
46. Bowler PG (2002) Wound pathophysiology, infection and therapeutic options. Ann Med 34(6):419–427
47. Hollander JE, Singer AJ (1999) Laceration management. Ann Emerg Med 34(3):356–367
48. Davison AM (2004) The incised wound. Essentials of autopsy practice: recent advances, topics and developments. Springer, London, pp 187–220
49. Downer M, Berry CE, Parker JB, Kameni L, Griffin M (2023) Current biomaterials for wound healing. Bioengineering 10(12):1378
50. Pinto A, Russo A, Reginelli A, Iacobellis F, Di Serafino M, Giovine S, Romano L (2019) Gunshot wounds: ballistics and imaging findings. In: Seminars in ultrasound, CT and MRI, vol 40. WB Saunders, pp 25–35
51. Kiran JVK, Fereed H, Jadhav SK (2023) Medicolegal evaluation of fatal firearm injury cases in Ethiopia: an autopsy study. J Indian Acad For Med 45(1):24–27
52. McKinley WO, Johns JS, Musgrove JJ (1999) Clinical presentations, medical complications, and functional outcomes of individuals with gunshot wound-induced spinal cord injury. Am J Phys Med Rehabil 78(2):102–107
53. Richards JS, Stover SL, Jaworski T (1990) Effect of bullet removal on subsequent pain in persons with spinal cord injury secondary to gunshot wound. J Neurosurg 73(3):401–404
54. Lamb CM, Garner JP (2014) Selective non-operative management of civilian gunshot wounds to the abdomen: a systematic review of the evidence. Injury 45(4):659–666

55. Maiden N (2009) Ballistics reviews: mechanisms of bullet wound trauma. For Sci Med Pathol 5:204–209
56. Lv X, Zhao N, Long S, Wang G, Ran X, Gao J et al (2024) 3D skin bioprinting as promising therapeutic strategy for radiation-associated skin injuries. Wound Repair Regener 32(3):217–228
57. Iddins CJ, Christensen DM, Parrillo SJ, Glassman ES, Goans RE (2014) Management of ionizing radiation injuries and illnesses, part 5: local radiation injury. J Osteop Med 114(11):840–848
58. Jarrett DG, Sedlak RG, Dickerson WE, Reeves GI (2007) Medical treatment of radiation injuries—current US status. Radiat Measur 42(6–7):1063–1074
59. Hegedus F, Schwartz RA (2019) Cutaneous radiation damage: updating a clinically challenging concern. Giornale Italiano di Dermatologia e Venereologia: Organo Ufficiale, SocietaItaliana di Dermatologia e Sifilografia 154(5):550–556
60. Donnelly EH, Nemhauser JB, Smith JM, Kazzi ZN, Farfan EB, Chang AS, Naeem SF (2010) Acute radiation syndrome: assessment and management. South Med J 103(6):541
61. Christensen DM, Iddins CJ, Sugarman SL (2014) Ionizing radiation injuries and illnesses. Emerg Med Clin 32(1):245–265
62. Barry CL (2021) Surgical wound infections. Phys Assis Clin 6(2):295–307. https://doi.org/10.1016/j.cpha.2020.11.003
63. Zhang H, Zhao J, Farzan R, Alizadeh Otaghvar H (2024) Risk predictions of surgical wound complications based on a machine learning algorithm: a systematic review. Int Wound J 21(1):e14665. https://doi.org/10.1111/iwj.14665
64. Frykberg RG, Banks J (2015) Challenges in the treatment of chronic wounds. Adv Wound Care 4(9):560–582. https://doi.org/10.1089/wound.2015.0635
65. Bhat KS, Vishwesh BN, Sahu M, Shukla VK (2014) A clinical study on the efficacy of Panchavalkala cream in Vrana Shodhana wsr to its action on microbial load and wound infection. Int Quart J Res Ayurveda 35(2):135–140
66. Langer V (2014) Leg ulcers: an Indian perspective. Indian Dermatol Online J 5(4):535
67. Chi YW, Raffetto JD (2015) Venous leg ulceration pathophysiology and evidence based treatment. Vascul Med 20(2):168–181. https://doi.org/10.1177/1358863X14568677
68. Alavi A, Sibbald RG, Phillips TJ, Miller OF, Margolis DJ, Marston W, Woo K, Romanelli M, Kirsner RS (2016) What's new: management of venous leg ulcers: treating venous leg ulcers. J Am Acad Dermatol 74(4):643–664. https://doi.org/10.1016/j.jaad.2015.03.059
69. Raffetto JD (2018) Pathophysiology of chronic venous disease and venous ulcers. Surg Clin N Am 98(2):337–347
70. Vivas A, Lev-Tov H, Kirsner RS (2016) Venous leg ulcers. Ann Intern Med 165(3):ITC17–ITC32. https://doi.org/10.7326/AITC201608020
71. Lal BK (2015) Venous ulcers of the lower extremity: definition, epidemiology, and economic and social burdens. Semin Vasc Surg 28(1):3–5. https://doi.org/10.1053/j.semvascsurg.2015.05.002
72. Rice JB, Desai U, Cummings AKG, Birnbaum HG, Skornicki M, Parsons N (2014) Burden of venous leg ulcers in the United States. J Med Econ 17(5):347–356. https://doi.org/10.3111/13696998.2014.903258
73. Meaume S, Couilliet D, Vin F (2005) Prognostic factors for venous ulcer healing in a non-selected population of ambulatory patients. J Wound Care 14(1):31–34
74. Mayrovitz HN, Wong S, Mancuso C (2023) Venous, arterial, and neuropathic leg ulcers with emphasis on the geriatric population. Cureus 15(4):e38123. https://doi.org/10.7759/cureus.38123
75. Kamal R, Awasthi A, Pundir M, Thakur S (2024) Healing the diabetic wound: unlocking the secrets of genes and pathways. Eur J Pharmacol 975:176645. https://doi.org/10.1016/j.ejphar.2024.176645
76. Kim J (2023) The pathophysiology of diabetic foot: a narrative review. J Yeungnam Med Sci 40(4):328

77. Pastar I, Balukoff NC, Sawaya AP, Vecin NM, Tomic-Canic M (2024) Physiology and pathophysiology of wound healing in diabetes. In: Veves A, Giurini JM, Schermerhorn ML (eds) The diabetic foot: medical and surgical management. Springer, Berlin, pp 109–134
78. Peng Z, Lavigne J-P (2023) Editorial: pathogenesis of diabetic foot ulcers. Front Pharmacol 14:201. https://doi.org/10.3389/fphar.2023.1218201
79. Patel M, Patel V, Shah U, Patel A (2024) Molecular pathology and therapeutics of the diabetic foot ulcer; comprehensive reviews. Arch Physiol Biochem 130(5):591–598. https://doi.org/10.1080/13813455.2023.2219863
80. Raja JM, Maturana MA, Kayali S, Khouzam A, Efeovbokhan N (2023) Diabetic foot ulcer: a comprehensive review of pathophysiology and management modalities. World J Clin Cases 11(8):1684–1693
81. Chan P, Zhang J, Zhang R (2024) A review of research status of self-healing hydrogels for treating a diabetic wound. Theor Nat Sci 45:232–238
82. Jain N, Singh Y, Nouri A, Garg U, Pandey M (2024) Assessment of healing capacity of glucose-responsive smart gels on the diabetic wound: a comprehensive review. J Drug Deliv Sci Technol 93:105403. https://doi.org/10.1016/j.jddst.2024.105403
83. Zhou Q, Li X, Gao N, Ling G, Zhang P (2024) A multimodal therapy for infected diabetic wounds based on glucose-responsive nanocomposite-integrated microneedles. J Mater Chem B 12(4):1007–1021. https://doi.org/10.1039/D3TB02609D
84. Sharma A, Dheer D, Puri V, Alsayari A, Wahab S, Kesharwani P (2024) Insights of biopolymeric blended formulations for diabetic wound healing. Int J Pharm 656:124099. https://doi.org/10.1016/j.ijpharm.2024.124099
85. Liu Z, Lu L (2024) New advances in the treatment of diabetic wounds. Int J Biol Life Sci 6(1):1
86. Huang Y, Zhou W, Du H (2024) Association between hypertension and pressure ulcer: a systematic review and meta-analysis. Int Wound J 21(3):e14829. https://doi.org/10.1111/iwj.14829
87. Soppi E (2010) Pressure ulcer—occurrence, pathophysiology and prevention. Duodecim Laaketieteellinen Aikakauskirja 126(3):261–268
88. Rubayi S (2015) Etiology and pathology of pressure ulcers. In: Rubayi S (ed) Reconstructive plastic surgery of pressure ulcers. Springer, Berlin, pp 5–17
89. Verhun A, Krasnyi M, Verhun O, Makahonov I, Ivanykovych T (2023) Pressure ulcers (bedsores, decubitus ulcers): strategy of complex treatment in a palliative care department, author's views from the standpoint of clinical experience. J Dermatovenereol Cosmetol 2(50):16–30
90. Gefen A, Ousey K (2020) COVID-19: pressure ulcers, pain and the cytokine storm. J Wound Care 29(10):540–542
91. Bansal C, Scott R, Stewart D, Cockerell CJ (2005) Decubitus ulcers: a review of the literature. Int J Dermatol 44(10):805–810. https://doi.org/10.1111/j.1365-4632.2005.02636.x
92. Baldan-Martin M, Martin-Rojas T, Corbacho-Alonso N, Lopez JA, Sastre-Oliva T, Gil-Dones F, Vazquez J, Arevalo JM, Mourino-Alvarez L, Barderas MG (2020) Comprehensive proteomic profiling of pressure ulcers in patients with spinal cord injury identifies a specific protein pattern of pathology. Adv Wound Care 9(5):277–294. https://doi.org/10.1089/wound.2019.0968
93. Franco YAC, Camacho JRZ, Arizala JFC, Morales-García WC (2023) Pressure ulcers from a nursing perspective. Interam J Health Sci 3:157–157
94. Qian W, Li X, Yang M, Liu C, Kong Y, Li Y, Wang T, Zhang Q (2022) Relationship between antibiotic resistance, biofilm formation, and Biofilm-specific resistance in *E. coli* isolates from Ningbo, China. Infect Drug Resist 15:2865–2878. https://doi.org/10.2147/IDR.S363652
95. Xue C, Friedman A, Sen CK (2009) A mathematical model of ischemic cutaneous wounds. Proceed Natl Acad Sci 106(39):16782–16787. https://doi.org/10.1073/pnas.0909115106
96. Bhattacharya S, Mishra RK (2015) Pressure ulcers: current understanding and newer modalities of treatment. Indian J Plast Surg Off Publ Assoc Plast Surg India 48(1):4–16. https://doi.org/10.4103/0970-0358.155260

97. Patil P, Panarelli NC (2019) Educational case: ischemic disorders of the gut in adult patients. Acad Pathol 6:2374289519888709. https://doi.org/10.1177/2374289519888709
98. Soto JA, Lucey BC (2016) Emergency radiology: the requisites. Elsevier, Amsterdam
99. Frank U, Nikol S, Belch J, Boc V, Brodmann M, Carpentier PH, Chraim A, Canning C, Dimakakos E, Gottsäter A, Heiss C, Mazzolai L, Madaric J, Olinic DM, Pécsvárady Z, Poredoš P, Quéré I, Roztocil K, Stanek A, Terlecki P (2019) ESVM guideline on peripheral arterial disease. Vasa 48(102):1–79. https://doi.org/10.1024/0301-1526/a000834
100. Singh VP, Verma V (2018) Assessment and documentation of injuries. Recent Adv For Med Toxicol Good Pract Guidel Curr Medicol 2:145
101. Smith J, Greaves I (2003) Crush injury and crush syndrome: a review. J Trauma Acute Care Surg 54(5):S226–S230

Mechanistic Insights into the Biology of Wound Healing

Deepyaman Das

Abstract The mechanism via which the body generates new tissue to substitute for destroyed tissue is referred to as wound healing. A multitude of growth factors and cell types participate in this complex process. This chapter initially delineates the several types of wounds according the nature of cut formed like—Abrasions, Lacerations, Bruises, Punctures, Burns, Avulsions, Firearm injury. It subsequently examines the mechanistic process of wound healing, highlighting the significant function of growth factors. Accordingly, the significance of growth factors i.e. PDGF, FGF, TGF-β, EGF and IGF in wound healing have been discussed. Lastly, this chapter deliberates about the state-of-the-art research techniques that are being devised to help with wound healing in process. Such systematic description of the process of wound healing will help in understanding the biology behind wound healing and help devise newer methodologies for wound healing in future.

Keywords Wound healing · Growth factors · Haemostasis

1 Introduction

A wound is any interruption in the skin's continuity. "Disruption of normal anatomic structures and function" is one definition of it. Wounds remain a challenging clinical concern in daily life, with both immediate and delayed effects often leading to serious complications [1]. The function of the skin barrier is severely compromised by wounds because the skin acts as a barrier against disease entry, radiation or heat stress, and physical and chemical hazards. A superficial skin wound is almost never sterile because the wound site is always under microbial pressure. Treatment of wound infections can vary from simple to nearly impossible, depending on the wound's surroundings, depth, and size in addition to the patient's immunocompetence. Enhancing and elucidating local antimicrobial defence mechanisms is a primary objective of research in wound-healing due to the substantial influence of

D. Das (✉)
Department of Zoology, Raiganj University, Raiganj, Uttar Dinajpur, India
e-mail: deepyaman.das@gmail.com

A. K. Mandal et al. (eds.), *Functionalized Nanoparticles Hydrogels for Wound Healing*, Smart Nanomaterials Technology, https://doi.org/10.1007/978-981-95-0720-7_4

microbial colonisation. The cornerstone of wound research involves elucidating the essential mechanisms of the healing procedure and subsequently regulating these mechanisms to enhance healing speed or avert negative outcomes such as infection or scarring. Although the natural healing process is essentially understood, the most recent molecular tools and methodologies must be used to characterise the sequence of activities at an elevated resolution in order to fully comprehend the extremely complex features involved [2].

Despite decades of research, the molecular pathways underlying skin wound healing remain unclear, and the majority of our expertise at the molecular level has come from studying different animal models. Different biological processes are always species specific because of the ways that anatomical and physiological distinctions across species impact healing mechanisms. Therefore, additional methods for directly studying human wound healing are required to prevent erroneous interpretations resulting from species difference [2]. Here, we outline the benefits and drawbacks of various methods for characterising human wound healing in a medical context. This chapter discusses on the biology behind wound healing and the modern research that are taking place in this field.

2 Types of Wounds

Lacerations

Deep cuts to the skin are called lacerations. Usually, a cut from a knife, glass fragments, or another sharp instrument causes them. Mild cuts can occasionally be healed with stitches, while severe wounds can need more. The wound should be cleansed and bandaged as soon as the bleeding stops [3].

Abrasions

Mild abrasions caused by rubbing or scraping the skin against a hard surface can usually be treated at home; however, abrasions near the eyes, showing signs of infection, or covering a large area of the body may necessitate medical treatment [4].

Contusions/Bruises

When blood vessels in the skin or internal organs burst, this kind of injury happens. A bruise recovers through devastation and removal of the blood that has been extravasated [5].

Punctures

Usually caused by knives, broken glass bottles, and tools, this kind of damage results from the penetration of a pointed or sharp item or weapon deeper than its length into the body. Some puncture wounds are minor and don't require extensive medical care, but if the wound is deep and challenging to clean, it becomes susceptible to bacteria and infection. These are tiny, deep holes in the skin that are made by a pointed item, such as a nail [5].

Burns

This is tissue damage brought on by the sun, heat, radiation, chemicals, or electricity. Although they are susceptible to infection, first-degree and the majority of second-degree burns may typically be treated at home. Any burn that is more severe than a second degree might be fatal and has to be treated by a professional right away [5].

Avulsions

Severe wounds that require the removal of multiple layers of skin, sometimes revealing visible muscle layers or fat underneath, are rarely treatable at home. Unless the injury is <0.25 inches by 0.25 inches, the patient should seek medical attention right away and take any removed skin to the medical centre [5].

Firearms Injury

Depending on the projectile, muzzle velocity, distance, firing angle, and body part involved, different firearms can cause different injuries. These wounds can be further classified as entry wounds, which occur when a gunshot enters the body through the skin, and exit wounds, which occur when the bullet leaves the body. In the entering wound, the skin edges are inverted, while in the exit wound, they are everted [6].

Figure 1 illustrates various types of wounds in human.

3 Wound Healing: A Magic Wand for Damaged Tissue

The repair of a wound is a standard biological reaction to tissue injury; However, it's a complex phenomenon that entails numerous distinct cell types, cytokines, mediators, and the vascular system. In order to stop bleeding, a series of inflammatory cells, including neutrophils, enter the body and release a variety of mediators and cytokines that aids first in angiogenesis secondly in thrombosis, and lastly in reepithelialisation. Extracellular components that will serve as scaffolding are then deposited by fibroblasts [5].

4 Steps in Wound Healing

The complex wound-healing process consists of four stages: haemostasis, inflammation, proliferation, and remodelling. Soon after an injury, haemostasis occurs, using clotting factors to prevent further blood loss from the site of the wound and to create the structural foundation for future granulation tissue creation [7]. Phagocytic cells that generate reactive oxygen species are involved in the subsequent inflammatory phase, which can last up to seven days in acute wounds but longer in persistent

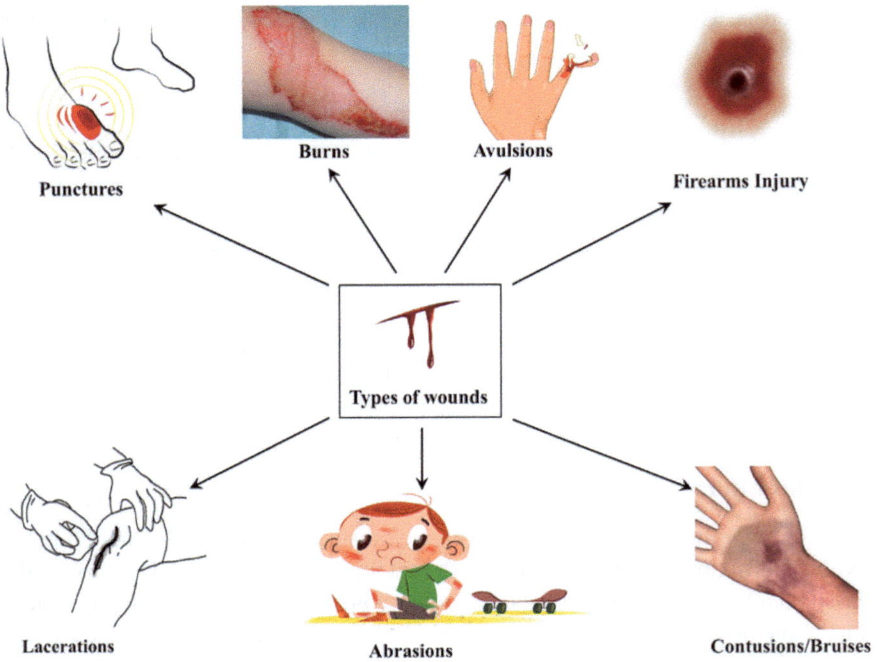

Fig. 1 Types of wounds

wounds. As inflammatory cells die, wound healing progresses to the proliferative phase, which is characterised by the formation of granulation tissue, angiogenesis (the creation of blood vessels), wound contraction, and the epithelialisation process. The final phase of remodelling, characterised by the formation of scar tissue, may occur across several months or years [7]. The steps of wound healing have been delineated in Fig. 2.

Haemostasis

The preliminary stage of wound healing, termed haemostasis, may last for as long as two days. The vessels of blood in the injured region constrict immediately upon sustaining a wound, thus lowering the blood flow. We call this vasoconstriction. The release of clotting factors at the injury site and their congealing with fibrin results in a thrombus, commonly known as a blood clot. The clot acts as a wall between the broken blood vessels to prevent blood loss [7].

Inflammation

The second stage of wound healing is known as the inflammatory phase. Phagocytic cells release reactive oxygen compounds that can persist for up to seven days in acute wounds and even longer in chronic wounds. To avoid infection, white blood cells and specific enzymes enter the wound area during this phase, removing germs and debris and creating a suitable environment for new tissue growth. The phase's

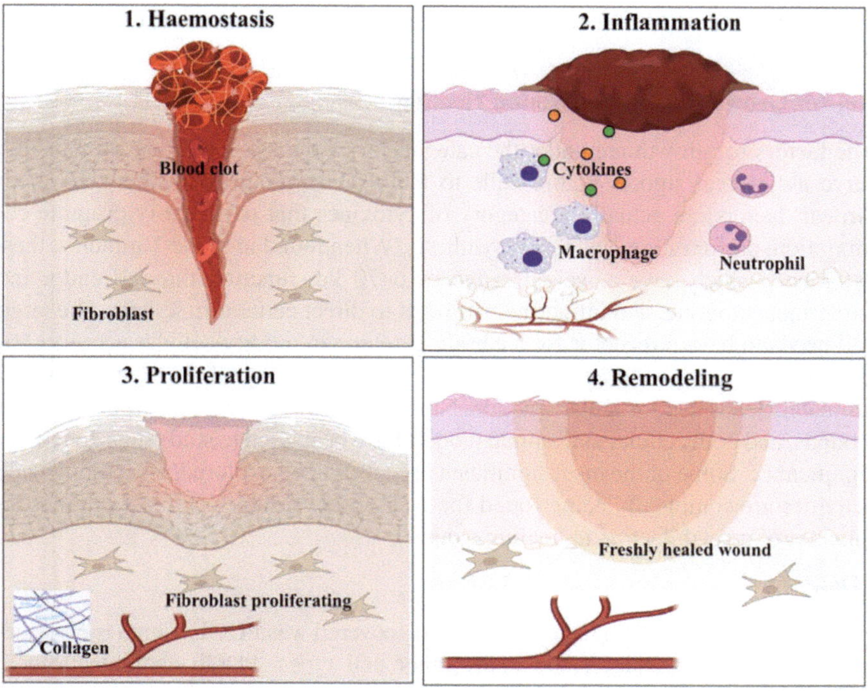

Fig. 2 Steps in wound healing

physical characteristics encompass pain, swelling, redness, and inflammation at the injury site [8].

Proliferation

The focus of the third phase of wound healing, known as the Proliferative Phase, is on filling and covering the wound. As inflammatory cells die, wound healing progresses to the proliferation phase, which is characterised by the formation of tissue granulation, angiogenesis (the creation of blood vessels), contraction of the wound, and the epithelialisation process. The new tissue usually looks red or pink due to the presence of inflammatory chemicals. The production of collagen proteins by fibroblasts, a kind of cell found in connective tissue, controls the rate of tissue regeneration. This stage of healing a wound could last anywhere from four days through three weeks or more [8].

Remodelling

Scar tissue formation is a hallmark of the final remodelling phase, also known as maturation. It may take months or years, depending on the location of the wound, its original severity, and available treatment choices. This stage involves the gradual increase in strength and flexibility of the regenerated tissue. Collagen synthesis helps to keep the skin supple and strong under stress. Collagen builds up in granulation tissue to form scarring tissue, that is 20% weaker yet more elastic than the skin before the injury [9].

Role of Growth Factors in Wound Healing

The factors for growth are under the category of cytokines, which are proteins that serve as internal signals within cells to facilitate communication between them. Growth factors are actually a category of cytokines that particularly promote cell migration, proliferation, and tissue synthesis. When found in modest amounts, these cytokines, which have a weight range of 6–70 kd, function through endocrine, paracrine, autocrine, and intracrine pathways to direct cellular functions. The target cell needs to have a receptor for a certain cytokine to modify cellular activity [10]. Following receptor binding, a sequence of intracellular signals is triggered, which ultimately leads to a particular reaction. Numerous growth factors have been found in wounds, and both cellular and animal models have been used to study their functional significance. Some of the most significant aspects will be explored, even though new activities are continually being found for each of the variables. Table 1 outlines the function of growth factors in healing wounds.

PDGF

One of the earliest growth factors that was discovered was PDGF, which is produced by platelets, macrophages, and many other cell types. PDGF chemotactic cells include smooth muscle cells, fibroblasts, and maybe neutrophils and monocytes. PDGF is a mitogen, meaning it promotes the growth of smooth muscle cells and fibroblasts. It has a significant impact on the extracellular matrix as well. The PDGF family of proteins includes at least two more proteins. A strong angiogenic factor, vascular endothelial growth factor (VEGF) shares similarities with PDGF [11].

FGF

There are now at least nine members of the fibroblast growth factor family. FGF is not only mitogenic for fibroblasts but also exhibits strong angiogenic properties. One of the more recently discovered fibroblast growth factors, which has been demonstrated to have strong stimulatory impact on the proliferation of keratinocyte, is keratinocyte growth factor (KGF or FGF-7) [12].

TGF-β

TGF-beta1 is the most well-characterized growth factor; it has been shown to have a variety of effects based on the environment and cell type. TGF-beta1 prevents the breakdown of collagen and is a very strong stimulator for collagen deposition. Research has indicated that inhibiting TGF-beta1 may reduce the formation of scars. It is interesting to note that TGF-beta3 may prevent scarring on its own. Additionally, TGF-beta1 plays a critical role in controlling inflammation. TGF-beta1-deficient animals perish from an excessive inflammatory reaction. In addition to being crucial for inhibiting the proliferation of numerous cell types, TGF-beta has been linked to the development of cancer when improperly controlled [13].

Table 1 Role of growth factors in wound healing

Name of the growth factor	Name	Source	Function in wound healing
Platelet-derived growth factor	PDGF	Platelets, macrophages, endothelial cells and injured cells	Chemotaxis, collagenase production, fibroblast proliferation
Recombinant human granulocyte macrophage colony stimulating factor	Rh-GM-CSF	Macrophages, mast cells, T cells, fibroblasts and endothelial cells	Heals Chronic Venous Ulcers. It stimulates the recruitment and activation of macrophages, which are essential for clearing debris and promoting angiogenesis
Recombinant human granulocyte colony-stimulating factor	Rh-G-CSF	Macrophages and endothelial cells	Stimulates the production of neutrophils for the initial inflammatory phase of wound healing. Çan heal second degree burns
			Can heal deep diabetic foot ulcers
Recombinant human platelet-derived growth factor-BB	Rh-PDGF-BB	Activated platelets	Can heal deep diabetic foot ulcers
			Can heal stage III or IV Pressure ulcers
Recombinant human vascular endothelial growth factor	Rh-VEGF	Keratinocytes, fibroblasts, macrophages, platelets, and endothelial cells	Can heal Diabetic foot ulcers
Recombinant human basic fibroblast growth factor	Rh-bFGF	Fibroblasts	Can heal stage III or IV pressure ulcers
Transforming growth factor beta	TGF-β	Macrophages, platelets, neutrophils, lymphocytes, fibroblasts, epithelial and endothelial cells, injured cells	Fibroblast proliferation, chemotaxis, collagen metabolism
Epidermal growth factor	EGF	Plasma, platelets, macrophages, epithelial cells	Epithelial cell proliferation, granulation tissue formation
Transforming growth factor alpha	TGF-α	Activated macrophages, platelets, epithelial cells, injured cells, macrophages, T lymphocytes	Epithelial cell proliferation, granulation tissue formation, fibroblast proliferation

(continued)

Table 1 (continued)

Name of the growth factor	Name	Source	Function in wound healing
Keratinocyte growth factor	KGF	Fibroblasts epithelial cell proliferation	IL-1 macrophages fibroblast proliferation
Fibroblast growth factor	FGF	Pituitary, macrophages, fibroblasts, endothelial cells	Fibroblast proliferation, matrix deposition, wound contraction angiogenesis
Insulin-like growth factor-1	IGF-1	Plasma, liver, fibroblasts	Synthesis of sulfated proteoglycans and collagen, fibroblast proliferation
Interferons	IFNs	Lymphocytes, fibroblasts	Inhibition of fibroblast proliferation and collagen synthesis
Insulin-like growth factor	IGF	Liver	Mitogen for fibroblasts, stimulates lymphocytes
Fibroblast growth factor 7	KGF or FGF-7	Mesenchymal cells like fibroblasts	Promotes the proliferation and migration of epithelial cells in the re-epithelialization phase of wound repair
Vascular endothelial growth factor	VEGF	Platelets, macrophages, keratinocytes, renal mesangial cells, and tumor cells	Heal by stimulating angiogenesis, collagen deposition, and epithelialization; endothelial cell migration
Transforming growth factor beta-1	TGF-β1	Platelets, monocytes, macrophages, endothelial cells, and fibroblasts	Stimulant for collagen deposition, inhibits collagen breakdown and regulation of inflammation
Transforming growth factor beta-3	TGF-β3	Fibroblasts	Associated with promoting scar-less wound healing by regulating the deposition of extracellular matrix proteins

EGF

According to research, EGF is essential for the repair of the gut mucosa and may aid in the prevention and management of ulcers. Though it may be more powerful than EGF, Transforming Growth Factor-alpha, is entirely distinct from TGF-beta, is structurally similar to EGF [14].

IGF-I and IGF-II

IGF-I has a significant impact in promoting development, particularly the minor growth traits of puberty. It is also crucial for encouraging the synthesis of proteins. Although IGF-I stimulates the growth of fibroblasts and other cell types, it typically necessitates the inclusion of another growth factor, such as FGF or PDGF. IGF-II has been demonstrated to have effects similar to those of IGF-I and can also aid in healing, despite being believed to be more active in foetal growth [15].

5 Modern Research in Wound Healing

Due to the diverse molecular and cellular pathways involved in different types of wounds, there is currently no single therapy technique that can effectively promote complete wound healing. In the realm of wound healing research, recent technological breakthroughs and progress have greatly enhanced our comprehension of evaluation and treatment methodologies. The advancement in this field has been completely transformed by the paradigm shift from dry dressings to moist wound dressings, stimuli-responsive wound dressings, growth factor-based therapy, tissue-engineered skin, bioengineered human skin substitutes, gene therapy, nanotherapeutics, and stem cell therapy [16]. Advancements in cutting-edge technology, such as platelet-rich plasma (PRP) therapy, 3D bioprinting, and ECM-based methods, have enabled personalized wound healing. Nevertheless, there are still challenges in accurately diagnosing and effectively treating wounds that do not heal. In order to promote effective wound healing, it is necessary to adopt more recent and innovative approaches that consider cost-effectiveness, benefit-to-risk, and efficacy [16].

Various cellular therapies for wound healing have been identified in the recent times which could advance would healing procedure. T-regs are a diverse fraction of CD4+ helper T-cells, which promote self-tolerance, reduce overreactive immunological responses, and inhibit autoimmunity to preserve immune tissue homeostasis [16]. T-regs could be combined with chimeric antigen receptors to create a new type of treatment for reducing rejection in human skin grafts, which could potentially help with wound healing.

Stem cells have also been implicated in tissue repair. In general, pluripotent stem cells (pscs) can be categorized into two groups: induced pluripotent stem cells (iPSCs) and embryonic stem cells (ESCs). The core cell mass of a blastocyst develops into ESCs, which are PSCs capable of differentiating into any of the three germ layers' tissues. [17]. However, iPSCs have replaced ESCs as their functional substitute due to the controversial ethical issues surrounding ESC research and its practical application in therapy. The process of immunoregulation, angiogenesis, cell proliferation, and extracellular matrix (ECM) deposition can be supported by iPSC-derived cells. Grafix, a cryopreserved amniotic membrane with a 3D-ECM scaffold comprising growth factors, fibroblasts, endothelial cells, and MSCs, is currently the only FDA-approved stem cell-based treatment for chronic wounds. Even in cases where diabetic foot ulcers do not respond to treatment, Grafix therapy is considered safe and effective, as demonstrated by two clinical trials. However, further investigation is needed to determine the efficacy of this method in promoting wound healing [17].

6 Conclusion

This chapter explores the different categories of wounds based on the shape of the cut. Then it delves into the process of wound healing, providing a brief overview of the role of growth factors in this intricate process. The future of wound healing is progressing towards more tailored, effective, and innovative treatments, integrating biological therapies, advanced technologies, and a comprehensive comprehension of the underlying mechanisms. As these advancements progress, the expectation is for quicker healing periods, reduced complications, and the preservation of tissue function with minimal scarring.

References

1. Gonzalez AC, Costa TF, Andrade ZA, Medrado ARAP (2016) Wound healing: a literature review. Ann Bras Dermatol 91:614–620. https://doi.org/10.1590/abd1806-4841.20164741
2. Nuutila K, Katayama S, Vuola J, Kankuri E (2014) Human wound-healing research: issues and perspectives for studies using wide-scale analytic platforms. Adv Wound Care 3:264–271. https://doi.org/10.1089/wound.2013.0502
3. Hollander JE, Singer AJ (1999) Laceration management. Ann Emerg Med 34:356–367. https://doi.org/10.1016/S0196-0644(99)70131-9
4. Shrestha R, Krishan K, Ishaq H, Kanchan T (2025) Abrasion. In: Treasure island (FL)
5. Abazari M, Ghaffari A, Rashidzadeh H, Badeleh SM, Maleki Y (2022) A systematic review on classification, identification, and healing process of burn wound healing. Int J Low Extrem Wounds 21:18–30. https://doi.org/10.1177/1534734620924857
6. Kaufman EJ, Richmond TS, Hoskins K (2023) Youth firearm injury: a review for pediatric critical care clinicians. Crit Care Clin 39:357–371. https://doi.org/10.1016/j.ccc.2022.09.010
7. Almadani YH, Vorstenbosch J, Davison PG, Murphy AM (2021) Wound healing: a comprehensive review. Semin Plast Surg 35:141–144. https://doi.org/10.1055/s-0041-1731791
8. Landén NX, Li D, Ståhle M (2016) Transition from inflammation to proliferation: a critical step during wound healing. Cell Mol Life Sci 73:3861–3885. https://doi.org/10.1007/s00018-016-2268-0
9. Chester D, Marrow EA, Daniele MA, Brown AC (2019) Wound healing and the host response in regenerative engineering. In: Narayan R (ed) Encyclopedia of biomedical engineering. Elsevier, Oxford, pp 1–12
10. Steed DL (1997) The role of growth factors in wound healing. Surg Clin North Am 77:575–586. https://doi.org/10.1016/S0039-6109(05)70569-7
11. Pierce GF, Mustoe TA, Altrock BW, Deuel TF, Thomason A (1991) Role of platelet-derived growth factor in wound healing. J Cell Biochem 45:319–326. https://doi.org/10.1002/jcb.240450403
12. Farooq M, Khan AW, Kim MS, Choi S (2021) The role of fibroblast growth factor (FGF) signaling in tissue repair and regeneration. Cells 10:242. https://doi.org/10.3390/cells10113242
13. Ramirez H, Patel SB, Pastar I (2014) The role of TGFβ signaling in wound epithelialization. Adv Wound Care 3:482–491. https://doi.org/10.1089/wound.2013.0466
14. Shin SH, Koh YG, Lee WG, Seok J, Park KY (2023) The use of epidermal growth factor in dermatological practice. Int Wound J 20:2414–2423. https://doi.org/10.1111/iwj.14075
15. Garoufalia Z et al (2021) Insulin-like growth factor-I and wound healing, a potential answer to non-healing wounds: a systematic review of the literature and future perspectives. Biomed Rep 15:66. https://doi.org/10.3892/br.2021.1442

16. Davis M, Hom D (2023) Current and future developments in wound healing. Facial Plast Surg 39:477–488. https://doi.org/10.1055/s-0043-1769936
17. Kolimi P, Narala S, Nyavanandi D, Youssef AAA, Dudhipala N (2022) Innovative treatment strategies to accelerate wound healing: trajectory and recent advancements. Cells 11:439. https://doi.org/10.3390/cells11152439

Metallic Nanoparticles for Wound Healing Management: A Promising Therapeutic Approach

Manab Deb Adhikari and Bipransh Kumar Tiwary

Abstract Wounds are a most important medical concern, and they are getting more frequent as a result of burns, surgeries, traumas, and long-term illnesses like diabetes. Chronic wounds pose a significant burden on healthcare systems worldwide, impacting ~6.5 million individuals yearly, resulting in substantial economic expenses for treating and managing wounds, necessitating innovative and efficient wound dressing materials and formulations. Recent progress in nanotechnology has resulted in the development of metal nanoparticles (MNPs) as a promising therapeutic approach for wound healing because of their unique physical, chemical, and optical characteristics. Nanoparticles can change the wound environment from one that is not healing to one that is healing through their antibacterial, anti-inflammatory, and angiogenic properties. Biocompatible metallic nanoparticle-loaded wound dressing provides a revolutionary way to successfully overcome the inherent disadvantages of standard wound dressing materials. This chapter explores the wound-healing process and its basic mechanisms. Documented the efficiency of metal nanoparticles, including silver, gold, zinc oxide, and copper, in enhancing wound healing through their antibacterial, anti-inflammatory, and tissue-regenerative properties. Additionally, the chapter discusses the challenges associated with the use of metal nanoparticles in wound healing and outlines future perspectives on their applications.

Keywords Wound · Chronic wounds · Wound healing · Metal nanoparticles · Silver nanoparticles · Gold nanoparticles · Copper nanoparticles · Zinc oxide nanoparticles

---

M. D. Adhikari
Department of Biotechnology, University of North Bengal, Darjeeling, Siliguri, West Bengal, India

B. K. Tiwary (✉)
Department of Microbiology, North Bengal St. Xavier's College, Rajganj, West Bengal, India
e-mail: bipra.tiwary@gmail.com

A. K. Mandal et al. (eds.), *Functionalized Nanoparticles Hydrogels for Wound Healing*, Smart Nanomaterials Technology, https://doi.org/10.1007/978-981-95-0720-7_5

1 Introduction

Wounds are sharp injuries to the skin induced by external stimuli or trauma, which cause cuts, punctures, or tears in the dermal layers, and promote the spread of infections. Wound healing is a complex process that involves many different types of specialized cells, including platelets, macrophages, fibroblasts, epithelial, and endothelial cells [1–3]. These cells communicate with one another and with the extracellular matrix. In addition to cellular connections, proteins and glycoproteins such as cytokines, chemokines, growth factors, inhibitors, and receptors all have an impact on healing [1, 4]. Each stage of wound healing has specific milestones that must be met in order for normal healing to continue. Wounds are often categorized into two kinds based on their healing time. Acute wounds are those that heal within a predictable time frame and restore anatomical and functional tissues. These wounds are often caused by traumatic tissue loss or surgical operations. Chronic wounds are those that do not heal within a predicted time frame, resulting in further complications and diseases. Chronic illnesses like diabetes, obesity, and vascular dysfunction have significantly grown recently, leading to a rise in the number of individuals with chronic wounds [5, 6]. Chronic wounds are a major concern, particularly among diabetic patients, and have a considerable impact on healthcare expenses. Bacterial biofilms, which are bacterial colonies encased in an extracellular polymeric substance that the bacteria produce for protection, are a characteristic of chronic wounds [7–9]. It has been demonstrated that wound-associated biofilms can cause inflammatory cytokines, reactive oxygen species, and apoptosis, contributing to persistent inflammation and preventing re-epithelialization. Moreover, because of their persistence and resistance to standard antibiotic treatment, microorganisms within biofilms pose a therapeutic challenge [10, 11].

Millions of people worldwide suffer from poor wound healing, increasing mortality rates and associated costs [12, 13]. The three main issues associated with wounds are: (i) the lack of a proper environment for cell migration, proliferation, and angiogenesis, (ii) microbial infection, and (iii) unstable and prolonged inflammation. Additionally, by developing a biofilm, pathogenic bacteria can cause healing delays, treatment issues due to antibiotic resistance, and potentially fatal outcomes. Burns, surgical incisions and cuts, bruises, and lacerations from trauma are the most common types of wounds. Most of these wounds heal within three months due to the organism's self-healing properties. However, in some situations, such as an uncontrolled infection, acute wounds can progress to chronic wounds that last months or even years [14, 15].

Wound healing has long been a challenging issue for clinicians, and novel materials and treatments are urgently required. The global advanced wound management products market reached \$10.7 billion in 2022 and is expected to reach \$15.3 billion by 2028, with a compound annual growth rate (CAGR) of 6.2% from 2023 to 2028 [16]. The ideal healing system should speed up wound closure, decrease infection, activate healing mechanisms, imitate extracellular matrix features, moisturize the wound, and minimize scar formation.

Nanotechnology has significantly developed in this century. Nanomaterials produced via the teamwork of many disciplines have been used in many areas of life, including agriculture, electronics, the environment, and medicine [17, 18]. Nanomaterials can promote a variety of cellular and molecular processes that aid in the wound microenvironment, including antibacterial, anti-inflammatory, and angiogenic actions, potentially shifting the environment from nonhealing to healing. Nanomaterials have recently been extensively used in wound healing due to their superior adsorption capacity, antibacterial characteristics, and medication loading [19–21].

Metal and metal oxide nanoparticles such as gold, silver, copper, titanium, zinc oxide, magnesium oxide, and their derivatives, are known to have antibacterial effects and are widely used in the design of novel wound dressings [22–24]. These nanoparticles efficiently inhibit infection and accelerate wound healing when added to scaffolds or wound dressings. The nanoparticles may directly contribute to wound healing through involvement in the repair process or serve as delivery systems. The combination of 3D printing and nanotechnology offers enormous transformational prospects in the field of wound rehabilitation. By incorporating AgNPs into 3D-printed structures, researchers were able to create wound dressings and scaffolds that were tailored to individual wounds while also increasing their antibacterial effectiveness [25]. The integration of silver nanoparticles with chitosan-based nanofibers exhibits significant potential for use in wound healing [26].

In this chapter, we begin with a concise overview of the wound healing process, highlighting antibacterial properties, angiogenesis, and immunomodulation mechanisms. Next, we have discussed how metal nanoparticles are used in wound healing, along with recent developments in innovative compositions and formulations for effective wound dressings that include gold, silver, copper, and zinc oxide nanoparticles. Finally, we have discussed the challenges, limitations, and future perspectives in this field.

2 Wound Healing: Processes and Mechanism

Wound healing is one of the most important biological processes which relies on and self-adjusts to intercourse of multiple external and internal variables. Various methods are used to classify wounds based on factors such as cleanliness, wound condition, and healing time. The CDC classifies wounds into four groups according to the degree of infection and cleanliness. (a) Clean wounds are those that are closed, and devoid of inflammation and infection. (b) Clean-contaminated, indicating a low level of contamination. These wounds require entrance into the respiratory, alimentary, vaginal, or urinary tracts, but only under specific conditions. (c) Infected: fresh, open, and free of inflammation and exudate. These types of wounds are incisions caused by acute or non-purulent inflammation, (d) Highly infected wounds: these wounds, which frequently result from improper treatment of traumatic wounds, include the most common pathogens. This is frequently caused by surgery or germs discovered in punctured organs [27;Gorvetzian et al. 2018).

A controlled cascade of biochemical reactions is triggered when the skin is injured in order to heal the damage. It is vital to anticipate the needs of injured tissue and meet the expectations for regeneration. Wound healing is facilitated by a complex network of cell types, chemokines, and growth factors that work together [28]. The wound-healing process is divided into four phases: haemostasis, inflammation, proliferation, and remodeling [29]. The schematic representation of the wound healing process is depicted in Fig. 1.

Hemostasis Phase

Within a few minutes of spraining, the very first phase, hemostasis is the stopping of blood leaking out of the injured area alongside making a barrier [30]. Hemostasis tries to keep the site disorder from losing sanguine supply and abnormal damage. Several myocytes come together and trigger vasoconstriction, primary and secondary hemostasis, all in a race to suppress blood circulation [1, 31]. First, the primary step starts with Platelets aggregating at the injury site and then releasing thrombin pro-coagulation which changes fibrinogen into fast-acting fibrin capturing the location in a dramatic, fibrous net instantly thereafter. As the final platelets make fibrous doom, pro-coagulate thrombin helps to slack blood inflow into the site and causes local inflammation [32] (Fig. 1).

Inflammatory Phase

After hemostasis, the inflammation phase starts characterized by the formation of inflammatory mediators such as prostaglandins and histamine. These chemicals increase vascular permeability and are responsible for a local vasodilation that provides an avenue through which the infiltrating immune cells reach the location of injury. Neutrophils are some of the first responders through phagocytosis kill invading pathogens [32]. Platelets add their share by releasing a variety of growth factors that help in the recruitment of immune cells, aid in tissue repair, and boost immunity against infections. Macrophages then take over to remove apoptotic cells, control the immune response, and release cytokines that help in the regeneration of tissue. Other immune cells, like mast cells and leukocytes also help in wound healing by regulating inflammation and preparing the site for tissue remodeling. Clinically, this phase is often associated with redness, heat, swelling, and pain—classic signs of an inflammatory response [34–36].

Proliferation Phase

It leads to the proliferative stage after the inflammatory phase, where the cellular activity shifts towards tissue regeneration. At this point, there's deposition of extra-cellular matrix (ECM) as well as generation of new connective tissue richly made up of collagen. Here, fibroblasts tend to differentiate to myofibroblasts [37]. The effect is contraction through forces applied which close the edges. In concert with these happenings, the newly established vascular plexi from angiogenesis assure sufficient blood supply of the cells by their growth to allow more oxygenation and nutrient influx [38]. Fibroblasts continuing in their process build the tissues' elasticity

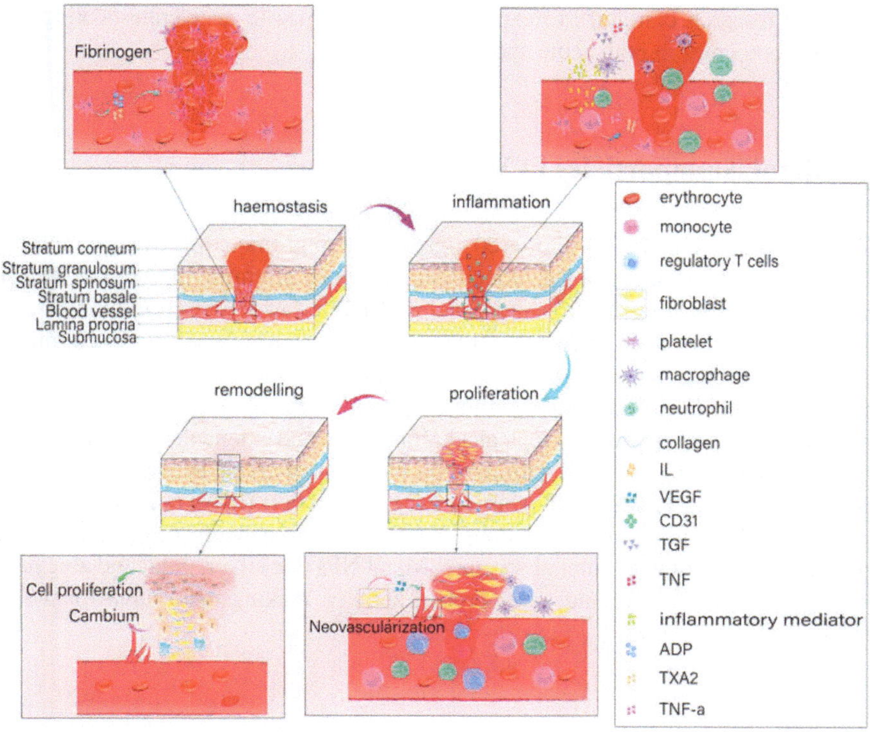

Fig. 1 The process of wound healing. (Copyright © 2024 Zhao et al. [33]. Published by Frontiers under the terms of the Creative Commons Attribution License (CC BY)

via elastin deposition. The concerted actions of both fibroblasts and myofibroblasts significantly enhance the closure of wounds and tissue integrity [39].

Maturation (Remodeling) Phase

During the maturation phase, structural refinement and tissue strengthening of newly formed tissue is completed. The process eliminates surplus fibroblasts and immune cells via apoptosis. Moreover, collagen cross-linking will also improve tensile strength within the repaired skin. Type III collagen, that initially was deposited at earlier stages will be progressively substituted by Type I collagen which improves strength and durability [40]. This period can last up to a few weeks or months, depending on the wound's severity and the healing capacity of the individual. By the maturation phase, the key objective is to restore the skin as much as possible in terms of function and structural anatomy to its original state.

Wound healing is a highly coordinated process involving multiple cellular and molecular mechanisms. Understanding the complex interplay between hemostasis, inflammation, proliferation, and maturation is essential for optimizing therapeutic interventions aimed at enhancing tissue repair and minimizing complications.

Ongoing research continues to explore novel strategies to accelerate healing and improve outcomes in both acute and chronic wounds.

3 Metal Nanoparticles in Wound Healing

Recently, nanotechnology has offered a great way to promote appropriate mobility during the various stages of the wound-healing process, improving both acute and chronic wound healing. Metal nanoparticles (NPs) are widely used in biomedicine because to their lots of advantages, including easy synthesis with predetermined shapes and sizes, superior surface to-volume ratio, surface functionalization, improved biocompatibility, and superior physical–chemical properties. These physiochemical properties promote topical administration and extend the half-life of therapeutic drugs when combined with the nanoparticles, resulting in fewer administrations and lower costs. Metal nanoparticles can target microRNAs in aberrant wound healing, such as infectious conditions, by encapsulating them, protecting charge groups and permitting cellular absorption. Metal NPs like silver, gold, and zinc have excellent wound healing and antimicrobial activity, making them ideal for use in dressings.

3.1 Gold Nanoparticles

Gold nanoparticles (AuNPs) have garnered significant attention in the realm of regenerative medicine particularly for wound healing due to their biocompatibility with and anti-inflammatory properties [41–44]. Among precious metal nanoparticles, AuNPs are considered one of the most promising materials because of their multifunctionality and relatively low cytotoxicity, although their high cost may limit widespread adoption. Researchers are continually exploring innovative approaches to harness the potential of AuNPs in wound care despite this limitation [45].

The unique properties of AuNPs make them a rational choice for wound healing applications. AuNPs exhibit anti-inflammatory effects promote angiogenesis and support regeneration of damaged tissue. Studies have demonstrated that AuNPs may enhance the secretion of cytokines such as interleukin-10 (IL-10) and interleukin-4 (IL-4), as well as growth factors like fibroblast growth factor (FGF) and T AuNPs can significantly increase collagen expression and vascular endothelial growth factor (VEGF) levels, thereby leading to rapid wound closure [46, 47]. For example, Chen et al., designed a combination of antioxidants with AuNPs which demonstrated significant efficacy in healing diabetic skin wounds. This combination not only regulated angiogenesis but also expressed anti-inflammatory and shortened the healing cycle of diabetic ulcer wounds [48]. Similarly, Kim et al. [46] it has been reported that hydrocolloid membranes coated with AuNPs accelerated wound healing by enhancing tissue regeneration. Unlike silver nanoparticles (AgNPs), AuNPs lack intrinsic antibacterial activity which is crucial in preventing wound infections. To address this limitation AuNPs are often hybridized with biomacromolecules such as chitosan gelatin, dextran polyvinyl alcohol (PVA), poly(lactic-co-glycolic acid) (PLGA) and poly(caprolactone). These combinations enhance their antibacterial properties and accelerate the process of wound healing [49–52]. For example, the supramolecular hydrogel wound dressing incorporating AuNPs (Au@hydrogel) has been designed and prepared which demonstrates excellent antibacterial activity and biocompatibility. In vitro studies confirmed that this dressing increased wound healing effectiveness while maintaining excellent cell compatibility making it safe for in vivo also [53]. In a study by Volkova et al. [54] AuNPs demonstrated antioxidative and antimicrobial properties, proving effective in regenerating damaged collagen tissues and promoting faster wound closure. Another example involves the development of an AuNPs-integrated hydrogel with epigallocatechin gallate and -lipoic acid which exhibited potent antioxidant and anti-inflammatory activities [45, 55, 56]. AuNPs contribute to wound healing by modulating inflammatory responses. They enhance the secretion of vascular endothelial growth factor (VEGF), tumor necrosis factor- alpha (TNF-), interleukin-8 (IL8) and interleukin-12 (IL-12) which collectively play a role in the resolution AuNPs' anti-inflammatory properties are particularly beneficial for chronic wounds where prolonged inflammation can inhibit the healing process [55, 56]. AuNPs also exhibit anti-angiogenic activity a property that might seem contradictory but is contextually beneficial. By regulating angiogenesis AuNPs ensure that the formation of new blood vessels occurs in a controlled manner, preventing excessive or abnormal vascularization which can compromise wound healing [48]. The efficacy of AuNPs has been validated through various in vivo studies. For instance, the topical application of AuNPs significantly increased wound healing by increasing collagen synthesis and VEGF expression in rats models [56]. The effects were observed alongside a significant reduction in inflammation and infection. Wound closure occurred four times faster in groups treated with AuNPs compared to controls. Another innovative approach involved the sun-assisted synthesis of AuNPs integrated into a hydrogel. This formulation exhibited uniform size and shape along with surface functionalization that enhanced its antibacterial and wound-healing capabilities [57]. AuNPs have shown promise in managing chronic wounds diabetic ulcers and post-surgical wounds [58]. Their limited cytotoxicity and practical functionalization potential with other biopolymers make them ideal candidates for next-generation wound dressings.

While significant strides have been made in understanding the role of AuNPs in wound healing, several challenges remain. The high production cost of AuNPs limits their widespread use, necessitating the development of cost-effective synthesis methods. Moreover, the long-term safety and potential toxicity of AuNPs in humans requires further investigation through rigorous clinical *trials.*

3.2 Silver Nanoparticles

Silver nanoparticles (AgNPs) have emerged as the most widely studied, utilized and commercialized nanomaterial to promote skin regeneration. Their unique ant bactericidal properties and ability to stimulate tissue repair make them highly effective for preventing wound infections and accelerating healing compared to conventional topical treatments [59]. AgNPs are actionable against a broad spectrum of microorganism populations and achieve efficacy at low concentrations while exhibiting minimal toxicity and avoiding skin color [60]. This unique combination of properties has propelled AgNPs to the forefront for innovative wound care solutions. AgNPs have therapeutic benefits due to their ability to lower mitochondrial membrane potential, initiate neutrophil apoptosis and reduce cytokine production [60]. This modulation of the inflammatory response helps to create an optimal microenvironment for wound healing. AgNPs also disrupt bacterial cell replication by oxidizing silver ions that interact with oxygen and hydrogen atoms in thiol groups to form disulfide bonds interfering with bacterial DNA replication and leading to cell death (Vijaayakumar et al. 2019), [59, 61]. In addition to their antimicrobial activity AgNPs promote tissue repair by inducing apoptosis in dysfunctional cells and regulating key gene pathways such as p53. These are highly versatile in applications such as dental treatment wound management, bone healing and catheter modification [62, 63]. AgNPs are shown to be anti-angiogenic and inhibit angiogenesis by affecting pathways like PI3K/AKT and HIF-1 activation thereby promoting collagen deposition. Such mechanisms improve their utility in controlling excessive vascularization often associated with chronic wounds [64, 65]. AgNPs have been designed to maximize their therapeutic benefits in a variety of wound dressings through the incorporation into nanofiber mats, hydrogel materials and semipermeable membranes [66]. For example, chitosan and gelatin are commonly used polymers to stabilize AgNPs synergistically enhancing their healing properties [67–69]. Recent advances including the synthesis of gelatin-silver nanocomposite hydrogel wound dressings through Schiff base bonding and lactose as a crosslinker studied by Sudhakar et al. [70]. However, the development of AgNP-coated nonwoven fabrics with coated with alginate, gelatin and carboxymethyl chitosan has been investigated for their antibacterial activity [71]. In a study by Akin in 2022 silver-based dressings were tested on patient's post-ostomy closure to evaluate their ability to prevent surgical site infections. The results show a significant reduction in infection rates with no infections observed in the group treated with silver dressings compared to a 26.7% infection rate in the control group. Such findings highlight the clinical significance of AgNP-infused dressings in addressing multidrug-resistant bacterial infections, particularly in high-risk surgical environments [72]. Several studies have investigated the efficacy of commercially available AgNP-containing dressings in wound care. For example, Carvalho et al. evaluating products like Silvercel, Mepilex Ag Aquacel Ag and Acticoat for partial thickness burn treatment. All silver-based dressings outperformed saline controls above all, with Silvercel and Acticoat emerging as most effective throughout different stages of wound healing [73]. Another notable advancement is the development of anti-biofilm silver hydrofiber (ABSH) dressings. Biofilm formation is a major independent risk factor for delayed wound healing and ABSH dressings have shown promise in reducing biofilm mass. Davis et al. In this study, used a porcine model to assess the performance of ABSH dressings for deep-partial thickness wounds. Results showed comparable healing duration with other proven silver dressing formulations while effectively addressing biofilm-related complications [11, 74]. AgNPs have also found utility in dental interventions and burn management, two areas characterized by high infection risks. Dental wound healing often occurs in a bacterial-rich environment and interventions can lead to bacteremia. In burns extensive tissue damage exacerbates the risk of severe infections in persons with asthma by exposure to intense heat that can cause bleeding and swelling. Beyond traditional applications AgNPs are being combined with different biomolecules, including amino acids proteins alkaloids, and phenols to enhance their therapeutic potential [66]. Such combinations aid in the reduction and stabilization of nanoparticles while amplifying their wound healthfulness capabilities. For wound management, AgNPs are highly potential despite their remarkable potential. However ongoing research continues to explore the greater implications of these molecules for wound management [75]. Key areas of focus include understanding nanoparticle formulations, understanding long-term effects and integrating AgNPs with novel biomaterials to develop next-generation wound care products.

Advances in AgNP-infused dressings and their integration with other materials have expanded their therapeutic applications, addressing challenges in chronic wound care surgical facility infections and biofilm-related complications. As research progresses AgNP-based technologies are poised to redefine the standards of care, serving patients safer and more effective solutions for their chronic illnesses.

3.3 Copper Nanoparticles

Copper nanoparticles (CuNPs) are increasingly acknowledged for antimicrobial and wound-healing properties. These nanoparticles offer considerable promise in reducing fungal loads at the wound site [76], augmenting immunological response by the production of interleukin-2 and speeding up the healing process. In 2008, copper was recognized as the first metallic antimicrobial agent by the United States Environmental Protection Agency (EPA) because of its effectiveness against microbial hindrances in wound healing [77].

Copper is an indispensable component in wound healing due to the simultaneous operation of cellular and molecular processes. This includes its function as a cofactor for the enzyme such as superoxide dismutase and cytochrome oxidase for oxidative stress management and cellular metabolism. Additionally, it augments the immunological response through stimulating interleukin-2 (IL-2) production [78]. Copper regulates the expressions of proteins and growth factors that are critical in wound healing, such as: platelet-derived growth factor (PDGF) [79, 80], vascular endothelial growth factor (VEGF) [81, 82], and transforming growth factor (TGF)-β [83]. These concerted actions aid in angiogenesis, extracellular matrix stabilization, and tissue regeneration. Copper ions released from nanoparticles are responsible for the regulation of fundamental wound-healing proteins, including the matrix metalloproteinases (MMPs) [84] vital to tissue remodeling. Lower concentrations of copper have been known to stimulate MMP activity and increase the expression levels of MMPs in fibroblasts, thus accelerating tissue repair and angiogenesis [84, 85]. The above unique properties of CuNPs have given rise to some extraordinary applications in the field of wound healing. For instance, Qiao et al. developed copper sulfide nano-dots with enhanced stability and slower Cu ion release. Compared with conventional CuNPs, they have improved safety, suitable for advancement towards wound healing, without raising toxicity issues compared to native copper [86]. Liu et al. synthesized ultra-small Cu5.4O nanoparticles which demonstrated better biocompatibility and promoted rapid wound healing in the in vivo study. The nanoparticles were introduced to some animal models using scratch wounds before applying any treatment. They showed complete recovery three days after application and were proven to be non-toxic to test animals [87]. The hybridization of CuNPs with biomacromolecules has provided a further boost in their therapeutic efficiency. For example, CuNPs were biosynthesized using *Pseudomonas aeruginosa* by Tiwari et al. [88], and they demonstrated enhanced wound healing in rats. The biosynthesized nanoparticles had greater biocompatibility and efficacy compared to native copper. Copper-chitosan nanocomposites also showed great promise for wound healing applications. CCNCs offered additional advantages, such as antifungal and antibacterial activity, attributed to the bio-compatibility of chitosan as reported by Gopal et al. [83]. In rat models, CCNCs modulated key processes in wound healing by enhancing VEGF and TGF-β1 expression while downregulating pro-inflammatory cytokines such as TNF-α and upregulating anti-inflammatory cytokines like IL-10. This regulation on cellular and molecular pathways considerably accelerated the blossoming of closure and tissue regeneration. The roles of copper in skin restoration are attributed to its ability to stabilize components of the extracellular matrix, such as fibrinogen and collagen. The copper nanoparticles will enhance the expression of integrins on the surface of the cell, thus favouring its adhesion to the extracellular matrix and encouraging tissue repair [82, 89]. On top of that, copper stimulates the expression of hypoxia-induced factor-1-alpha (HIF-1α) [81], a key regulator of angiogenesis under hypoxic conditions. This stimulates the production of VEGF, an important angiogenic factor, responsible for the formation of new blood vessels to provide nutrients to the wound site [90]. New hydrogels loaded with CuNPs contain useful therapeutic agents with an important formulation-protective effect. For example, CuNPs loaded with epigallocatechin gallate and α-lipoic acid demonstrate great anti-inflammatory and antioxidant properties to execute a pronounced effect in improving healing outcomes. The gelatinous gels impregnated with CuNPs and loaded with PHT have developed into good platforms, with efficacy surpassing that of standalone treatments with either PHT or CuNPs. Further, these formulations promote procollagen synthesis and inhibit various inflammatory pathways, including the JAK3 pathway, thus promoting wound healing [91]. CuNPs-induced accelerated healing on acute full-thickness defects were reported by Alizadeh et al. It was shown that CuNPs enhanced skin cell migration, proliferation, neovascularization, and wound closure. Besides the dual antibacterial and pro-angiogenic properties, CuNPs showed higher benefits in reducing bacteremia and promoting tissue repair. Biochemical analyses of serum and liver tissue conducted on animal models treated with CuNPs showed no sign of significant accumulation or adverse effects. The present findings demonstrate that CuNP-based therapies can safely accelerate the repair of skin wounds [92].

Copper nanoparticles seem most fascinating and promising as a course of treatment for wound healing, while they act in multifaceted manners in angiogenesis, stabilization of the extracellular matrix, and modulation of the immune response. All these attributes augment the expression of VEGF, demonstrate regulation of cytokines, and encourage tissue regeneration to make them valuable to modern wound-care therapy. Embedding CuNPs in biomaterial-based formulations such as hydrogels and nanocomposites further broadens their therapeutic potential but holds minimal concern for safety issues.

3.4 Zinc Oxide Nanoparticles

Zinc oxide nanoparticles (ZnONPs) represent cutting-edge materials with the potential to accelerate healing. ZnONPs, with their unique properties and biocompatibility, have become the essential in modern wound management approaches. The roles of zinc in the healing process encompass antibacterial, anti-inflammatory, skin rebirth, blood vessel generation, and tissue reconstruction [93]. ZnONPs are less toxic than other antimicrobial agents like gold or silver nanoparticles while still providing zinc, required to carry out these functions [94, 95]. Other biological functions include the facilitation of keratinocyte mobility, fibroblast mitosis, and blood vessel formation, which all stand to benefit tissue repair. The generation of reactive oxygen species (ROS), especially hydrogen peroxide (H_2O_2), is one of the key mechanisms through which ZnONPs boost healing. ROS proved important in the signaling processes in cells, which, in turn, trigger fibroblast migration, adhesion, and tissue repair by a mechanism involving growth factors. Controlled ROS production facilitates a wound-healing process, during which biocompatibility is maintained [96, 97].

There is growing interest in some of the approaches to the synthesis of ZnONPs, namely, environmentally friendly ones based on the use of *Calendula officinalis* flowers. Green synthesis methods favourably reduce environmental impact and produce rich antioxidant nanoparticles that exhibit excellent anti-inflammatory and wound-healing properties [98]. These nanoparticles are very effective in promoting keratinocyte mobility, angiogenesis, and re-epithelialization, which recommend themselves in biomedicine. ZnONPs in various matrices for the preparation of wound dressings have been developed to enhance their antimicrobial activity and wound-healing performance. Among the advanced materials studied are electrospun membranes [99], hydrogels, and nanocomposites [100]. For example, Augustine et al. [96] employed electrospinning to conjugate polycaprolactone with ZnONPs, leading to an increase in ROS production and fibroblast migration in guinea pigs for a considerably accelerated healing process. Hydrogel systems that have incorporated ZnONPs into various biocompatible polymer matrices, including chitosan and poly(vinyl alcohol, have shown improved permeability, bacterial inhibition, and wound healing capability both in vitro and in vivo [101, 102]. Similarly, nanocomposite formulations, including alginate-gum/acacia and collagen-ZnONPs blend have been effective in providing mechanical support, limiting microbial infections, and aiding regeneration [103]. For instance, cellulose acetate nanofibers doped with ZnONPs and graphene oxide (GO) exhibit promise with respect to improved surface roughness and tensile strength and adhesion of human fibroblasts [104]. Thus, these hybrid materials generally not only improve mechanical performance but also promote cell proliferation in both acute and chronic wounds. It is also known from the studies that ZnONPs incorporated in nanofibrous membranes based on, i.e., poly (lactide-co-glycolic acid) or silk fibroin, exhibit a strong antimicrobial action favoring vascularization and tissue remodeling [105].

Zinc oxide nanoparticles are a great leap in wound healing. As the study continues, the integration of ZnONPs with innovative biomaterials, combined with green synthesis techniques, is thought to yield more effective and sustainable alternatives for wound-healing challenges.

4 Limitations of Metal Nanoparticles in Wound Healing Application

The idea of nanomedicine first appeared about 35 years ago, and as technology advanced in this field, the concept of "nanotoxicity" brought up moral questions about the usage of nanomaterials. Although nanomedicine has numerous benefits over conventional drugs, there are several drawbacks and difficulties that must be resolved (Fig. 2). Metal and metal oxide nanoparticles hold potential for a variety of applications, including diagnosis, treatment, theranostics, sensors, and cosmetics. The usage of each nanoparticle, particularly in medicine, requires fresh understanding and study since the properties of the nanomaterials differ from those of their bulk counterparts. One of the key advantages of MNPs is their high antibacterial activity. Several investigations have shown that MNPs are effective against various bacteria, fungi, and viruses [106–108]. This broad-spectrum antibacterial effect is very useful in preventing and treating infections, and can help in wound healing.

Although employing NPs in therapy or healing has several benefits, there is a higher chance of cytotoxicity and/or genotoxicity. Additionally, the size, shape, and concentration of the NPs affect this risk [110–112]. A careful balance between cytotoxicity and healing potential must be maintained while studying the use of metal nanoparticles for wound healing. Since the skin serves as the body's primary barrier, wounds that breach it expose a person to harmful compounds. Because NPs used

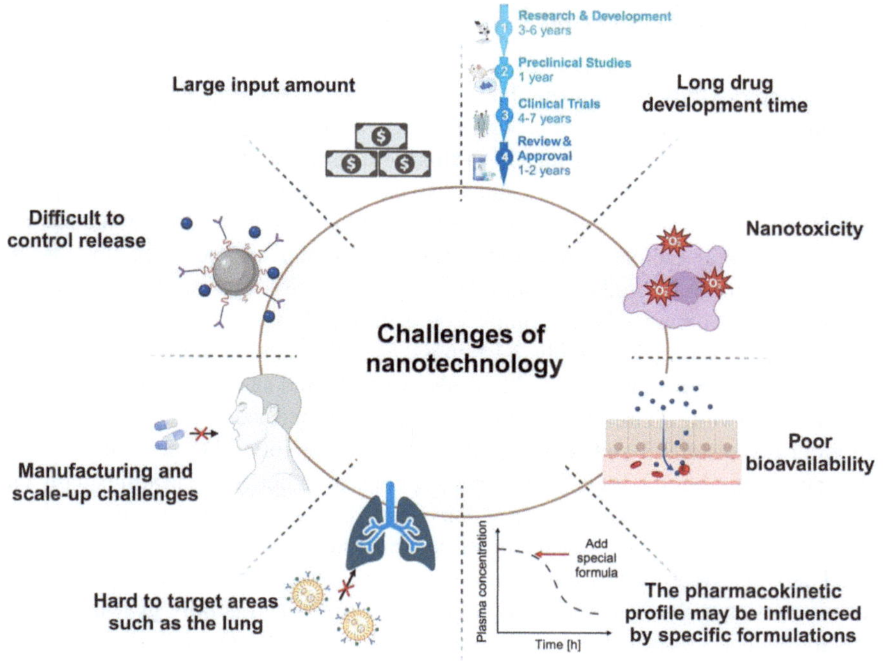

Fig. 2 Challenges of nanotechnology. (Copyright © Huang et al. [109], Published by Springer Nature under the terms of the Creative Commons Attribution 4.0 International License, http://creativecommons.org/licenses/by/4.0/)

in wound healing come into direct touch with the wound tissue, their biological safety must be assessed before application. There is no solid evidence that long-term exposure to NPs via the skin and their deposition will have major implications, which is critical to verify in future extended exposure studies. Skin irritation and allergy are the most frequently reported transdermal noxious effects of NPs. Several metal nanoparticles have been proven to cause liver damage, kidney failure, and neurological diseases, as well as impact immunological function [113, 114]. It was also reported that fibroblasts and keratinocytes exposed to NPs also showed oxidative stress, autophagy, and programmed cell death [115, 116]. Another issue is that NPs might cause multisystem problems by spreading throughout the body and to different organs. After entering the bloodstream, the amount of NPs drops dramatically compared to their original concentration and may be partially excreted in urine and feces. Furthermore, the size, shape, concentration inside the hydrogel, and surface charge are some of the variables that affect NP activity [117]. The release and uptake of NPs are another issue. While excessive release might cause toxicity, accumulation in nearby tissues or organs, or other undesirable outcomes, poor release or absorption restricts the effectiveness and quality of treatment given [118–120].

5 Future Directions and Perspectives

Nanotechnology-based platforms have emerged as an exciting new research area with numerous applications in wound healing in the recent decade, as evidenced by the significant number of papers in this field. Several studies have shown that inorganic and metal nanoparticles, such as magnetic, silica silver, and nanographene oxide scaffolds, as well as copper and titanium dioxide, can promote wound healing. However, several factors hinder the rapid development of therapies for chronic wounds, including limited understanding of pathophysiology, variability in patient populations and comorbidities, complex clinical trial design and related costs, alternative clinical, and lack of public awareness. Metallic nanoparticles can be used as nanocarriers to release bioactive chemicals precisely and sustainably in tissue engineering, a growing field. Nanoparticles have a high surface area to volume ratio, allowing for sustained and regulated release of therapeutic chemicals in wounds. The capacity to deliver drugs, genes, and peptides at higher concentrations with less side effects than with traditional drug delivery techniques is one of the main advantages of using NPs, which boosts the effectiveness of the treatment. Nanomaterial-based delivery systems play a crucial role in tissue engineering by releasing growth factors continuously. Nanotechnology-based dressings can supply growth factors, pro-angiogenesis chemicals, and antimicrobials, potentially revolutionizing wound care. While nanoparticles have significant promise for wound healing, various difficulties prevent their clinical application. Challenges include production costs, stability, biocompatibility, toxicity, and regulatory requirements. Improvements in characterization methods, optimization of formulations, and thorough testing are essential to facilitate the clinical application of nanomaterials in wound healing. The high cost of producing nanomaterials may limit their widespread usage in healthcare settings. Ensuring consistent quality and reproducibility of wound dressings made from nanomaterials can be challenging because of different production methods and material characteristics. Nanotechnology can help regenerate hair follicles, regulate paraesthesia, and improve aberrant pigmentation. New technologies, such as electronic skin and nanotechnology, offer innovative solutions for wound healing and paraesthesia rehabilitation.

More research and study are needed to understand how nanotechnology-based therapeutics can be used in wound care management. Future research should optimize the manufacturing and functionalization of these materials for better biocompatibility, efficacy, and patient-specific adaptation. The current capabilities in advanced manufacturing and development of wound dressing, combined with knowledge of chronic wounds, molecular pathology, and phenotype-genotype characteristics, are expected to drive the blueprint of future wound-healing nanomaterials.

6 Conclusions

Acute and chronic wounds affect millions of people annually, hence efficient and novel wound dressing materials have to be developed and evaluated for their effectiveness in wound care management. An ideal dressing must maintain a moist environment and pH equilibrium, facilitate oxygen exchange, sequester proteases, enhance growth factor activity, prevent infection, support autolytic debridement, and encourage granulation tissue formation and re-epithelialization. In the field of wound care, nanotechnology has become a game-changer, transforming the biomedical dressing market. Its use has exceptional effectiveness in promoting wound healing and preventing bacterial infections, marking a substantial improvement in wound care management. By improving mechanical strength and regulating medication release, the incorporation of nanotechnology into dressings has made customized therapeutic interventions possible. Nanomaterials have been used to help wound healing process at various stages. Nanoparticles' unique physicochemical and biological characteristics, as well as their extremely specific surface area, make them ideal for delivering and releasing therapeutic compounds in wound dressings over time. NPs can be used to create a wound dressing that targets microbial infections using natural cues including pH, temperature, enzymes, and toxins released by bacteria.

References

1. Rodrigues M, Kosaric N, Bonham CA, Gurtner GC (2019) Wound healing: a cellular perspective. Physiol Rev 99(1):665–706
2. Velnar T, Bailey T, Smrkolj V (2009) The wound healing process: an overview of the cellular and molecular mechanisms. J Int Med Res 37(5):1528–1542
3. Wilkinson HN, Hardman MJ (2020) Wound healing: cellular mechanisms and pathological outcomes. Open Biol 10(9):200223
4. Werner S, Grose R (2003) Regulation of wound healing by growth factors and cytokines. Physiological Rev 83(3):835–870. https://doi.org/10.1152/physrev.2003.83.3.835
5. Falcone M, De Angelis B, Pea F, Scalise A, Stefani S, Tasinato R et al (2021) Challenges in the management of chronic wound infections. J Glob Antimicrob Resist 26:140–147
6. Sen CK (2021) Human wound and its burden: updated 2020 compendium of estimates. Adv Wound Care 10(5):281–292
7. Clinton A, Carter T (2015) Chronic wound biofilms: pathogenesis and potential therapies. Lab Med 46(4):277–284
8. Goswami AG, Basu S, Banerjee T, Shukla VK (2023) Biofilm and wound healing: from bench to bedside. Eur J Med Res 28(1):157
9. Percival SL, McCarty SM, Lipsky B (2015) Biofilms and wounds: an overview of the evidence. Adv Wound Care 4(7):373–381
10. Kunnath AP, Suodha Suoodh M, Chellappan DK, Chellian J, Palaniveloo K (2024) Bacterial persister cells and development of antibiotic resistance in chronic infections: an update. Br J Biomed Sci 81:12958
11. Metcalf DG, Bowler PG (2013) Biofilm delays wound healing: a review of the evidence. Burns Trauma 1(1):2321–3868

12. Järbrink K, Ni G, Sönnergren H, Schmidtchen A, Pang C, Bajpai R, Car J (2017) The humanistic and economic burden of chronic wounds: a protocol for a systematic review. Syst Rev 6:1–7
13. Lindholm C, Searle R (2016) Wound management for the 21st century: combining effectiveness and efficiency. Int Wound J 13:5–15
14. Hurlow J, Bowler PG (2022) Acute and chronic wound infections: microbiological, immunological, clinical and therapeutic distinctions. J Wound Care 31(5):436–445
15. Schreml S, Szeimies RM, Prantl L, Landthaler M, Babilas P (2010) Wound healing in the 21st century. J Am Acad Dermatol 63(5):866–881
16. Elder M (2023) Markets for advanced wound management technologies. BCC Research, Wellesley
17. Bayda S, Adeel M, Tuccinardi T, Cordani M, Rizzolio F (2019) The history of nanoscience and nanotechnology: from chemical–physical applications to nanomedicine. Molecules 25(1):112
18. Malik S, Muhammad K, Waheed Y (2023) Nanotechnology: a revolution in modern industry. Molecules 28(2):661
19. Hamdan S, Pastar I, Drakulich S, Dikici E, Tomic-Canic M, Deo S, Daunert S (2017) Nanotechnology-driven therapeutic interventions in wound healing: potential uses and applications. ACS Central Sci 3(3):163–175
20. Kushwaha A, Goswami L, Kim BS (2022) Nanomaterial-based therapy for wound healing. Nanomaterials 12(4):618
21. Naskar A, Kim KS (2020) Recent advances in nanomaterial-based wound-healing therapeutics. Pharmaceutics 12(6):499
22. Gowda BHJ, Mohanto S, Singh A, Bhunia A, Abdelgawad MA, Ghosh S et al (2023) Nanoparticle-based therapeutic approaches for wound healing: a review of the state-of-the-art. Mater Today Chem 27:101319
23. Mendes C, Thirupathi A, Corrêa ME, Gu Y, Silveira PC (2022) The use of metallic nanoparticles in wound healing: new perspectives. Int J Mol Sci 23(23):15376
24. Pormohammad A, Monych NK, Ghosh S, Turner DL, Turner RJ (2021) Nanomaterials in wound healing and infection control. Antibiotics 10(5):473
25. Astaneh ME, Fereydouni N (2024) Silver nanoparticles in 3D printing: a new frontier in wound healing. ACS Omega 9(40):41107–41129
26. Chowdhury MFM, Khan MN, Rahman MM (2024) Metal nanoparticles incorporated chitosan-based electrospun nanofibre mats for wound dressing applications: a review. Int J Biol Macromol 13:137352
27. Onyekwelu I, Yakkanti R, Protzer L, Pinkston CM, Tucker C, Seligson D (2017) Surgical wound classification and surgical site infections in the orthopaedic patient. JAAOS: Glob Res Rev 1(3):e022. https://doi.org/10.5435/JAAOSGlobal-D-17-00022
28. Desai N, Rana D, Salave S, Gupta R, Patel P, Karunakaran B et al (2023) Chitosan: a potential biopolymer in drug delivery and biomedical applications. Pharmaceutics 15(4):1313
29. Pang Q, Jiang Z, Wu K, Hou R, Zhu Y (2023) Nanomaterials-based wound dressing for advanced management of infected wound. Antibiotics 12(2):351
30. Zhou Y, Li M, Gao W, Li X, Long L, Hou X et al (2023) Microstructure-united heterogeneous sodium alginate doped injectable hydrogel for stable hemostasis in dynamic mechanical environments. Int J Biol Macromol 248:125877
31. Matter MT, Probst S, Läuchli S, Herrmann IK (2020) Uniting drug and delivery: metal oxide hybrid nanotherapeutics for skin wound care. Pharmaceutics 12(8):780
32. Robson MC (2001) Wound healing; biologic features and approaches to maximize healing trajectories. Curr Probl Surg 38:61–140
33. Zhao Q, Leng C, Lau M, Choi K, Wang R, Zeng Y et al (2024) Precise healing of oral and maxillofacial wounds: tissue engineering strategies and their associated mechanisms. Front Bioeng Biotechnol 12:1375784
34. Hunt TK, Hopf H, Hussain Z (2000) Physiology of wound healing. Adv Skin Wound Care 13:6

35. Koh TJ, DiPietro LA (2011) Inflammation and wound healing: the role of the macrophage. Exp Rev Mol Med 13:e23
36. Noli C, Miolo A (2001) The mast cell in wound healing. Vet Dermatol 12(6):303–313
37. Brown GL, Curtsinger LJ, White M, Mitchell RO, Pietsch J, Nordquist R et al (1988) Acceleration of tensile strength of incisions treated with EGF and TGF-β. Ann Surg 208(6):788–794
38. Rodero MP, Khosrotehrani K (2010) Skin wound healing modulation by macrophages. Int J Clin Exper Pathol 3(7):643
39. Lu Z, Li J, Chen B, Feng J, Hu Q, Jin Y, Fu Z (2023) Mitochondria targeted nanoparticles potentiate tumor chemo-phototherapy by toxic oxidative stress mediated oxeiptosis. Macromol Biosci 23(11):2300151
40. Lee J, Kang S, Park H, Sun JG, Kim EC, Shim G (2023) Nanoparticles for lymph node-directed delivery. Pharmaceutics 15(2):565
41. Khan AU, Yuan Q, Wei Y, Khan GM, Khan ZUH, Khan S et al (2016) Photocatalytic and antibacterial response of biosynthesized gold nanoparticles. J Photochem Photobiol B Biol 162:273–277
42. Soliman WE, Elsewedy HS, Younis NS, Shinu P, Elsawy LE, Ramadan HA (2022) Evaluating antimicrobial activity and wound healing effect of rod-shaped nanoparticles. Polymers 14(13):2637
43. Wei SC, Chang L, Huang CC, Chang HT (2019) Dual-functional gold nanoparticles with antimicrobial and proangiogenic activities improve the healing of multidrug-resistant bacteria-infected wounds in diabetic mice. Biomater Sci 7(11):4482–4490
44. Zhang Q, Li M, Luo B, Luo Y, Jiang H, Chen C et al (2021) In situ growth gold nanoparticles in three-dimensional sugarcane membrane for flow catalytical and antibacterial application. J Hazard Mater 402:123445
45. Pivodová V, Franková J, Galandáková A, Ulrichová J (2015) In vitro AuNPs' cytotoxicity and their effect on wound healing. Nanobiomedicine 2:7. https://doi.org/10.210.5772/61132
46. Kim JE, Lee J, Jang M, Kwak MH, Go J, Kho EK et al (2015) Accelerated healing of cutaneous wounds using phytochemically stabilized gold nanoparticle deposited hydrocolloid membranes. Biomater Sci 3(3):509–519
47. Mendes C, dos Santos-Haupenthal DP, Zaccaron RP, de Bem-Silveira G, Corrêa MEAB, de Roch-Casagrande L et al (2020) Effects of the association between photobiomodulation and hyaluronic acid linked gold nanoparticles in wound healing. ACS Biomater Sci Eng 6(9):5132–5144
48. Chen SA, Chen HM, Yao YD, Hung CF, Tu CS, Liang YJ (2012) Topical treatment with anti-oxidants and Au nanoparticles promote healing of diabetic wound through receptor for advance glycation end-products. Eur J Pharm Sci 47(5):875–883
49. Arafa MG, El-Kased RF, Elmazar MM (2018) Thermoresponsive gels containing gold nanoparticles as smart antibacterial and wound healing agents. Sci Rep 8(1):13674
50. Nanda SS, Wang T, Hossain MI, Yoon HY, Selvan ST, Kim K, Yi DK (2022) Gold-nanorod-based scaffolds for wound-healing applications. ACS Appl Nano Mater 5(6):8640–8648
51. Xia C, Ren B, Liu N, Zheng Y (2022) A Feasible strategy of fabricating of gold-encapsulated dextran/polyvinyl alcohol nanoparticles for the treatment and care of wound healing. J Clust Sci 33(5):2179–2187
52. Boomi P, Ganesan R, Prabu Poorani G, Jegatheeswaran S, Balakumar C, Gurumallesh Prabu H, Anand K, Marimuthu Prabhu N, Jeyakanthan J, Saravanan M (2020) Phyto-engineered gold nanoparticles (AuNPs) with potential antibacterial, antioxidant, and wound healing activities under in vitro and in vivo conditions. Int. J. Nanomed 15:7553–7568. https://doi.org/10.10.2147/IJN.S257499
53. Chen W, Chu R, Li H, Hua T, Chen H, Li R et al (2023) A novel wound dressing based on a gold nanoparticle self-assembled hydrogel to promote wound healing. Mater Adv 4(14):2918–2925
54. Volkova N, Yukhta M, Pavlovich O, Goltsev A (2016) Application of cryopreserved fibroblast culture with Au nanoparticles to treat burns. Nanoscale Res Lett 11(1). https://doi.org/10.1186/s11671-016-1242-y

55. Leu JG, Chen SA, Chen HM, Wu WM, Hung CF, Yao YD et al (2012) The effects of gold nanoparticles in wound healing with antioxidant epigallocatechin gallate and α-lipoic acid. Nanomed Nanotechnol Biol Med 8(5):767–775
56. Ovais M, Ahmad I, Khalil AT, Mukherjee S, Javed R, Ayaz M et al (2018) Wound healing applications of biogenic colloidal silver and gold nanoparticles: recent trends and future prospects. Appl Microbiol Biotechnol 102:4305–4318
57. Huang X, Xu L, Yu X, Li Y, Huang Z, Xu R et al (2023) Near-infrared light-responsive multifunctional hydrogel releasing peptide-functionalized gold nanorods sequentially for diabetic wound healing. J Coll Interf Sci 639:369–384
58. Manisha N, Gaur P, Goel R, Lata K, Mishra R (2024) Understanding diabetic wounds: a review of mechanisms, pathophysiology, and multimodal management strategies. Curr Rev Clin Exper Pharmacol 12:965
59. Singh M, Thakur V, Kumar V, Raj M, Gupta S, Devi N et al (2022) Silver nanoparticles and its mechanistic insight for chronic wound healing: review on recent progress. Molecules 27(17):5587
60. Franková J, Pivodová V, Vágnerová H, Juráňová J, Ulrichová J (2016) Effects of silver nanoparticles on primary cell cultures of fibroblasts and keratinocytes in a wound-healing model. J Appl Biomaterials Funct Mater 14(2):137–142
61. Rigo C, Ferroni L, Tocco I, Roman M, Munivrana I, Gardin C et al (2013) Active silver nanoparticles for wound healing. Int J Mol Sci 14(3):4817–4840
62. Farah MA, Ali MA, Chen SM, Li Y, Al-Hemaid FM, Abou-Tarboush FM et al (2016) Silver nanoparticles synthesized from *Adenium obesum* leaf extract induced DNA damage, apoptosis and autophagy via generation of reactive oxygen species. Coll Surf B Biointerf 141:158–169
63. Wang ZX, Chen CY, Wang Y, Li FXZ, Huang J, Luo ZW et al (2019) Ångstrom-scale silver particles as a promising agent for low-toxicity broad-spectrum potent anticancer therapy. Adv Funct Mater 29(23):1808556
64. Gurunathan S, Lee KJ, Kalishwaralal K, Sheikpranbabu S, Vaidyanathan R, Eom SH (2009) Antiangiogenic properties of silver nanoparticles. Biomaterials 30(31):6341–6350
65. Yang T, Yao Q, Cao F, Liu Q, Liu B, Wang XH (2016) Silver nanoparticles inhibit the function of hypoxia-inducible factor-1 and target genes: insight into the cytotoxicity and antiangiogenesis. Int J Nanomed 15:6679–6692
66. Mihai MM, Dima MB, Dima B, Holban AM (2019) Nanomaterials for wound healing and infection control. Materials 12:2176
67. Dotto D, Scatolini M, Pugini S, Vercik L, Melo M, Vercik A, Rigo E (2021) A novel antibacterial and biocompatible wound cover made of gelatin/chitosan with silver nanoparticles of green synthesis. Mater Res Exp 8(11):115402
68. Setiyorini Y, Anggraeni A, Pintowantoro S (2022) In-vivo study of nano chitosan as therapeutic agent for toxic metal implant. Results Eng 13:100352
69. Ye H, Cheng J, Yu K (2019) In situ reduction of silver nanoparticles by gelatin to obtain porous silver nanoparticle/chitosan composites with enhanced antimicrobial and wound-healing activity. Int J Biol Macromol 121:633–642
70. Sudhakar K, Won SY, Han SS (2022) Gelatin stabilized silver nanoparticles for wound healing applications. Mater Lett 325:132851. https://doi.org/10.1016/j.matlet.2022.132851
71. Montaser AS, Jlassi K, Ramadan MA, Sleem AA, Attia MF (2021) Alginate, gelatin, and carboxymethyl cellulose coated nonwoven fabrics containing antimicrobial AgNPs for skin wound healing in rats. Int J Biol Macromol 173:203–210
72. Akin T, Kendirci M, Akgün AE, Cetinkaya E, Er S, Akin M, Yasti AC (2022) Applying a silver-containing dressing to the incision site and its effect on the development of surgical site infection after ostomy closure: a prospective randomized clinical pilot study. Wound Manag Prevent 68(4):34–43
73. Carvalho CDS, Bernardes MJC, Gonçalves RC, Vilela MS, Silva MVMD, Oliveira VDS et al (2022) Treatment of experimentally induced partial-thickness burns in rats with different silver-impregnated dressings. Acta Cir Bras 37(8):e370801

74. Davis SC, Li J, Gil J, Valdes J, Solis M, Higa A, Bowler P (2018) The wound-healing effects of a next-generation anti-biofilm silver Hydrofiber wound dressing on deep partial-thickness wounds using a porcine model. Int Wound J 15(5):834–839
75. ClinicalTrials.gov (2022) https://clinicaltrials.gov/ct2/results?cond=silver+nanoparticles. Accessed on 1 Oct 2022
76. El-Abeid SE, Ahmed Y, Daròs JA, Mohamed MA (2020) Reduced graphene oxide nanosheet-decorated copper oxide nanoparticles: a potent antifungal nanocomposite against fusarium root rot and wilt diseases of tomato and pepper plants. Nanomaterials 10(5):1001
77. Vincent M, Duval RE, Hartemann P, Engels-Deutsch M (2018) Contact killing and antimicrobial properties of copper. J Appl Microbiol 124(5):1032–1046
78. Hopkins RG, Failla ML (1997) Copper deficiency reduces interleukin-2 (IL-2) production and IL-2 mRNA in human T-lymphocytes. J Nutr 127(2):257–262
79. Ashino T, Sudhahar V, Urao N, Oshikawa J, Chen GF, Wang H et al (2010) Unexpected role of the copper transporter ATP7A in PDGF-induced vascular smooth muscle cell migration. Circul Res 107(6):787–799
80. Tsai CY, Finley JC, Ali SS, Patel HH, Howell SB (2012) Copper influx transporter 1 is required for FGF, PDGF and EGF-induced MAPK signaling. Biochem Pharmacol 84(8):1007–1013
81. Feng W, Ye F, Xue W, Zhou Z, Kang YJ (2009) Copper regulation of hypoxia-inducible factor-1 activity. Mol Pharmacol 75(1):174–182
82. Sen CK, Khanna S, Venojarvi M, Trikha P, Ellison EC, Hunt TK, Roy S (2002) Copper-induced vascular endothelial growth factor expression and wound healing. Am J Physiol Heart Circul Physiol 282(5):H1821–H1827
83. Gopal A, Kant V, Gopalakrishnan A, Tandan SK, Kumar D (2014) Chitosan-based copper nanocomposite accelerates healing in excision wound model in rats. Eur J Pharmacol 731:8–19
84. Philips N, Hwang H, Chauhan S, Leonardi D, Gonzalez S (2010) Stimulation of cell proliferation and expression of matrixmetalloproteinase-1 and interluekin-8 genes in dermal fibroblasts by copper. Connect Tissue Res 51(3):224–229
85. Michopoulou A, Rousselle P (2015) How do epidermal matrix metalloproteinases support re-epithelialization during skin healing? Eur J Dermatol 25:33–42
86. Qiao Y, Ping Y, Zhang H, Zhou B, Liu F, Yu Y et al (2019) Laser-activatable CuS nanodots to treat multidrug-resistant bacteria and release copper ion to accelerate healing of infected chronic nonhealing wounds. ACS Appl Mater Interf 11(4):3809–3822
87. Liu T, Xiao B, Xiang F, Tan J, Chen Z, Zhang X et al (2020) Ultrasmall copper-based nanoparticles for reactive oxygen species scavenging and alleviation of inflammation related diseases. Nat Commun 11(1):2788
88. Tiwari M, Narayanan K, Thakar MB, Jagani HV, Venkata-Rao J (2014) Biosynthesis and wound healing activity of copper nanoparticles. IET Nanobiotechnol 8(4):230–237
89. Tenaud I, Sainte-Marie I, Jumbou O, Litoux P, Dreno B (1999) In vitro modulation of keratinocyte wound healing integrins by zinc, copper and manganese. Br J Dermatol 140(1):26–34
90. Das A, Sudhahar V, Chen GF, Kim HW, Youn SW, Finney L et al (2016) Endothelial antioxidant-1: a key mediator of copper-dependent wound healing in vivo. Sci Rep 6(1):33783
91. Saddik MS, Alsharif FM, El-Mokhtar MA, Al-Hakkani MF, El-Mahdy MM, Farghaly HS, Abou-Taleb HA (2020) Biosynthesis, characterization, and wound-healing activity of phenytoin-loaded copper nanoparticles. AAPS PharmSciTech 21:1–12
92. Alizadeh S, Seyedalipour B, Shafieyan S, Kheime A, Mohammadi P, Aghdami N (2019) Copper nanoparticles promote rapid wound healing in acute full thickness defect via acceleration of skin cell migration, proliferation, and neovascularization. Biochem Biophys Res Commun 517(4):684–690
93. Lin PH, Sermersheim M, Li H, Lee PH, Steinberg SM, Ma J (2017) Zinc in wound healing modulation. Nutrients 10(1):16
94. Gupta M, Mahajan VK, Mehta KS, Chauhan PS (2014) Zinc therapy in dermatology: a review. Dermatol Res Pract 2014(1):709152

95. Shalaby MA, Anwar MM, Saeed H (2022) Nanomaterials for application in wound healing: current state-of-the-art and future perspectives. J Polym Res 29(3):91
96. Augustine R, Dominic EA, Reju I, Kaimal B, Kalarikkal N, Thomas S (2014) Investigation of angiogenesis and its mechanism using zinc oxide nanoparticle-loaded electrospun tissue engineering scaffolds. RSC Adv 4(93):51528–51536
97. Sharma DK, Sharma KK, Kumar V, Sharma A (2016) Effect of Ce doping on the structural, optical and magnetic properties of ZnO nanoparticles. J Mater Sci Mater Electr 27:10330–10335
98. Aydin-Acar C, Gencer MA, Pehlivanoglu S, Yesilot S, Donmez S (2024) Green and eco-friendly biosynthesis of zinc oxide nanoparticles using *C. officinalis* flower extract: wound healing potential and antioxidant activity. Int Wound J 21(1):e14413
99. Dobero A, Scarfi S, Pozzolini M, Vicini S, Alloisio M, Castellano M (2019) Alginate-based electrospun membranes containing ZnO nanoparticles as potential wound healing patches: biological, mechanical, and physicochemical characterization. ACS Appl Mater Interf 12(3):3371–3381
100. Masud RA, Islam MS, Haque P, Khan MNI, Shahruzzaman M, Khan M et al (2020) Preparation of novel chitosan/poly (ethylene glycol)/ZnO bionanocomposite for wound healing application: effect of gentamicin loading. Materialia 12:100785
101. Melnikova N, Knyazev A, Nikolskiy V, Peretyagin P, Belyaeva K, Nazarova N et al (2021) Wound healing composite materials of bacterial cellulose and zinc oxide nanoparticles with immobilized betulin diphosphate. Nanomaterials 11(3):713
102. Khorasani MT, Joorabloo A, Adeli H, Milan PB, Amoupour M (2021) Enhanced antimicrobial and full-thickness wound healing efficiency of hydrogels loaded with heparinized ZnO nanoparticles: In vitro and in vivo evaluation. Int J Biol Macromol 166:200–212. https://doi.org/10.10.1016/j.ijbiomac.2020.10.142
103. Manuja A, Raguvaran R, Kumar B, Kalia A, Tripathi BN (2020) Accelerated healing of full thickness excised skin wound in rabbits using single application of alginate/acacia based nanocomposites of ZnO nanoparticles. Int J Biol Macromol 155:823–833
104. Aly AA, Ahmed MK (2021) Nanofibers of cellulose acetate containing ZnO nanoparticles/graphene oxide for wound healing applications. Int J Pharm 598:120325
105. Huang K, Jinzhong Z, Zhu T, Morsi Y, Aldalbahi A, El-Newehy M et al (2021) Exploration of the antibacterial and wound healing potential of a PLGA/silk fibroin based electrospun membrane loaded with zinc oxide nanoparticles. J Mater Chem B 9(5):1452–1465
106. Hussain FS, Abro NQ, Ahmed N, Memon SQ, Memon N (2022) Nano-antivirals: a comprehensive review. Front Nanotechnol 4:1064615
107. Sánchez-López E, Gomes D, Esteruelas G, Bonilla L, Lopez-Machado AL, Galindo R et al (2020) Metal-based nanoparticles as antimicrobial agents: an overview. Nanomaterials 10(2):292
108. Wahab S, Salman A, Khan Z, Khan S, Krishnaraj C, Yun SI (2023) Metallic nanoparticles: a promising arsenal against antimicrobial resistance—unraveling mechanisms and enhancing medication efficacy. Int J Mol Sci 24(19):14897
109. Huang Y, Guo X, Wu Y, Chen X, Feng L, Xie N, Shen G (2024) Nanotechnology's frontier in combatting infectious and inflammatory diseases: prevention and treatment. Sig Transd Targ Ther 9(1):34
110. Abbasi R, Shineh G, Mobaraki M, Doughty S, Tayebi L (2023) Structural parameters of nanoparticles affecting their toxicity for biomedical applications: a review. J Nanopart Res 25(3):43
111. Egbuna C, Parmar VK, Jeevanandam J, Ezzat SM, Patrick-Iwuanyanwu KC, Adetunji CO et al (2021) Toxicity of nanoparticles in biomedical application: nanotoxicology. J Toxicol 2021(1):9954443
112. Manuja A, Kumar B, Kumar R, Chhabra D, Ghosh M, Manuja M et al (2021) Metal/metal oxide nanoparticles: toxicity concerns associated with their physical state and remediation for biomedical applications. Toxicol Rep 8:1970–1978

113. Bi J, Mo C, Li S, Huang M, Lin Y, Yuan P et al (2023) Immunotoxicity of metal and metal oxide nanoparticles: from toxic mechanisms to metabolism and outcomes. Biomater Sci 11(12):4151–4183
114. Yao Y, Zang Y, Qu J, Tang M, Zhang T (2019) The toxicity of metallic nanoparticles on liver: the subcellular damages, mechanisms, and outcomes. Int J Nanomed 34:8787–8804
115. Florance I, Cordani M, Pashootan P, Moosavi MA, Zarrabi A, Chandrasekaran N (2024) The impact of nanomaterials on autophagy across health and disease conditions. Cell Mol Life Sci 81(1):184
116. Petrache Voicu SN, Dinu D, Sima C, Hermenean A, Ardelean A, Codrici E et al (2015) Silica nanoparticles induce oxidative stress and autophagy but not apoptosis in the MRC-5 cell line. Int J Mol Sci 16(12):29398–29416
117. Khan I, Saeed K, Khan I (2019) Nanoparticles: properties, applications and toxicities. Arab J Chem 12(7):908–931
118. Hadrup N, Lam HR (2014) Oral toxicity of silver ions, silver nanoparticles and colloidal silver: a review. Regulat Toxicol Pharmacol 68(1):1–7
119. Innes ME, Umraw N, Fish JS, Gomez M, Cartotto RC (2001) The use of silver coated dressings on donor site wounds: a prospective, controlled matched pair study. Burns 27(6):621–627
120. Poon VK, Burd A (2004) In vitro cytotoxity of silver: implication for clinical wound care. Burns 30(2):140–147

Ceramic Nanofibre-Based Materials for Chronic Wounds: Emphasis on Silica Based Material

Rajrupa Bhattacharjee, Lopamudra Bhattacharjee, and Rama Ranjan Bhattacharjee

Abstract Chronic wound management remains a significant clinical challenge due to the limitations of traditional treatments such as skin grafts. The present chapter explores the potential of ceramic nanofibre-based materials, particularly silica (SiO_2), as advanced scaffolds for chronic wound healing. Silica nanofibres, fabricated via electrospinning and sol–gel techniques, exhibit exceptional biocompatibility, high surface-area-to-volume ratio, and tunable physicochemical properties that mimic the extracellular matrix (ECM) of skin tissue. The chapter highlights key synthesis methods, including self-assembly, phase separation, and polymer-assisted electrospinning, while addressing challenges such as fibre uniformity, mechanical stability, and controlled degradation. Characterization techniques such as scanning electron microscopy (SEM), Fourier-transform infrared spectroscopy (FTIR), and swelling tests are discussed to evaluate the structural and functional properties of silica nanomats. Additionally, the chapter emphasizes the material's dynamic wettability, which transitions from hydrophobic to hydrophilic under ambient conditions, enhancing its suitability for wound dressings. Future directions include optimizing mechanical properties, integrating bioactive agents, and scaling production for clinical applications. Silica-based nanofibrous ceramics represent a promising platform for regenerative medicine, offering multifunctional solutions for chronic wound care and beyond.

Keywords Silica nanofibres · Chronic wounds · Electrospinning · Sol–gel · Biomaterials · Tissue engineering · Wound healing scaffolds

---

R. Bhattacharjee
Department of Chemical Engineering, Heritage Institute of Technology, Kolkata, India

L. Bhattacharjee
MASOJA Techno Solution Pvt Ltd.,, Kolkata, India

R. R. Bhattacharjee (✉)
Dean Research and Innovation and Department of Chemistry, The Assam Royal Global University, Guwahati, India
e-mail: rama.rb@snuniv.ac.in

Department of Chemistry, Sister Nivedita University, New Town, Kolkata 700156, West Bengal, India

A. K. Mandal et al. (eds.), *Functionalized Nanoparticles Hydrogels for Wound Healing*, Smart Nanomaterials Technology, https://doi.org/10.1007/978-981-95-0720-7_6

1 Introduction

It is necessary to understand cellular responses to sophisticated structures, dimensions and spatial arrangements. A material that is compatible with live tissues and can sustain in the specific cellular environment instead of getting rejected by the immune system is a biomaterial [1]. Biomaterials can be in the form of scaffolds, implants, wound cover, prosthetics etc. Such materials must mimic the structure and nature of the native tissue, for which it is to be used. However, it is not possible to fabricate a material in its pristine form that has all the qualities of a bioengineered scaffold. Recent advancements of doping and coating these materials with bioactive compounds which have antimicrobial and anti-inflammatory properties have resulted in better interaction of the scaffold with the surrounding cells [2]. A variety of soft and hard materials in the form of hydrogels, sintered pellets, fibrous mats, porous templates etc. have been studied for their application in various tissue engineering products [3]. Nanosizing these materials tend to improve their properties in cellular environments by increasing their surface to volume ratio, bioactivity and drug delivering ability outside many other unique properties.

While skin has a natural self-healing ability, management of chronic burns, cuts, lacerations and secondary infections on the skin remain challenging clinical problems [4]. Wound healing progresses through a sequential, highly organized series of events: haemostasis, inflammation, angiogenesis, proliferation, and re-epithelialization. Skin wounds a few millimeters deep into the epithelial layers are quickly repaired via re-epithelialization, wounds as deep as the dermal and hypodermal layers required special treatment. The dermis, comprising of a thick connective tissue, sandwiched in the middle of the epidermis and the hypodermis. When the wound is deeper than the epidermis (>0.2 mm), the cells differentiate along the periphery of the wound forming a scar. In most cases (chronic wounds, diabetic, microbial infestation, genetic abnormalities etc.) the natural healing process is unable to fully recover the portion of damaged tissue. Skin autografts and allografts are the preferred medicinal practices performed on such injuries. However, the increase in demand and low availability of such grafts has directed the need to create materials that could mimic the properties of the native ECM of the skin [5]. These materials are to act as scaffolds and provide skeletal support to the cells at the site of injury and nurture their development and proliferation, accelerating the wound healing process.

2 Nanofibre

Nanofibres, with diameters ranging from 1 to 100 nm (10^{-9} m), represent a revolutionary class of materials prized for their exceptional porosity, high surface-area-to-volume ratio, and tunable physicochemical properties [6]. Inspired by natural fibrous structures—such as spider silk, collagen networks, and plant cellulose—researchers have sought to replicate their remarkable mechanical strength, flexibility, and bioactive functionalities in synthetic systems. These biomimetic nanofibres are fabricated through advanced techniques like electrospinning, phase separation, and self-assembly [7], enabling precise control over fibre morphology and composition. Their unique structural characteristics make them ideal for diverse applications, including wound dressings, tissue engineering scaffolds, drug delivery systems and wearable biosensors [8]. By emulating nature's design principles, scientists continue to expand the frontiers of nanofibre technology for next-generation biomedical and industrial innovations.

3 Properties of Nanofibre

Polymeric and ceramic nanofibres exhibit unique properties that make them highly advantageous for various applications [9]. One of their most notable characteristics is their high surface area-to-volume ratio, which enhances their reactivity and makes them ideal for applications like filtration, catalysis, and drug delivery. Additionally, nanofibres possess inherent porosity, allowing for efficient fluid permeability and increased interaction with surrounding media [10]. Their capability for mass transport further enhances their functionality, enabling effective diffusion of gases, liquids, or even nanoparticles through their fibrous networks. The physical characteristics of nanofibres, such as tensile strength, flexibility, and thermal stability, vary significantly depending on the polymer or ceramic material used in their synthesis [11], allowing for tailored properties to meet specific industrial or biomedical needs. These combined attributes make polymeric and ceramic nanofibres versatile materials in advanced technological and scientific fields.

4 Application of Nanofibre in Modern Technology

Nanofibres have emerged as a revolutionary material with diverse applications across multiple fields [12]. In biomedicine, they play a crucial role in drug delivery, gene therapy, tissue engineering, cancer treatment, and regenerative medicine due to their high surface area, biocompatibility, and ability to mimic extracellular matrices [13]. Additionally, nanofibres enhance wound healing by promoting cell proliferation and acting as scaffolds for tissue regeneration [14]. Beyond healthcare, they are transforming energy storage technologies, where their use in supercapacitors and advanced batteries improves charge storage capacity and efficiency [15]. Furthermore, nanofibres are integral to sensor technology, where their high sensitivity and responsiveness enable the detection of chemical, biological, and environmental changes [16]. These multifaceted applications highlight the immense potential of nanofibres in advancing modern science and technology.

5 Manufacturing of Nanofibres

Natural, synthetic, or hybrid materials can be used to create nanofibres [17]. Hybrid materials in the form of nanofibrous mats have been researched and tested for their scaffolding properties. Polymeric materials have garnered significant attention in research due to their exceptional biomechanical compatibility, biocompatibility and low cost, making them promising candidates for various applications [18]. However, they show poor stimulated angiogenesis, cell adhesion and proliferation which limits bioactivity and overall repair capacity. Bio ceramics are advanced ceramic materials, either amorphous or crystalline, developed for safe and effective use in human body, facilitating tissue repair and organ substitution [19]. Although their mechanical hardness and brittle nature makes them ideal for hard tissue implants, they offer great potential for soft tissue regeneration due to their outstanding biological compatibility, capacity to release bioactive/therapeutic ions, regulate the microenvironment, and promote intercellular communication [20].

Bio ceramics: to regulate stem cells and their microenvironment for tissue regeneration.

Introducing active ionic dopants and antimicrobial coatings allow polymers and nanoscale bio ceramics to exhibit exceptional cell repair at the site of the wound [21]. Nanofibrous materials make up ideal scaffolds due to their outstanding ability of mimicking the ECM of the skin tissue. These materials have a high surface to volume ratio, micro/mesoporous structure allowing sufficient exchange of gases, fluids and wound exudates, a structure that provides cell mobility for cellular infiltration into the scaffold and micro roughness that allows fibroblasts and epithelial cells to adhere to the fibres [22].

It is difficult to synthesize standalone ceramic nano structures, however there are a few low-cost techniques like self-assembly, phase separation, freeze drying followed by higher yielding techniques like melt spinning, blow spinning, centrifugal jet spinning and polymer assisted electrospinning [23]. Alongside an examination of relevant techniques, this chapter will provide an overview of the various nanostructured bio ceramic materials that have been studied for their scaffolding capabilities [24].

6 Methods of Preparation

Self-assembly

Self-assembly is a bottom-up approach in which natural or synthetic nanofibres are formed through the spontaneous organization of individual macromolecules driven by covalent or non-covalent interactions. This process results in highly ordered and stable nanostructures, including nanofibres, with precise control over morphology and functionality.

Phase Separation

In this method, a ceramic precursor-containing gel is immersed in distilled water, leading to phase separation due to the inherent incompatibility between the aqueous and precursor phases. This spontaneous segregation induces the formation of well-defined nanostructures, including nanofibres, with tunable porosity and architecture.

Polymer-Assisted Freeze-Drying

This technique involves a two-step process: rapid freezing of a ceramic precursor solution in liquid nitrogen, followed by sublimation under controlled conditions. The directional growth of ice crystals during freezing dictates the scaffold's microstructure, yielding highly porous, unidirectional frameworks that serve as templates for composite nanofibre growth. The resulting materials exhibit tailored pore morphology and enhanced mechanical properties.

Melt Spinning

Melt spinning involves extruding a ceramic melt through a micron-sized orifice at elevated temperatures. External forces—such as gravity, air drag, centrifugal motion, or electrostatic charges—stretch and attenuate the molten material into continuous nanofibres. This method is particularly advantageous for producing high-strength, thermally stable nanofibres in scalable quantities.

Blow Spinning/Solution Blow Spinning

In this approach, a gelated ceramic sol is extruded through a fine nozzle and subjected to a high-velocity gas stream (typically air or nitrogen). The resulting shear forces elongate and thin the sol into ultrafine, thread-like structures with diameters in the nanometer range. This technique enables rapid, scalable production of nanofibres with controlled diameters and alignment, suitable for various advanced applications.

These fabrication methods highlight the versatility of nanofibre synthesis, allowing for precise control over structural and functional properties to meet diverse technological demands.

7 Electrospinning

The technique has gained popularity due to its versatile nature and quality fibre production. Upscaling the process of fabricating 3D nanofibrous material while retaining consistent morphological traits, at lower costs and commercial scales. These advantages give electrospinning the edge over other conventional methods of NF synthesis like template synthesis, self-assembly, polymerization, and sonochemical synthesis [6]. It is a voltage-driven electro-hydrodynamic process that relies on various parameters [9]. The fibre diameter ranges from sub-micro to nanoscale, which is obtained via adjusting voltage and flow rates of solution at the electrically charged spinneret. The solution at the tip of the spinneret experiences electrostatic polarization forces due to an applied electric field, creating a stable Taylor cone. When the polarizing field overcomes the force of surface tension, the repulsive interaction forms a charged jet of fibres that elongates on its way to the oppositely charged collector drum (collector may be a flat plate, rotary drum, disc, or mesh, the solvent evaporates, and fibres are deposited [7]. Continuous deposition on the collector (covered with an aluminum foil) results in the formation of a mat. Electrospinning has gained significant popularity in the last few decades, creating opportunities in biomedical and energy-related fields by producing NF mats with high porosity (>80%), pore size distribution (0.5–1.2 μm), and pore interconnectivity [3]. Today it is widely used for fabricating water [8] and air filtration [10] membranes, electrodes and separators for batteries [15], and in tissue engineering as scaffolds [5].

8 Polymeric Nanomats

Chain entanglement chemistry must be held of great importance when preparing electrospinnable solutions. Polymers are compounds with high molecular weights; thus, they can be prepared simply using a polymer precursor and a suitable solvent. Synthetic polymers like PLGA, PLU, PCL, PVA, and PVDF are often used for their excellent structural properties, low cost, and cytotoxicity [12]. Natural polymers like silk, collagen, and chitosan are known for their biocompatibility [17]. These are often used together in different concentrations while fabricating scar scaffolds [4]. While it is simpler to electrospin synthetic polymers with high chain entanglements, they lack a bioactive surface. Hence, they are modified with bioactive dopants and coatings.

9 Tackling Troubles with Ceramic Fibres

Ceramic solutions must be prepared such that there are sufficient interactions between the linear chains and neighboring chains. A viscoelastic solution ensures the existence of such entanglements. Generally, the ceramic sol has a volatile polymer binder added to enhance its viscoelasticity; this polymer is later removed through calcination [9]. Nevertheless, thermal stresses due to calcination create microcracks and fractures, resulting in mechanically weaker fibres [11, 19].

One can modify the synthesis of the ceramic sol such that the concentration of the ceramic chain network is optimal by controlling the degree of crosslinking and solvent evaporation [7]. Electrospun ceramic mats are presently being used for fabrication of filtration membranes, catalysts, sensors, biomaterials, and energy harvesting systems [3]. Materials classified as bioceramics include alumina, zirconia, calcium phosphates, titania, silica-based glasses, or glass ceramics and pyrolytic carbons, which have properties like high surface-to-volume ratio and aspect ratio [1, 2]. The presence of surface-active groups like -COOH, -NH2, and -OH allows adhesion of serum glycoproteins like fibronectins [18].

10 Silica Sol Synthesis

The sol–gel synthesis of spinnable silica sol with TEOS precursor is an acid-catalyzed open-system condensation reaction, as mentioned by [9]. The process involves the following compounds (Fig. 1):

- TEOS (tetraethyl orthosilicate) (98 wt.%)
- Ethanol (99 wt.%)
- DI Water (deionized water)
- HCl (hydrochloric acid) (36–38 wt.%), in the suitable molar ratio.

Initially, TEOS mixed with ethanol has to be stirred for a certain period. To the TEOS-ethanol mixture, a solution containing DI water and hydrochloric acid is added dropwise under vigorous stirring at 350–400 rpm. The solution, kept around room temperature under stirring for 10–15 min, undergoes the first phase exothermal hydrolysis (Fig. 2). The solution is then kept in a water bath/oil bath at 80 °C (endothermic condensation) for 2–3 h or more, depending on the volume of the sample, until the volume reduces to 3/8th of the initial [9]. The sol, when cooled to room temperature (open system), turns moderately viscous (100–180 mPa·s) and must be sealed for electrospinning. This step is the second-phase condensation reaction, where acid-catalyzed linear chains of SiO_2 form, increasing molecular weight, producing sufficient chain entanglements, and optimizing solvent evaporation to form a viscous, spinnable sol [20].

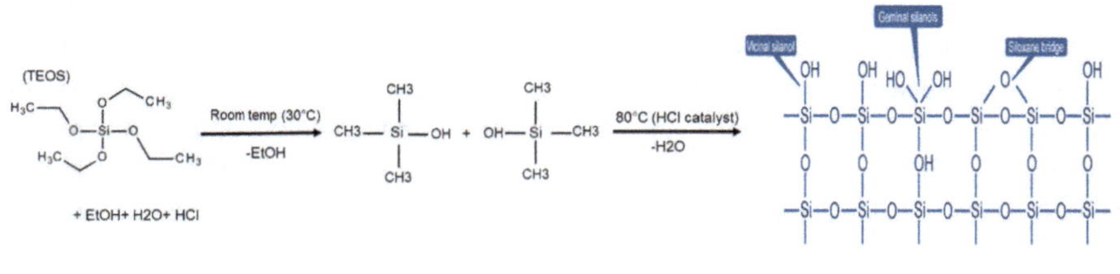

Fig. 1 First phase exothermal hydrolysis followed by second phase acid catalysed condensation polymerization at 80 °C and final phase loosely crosslinked silica

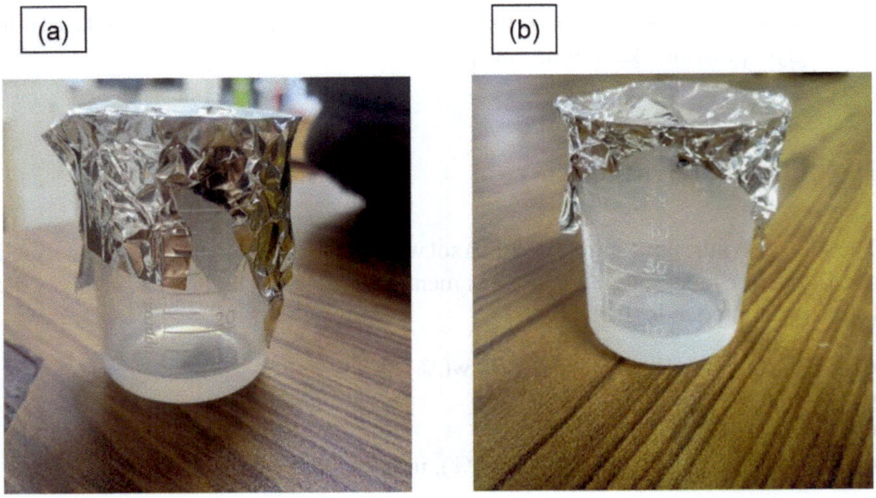

Fig. 2 **a** Homogenous clear sol of silica for prepared in open system, **b** turbid sol for sample sol made in closed system (premature gelation)

11 Limitations in Open System and Lack of Control

The ethanol concentration in the sol must be carefully optimized such that complete evaporation occurs during the stirring period, which coincides with the crosslinking reaction time at 80 °C [9, 24]. However, precise control of ethanol evaporation is challenging in open systems, and excess ethanol can lead to the formation of insoluble SiO_2 crystals [7, 23]. Such crystalline sols.

exhibit high surface tension and reduced viscoelasticity, often resulting in beaded, non-uniform fibres or electrospraying rather than continuous electrospinning. The degree of crosslinking plays an equally critical role, as acid-catalyzed sol–gel reactions initially produce linear chains that progressively aggregate into three-dimensional networks, ultimately forming spherical SiO_2 particles [9]. The resulting solution remains spinnable for only a few hours under sealed conditions before transitioning into a transparent gel. Maintaining the system at an optimal pH of approximately 2 is essential to promote loosely crosslinked structures by slowing the condensation polymerization rate and preventing the irreversible growth of dense clusters, a transition often signaled by the onset of turbidity [9]. In closed systems, matching the viscosities achieved in open systems requires extended condensation reactions, which risk premature gelation due to the formation of an interconnected SiO_2 network. To mitigate this, ethanol evaporation should be carefully controlled using a rotary evaporator [9]. Furthermore, maintaining the molar ratio of water to TEOS (r ratio) between 1.5 and 2 is crucial for ensuring spinnability, though achieving precise control of this ratio remains experimentally challenging [7].

12 Electrospinning Parameters

The typical electrospinning setup is capable of operating at voltages up to 30 kV and is equipped with two precision pumps [6]. Pump 1, connected to the plunger, controls the flow rate of the polymer solution through its X-directional motion (X-scan), ensuring consistent delivery to the spinneret tip. The Y-scan, on the other hand, corresponds to the Y-directional movement of the horizontally aligned rotary drum/flat plate collector, positioned perpendicular to the syringe tip [9]. When an electric field is applied, a potential gradient is established between the spinneret and the grounded aluminium foil-covered collector [7]. At the spinneret tip, the conductive solution is subjected to competing forces, including surface tension, gravity, and electrostatic repulsion [20].

For fibre formation to occur, the solution must overcome surface tension, transitioning from a spherical droplet to a Taylor cone, after which a stable polymer jet is ejected, yielding ultrafine fibres [25]. Maintaining optimal voltage and flow rates is critical to preserving the stability of the Taylor cone and the ensuing jet [24]. For example, Spinnability tests of silica sol conducted using a rotary type, had parameters adjusted as follows: applied voltage (18–24 kV, flow rate 1.6–3 mL/h)-with lower flow rates preferred for finer gauge needles (24-gauge-and tip-to-collector distance 10–15 cm) [9]. Given the brittle nature of the resulting fibres, a shorter collection distance (10–12 cm), and the drum rotation speed was maintained at 50–150 rpm to ensure uniform deposition [3]. Field-emission scanning electron microscopy (FESEM) or W-scanning electron microscopy (WSEM) analysis of the collected fibres, deposited as thin sheets on aluminium foil, confirms the formation of continuous fibres with diameters in nanometre range [14]. Fibre collection will commence immediately upon achieving a stable Taylor cone and continued to ensure sufficient mat formation [6].

13 Stability and the Taylor Cone

The transition of a homogeneous sol into a gel during electrospinning plays a critical role in determining the stability of the Taylor cone and the resulting fibre morphology. As the sol is subjected to an electric field, it undergoes progressive gelation, wherein the elongation and stretching of the partially gelated sol—coupled with solvent evaporation—lead to complete solidification into a fibrous structure. However, this process is highly sensitive to the kinetics of gelation and the rheological properties of the sol (Fig. 3).

A key challenge arises when premature gelation occurs at the spinneret tip, causing accumulation of crosslinked material that disrupts the formation of a stable Taylor cone. This is particularly problematic in systems with highly crosslinked polymer chains, where rapid gelation can lead to clogging, necessitating frequent tip

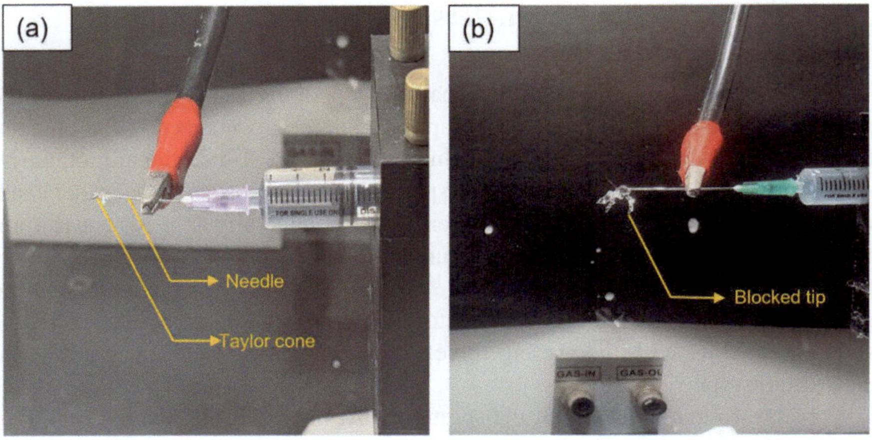

Fig. 3 **a** Stable Taylor cone at the tip of the needle and **b** blocked tip due to sol improper curing time

cleaning to maintain continuous electrospinning. The viscoelasticity of the sol, evaporation rate of the solvent, and applied electric field strength must be carefully balanced to prevent such instabilities.

The equilibrium of forces governing Taylor cone dynamics can be conceptually represented through a schematic diagram (Fig. 4). In this balanced system, electrostatic stretching forces counteract the solution's inherent surface tension. Viscoelastic forces arising from polymer chain entanglement modulate the cone's deformation behaviour, while gravitational influences remain negligible relative to the dominant electric forces. Hydrodynamic pressure from the syringe pump provides the necessary material supply, completing the force balance required for stable cone formation. When these forces maintain equilibrium, continuous jet ejection occurs; however, instability arises when gelation kinetics outpace stretching or when solvent evaporation occurs too rapidly, manifesting as bead defects or fibre discontinuities.

Fig. 4 Forces acting on Taylor cone

```
    [Syringe Tip]
      |
      • (Taylor Cone)
     /|\
    / | \
   ▲  ▲  ▲
   |  |  |
Electrostatic  Surface   Viscoelastic
 Force (↑)   Tension (↔)  Force (□)

Gravity (↓)
Hydrodynamic Pressure (→)

~~~ Jet ~~~> [Collector Drum]
(Solvent Evaporation: "..." dots)
```

14 Silica Nanomat

Silica (SiO_2) is known for it's abundance in nature. It's chemical inertness and biocompatibility have created a growing demand in biomedical applications. Once synthesized in nanoscale structures, silica tends to show better physiochemical properties as compared to bulk silica. Recent advancements have employed silica nanoparticles for drug delivery, improved fibrous scaffolds in combination with synthetic polymers, for bioactive glasses and many more (Fig. 5).

Nevertheless, once the techniques of synthesis are optimized, ceramic silica will be able to compete with synthetic polymers in the market as a nanofibrous skin scaffold.

The aim to produce flawless nanomat is to synthesize a silica sol that would form reproducible beadless silica nanofibres when electrospun with suitable parameters. A flexible mat is generated when silica solution (r ratio (HCl: TEOS) of 0.05 and suitable amount of solvent evaporation in open system for 2–2.5 h followed by cooling at room temperature) was electrospun at 22–24 kV, flow rates of 2–2.5 ml/h at tip-to-collector distance of 10 cm. Scanning electron microscopy (SEM) indicated the formation of defect free yet non-uniform (200–800 nm) and non-woven fibres. The mat was stored under sealed and unsealed ambient temperature and humidity (50–60%). This resulted in change in surface properties of the mat as indicated by contact angle and FT-IR tests. Initial experiments of contact angle measurements performed on the prepared mat indicated hydrophobic surface (100–125° angle of surface with DI water drop). The same material upon storage for a week at relative humidities > 50% turns the mat into a highly hydrophilic material. Swelling experiment performed on the mat in phosphate-buffered saline (PBS) solution indicated that the mats had almost no saturation period and degraded at a steady rate. These results concluded that it is indeed possible to synthesize pristine silica mat without a polymeric template. Improving swelling property of the material in future can serve as a beneficial addition to the application of the mat as wound cover.

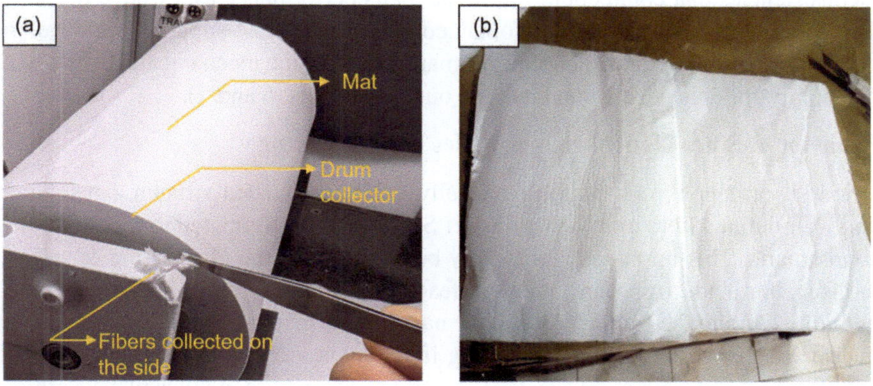

Fig. 5 Fibre deposition and collection for sample mat **a, b**

15 Nanomat as Wound Healing Scaffold

It is a hypothesis that the mat produced can act as a bandage material or an implant at the site of the injury and support the proliferating cells by mimicking the supportive tissue of the ECM. It is required to generate uniform beadless fibres with diameters < 700 nm and significantly high inter fibre pore connectivity. These properties are visible under SEM and ensures a high surface to volume ratio that positively impact cellular infiltration and proliferation through the scaffold material. It must keep the site of the wound moist and allow exchange of gases with the environment and the cells. The surface wettability depicts the adhesion of proteins which may be hydrophobic or hydrophilic in nature. The swelling tests must be such that the scaffold absorbs wound exudates without degrading too early.

16 Characteristics of Silica Nanomat

The following characterizations are done to understand the properties of nanomats.

Scanning electron microscopy

Tungsten scanning electron microscopy (W-SEM) serves as a valuable tool for characterizing ceramic nanofibrous mats designed for wound healing applications. The technique provides high-resolution visualization of critical structural features including fibre diameter uniformity, surface roughness, and three-dimensional porous architecture—all essential parameters that influence cellular interactions and exudate management in wound care. Secondary electron imaging clearly reveals the interconnected pore network that facilitates nutrient transport and cell migration while maintaining mechanical integrity. EDS elemental mapping confirms the presence of bioactive ceramic components like silica or calcium phosphate that promote tissue regeneration. W-SEM analysis also enables quality control by detecting structural defects or inconsistencies that could compromise the mat's performance as a wound dressing. The detailed microstructural data obtained through W-SEM supports the development of optimized ceramic nanofibrous mats that combine mechanical stability with bioactivity for enhanced wound healing outcomes (Figs. 6 and 7).

Behavior of Silica Nanofibrous Samples During Drying

Dried silica nanofibrous samples typically exhibit enhanced structural resolution and well-defined fibre morphology under SEM imaging compared to their hydrated counterparts. This improved clarity may be attributed to the elimination of surface moisture that tends to obscure nanoscale features in samples stored under high relative humidity. During the drying process, the nanofibrous mats often undergo significant interfacial changes, sometimes resulting in complete detachment from underlying substrates such as aluminum sheets. The hygroscopic nature of silica plays a critical role in this behavior, as atmospheric moisture induces a gradual transformation from

Fig. 6 The samples were sputter coated (Au coated) for 60 s and analysed with SEM (CARL ZEISS EVO 50) fitted with W-filament at an accelerating voltage of 20 kV

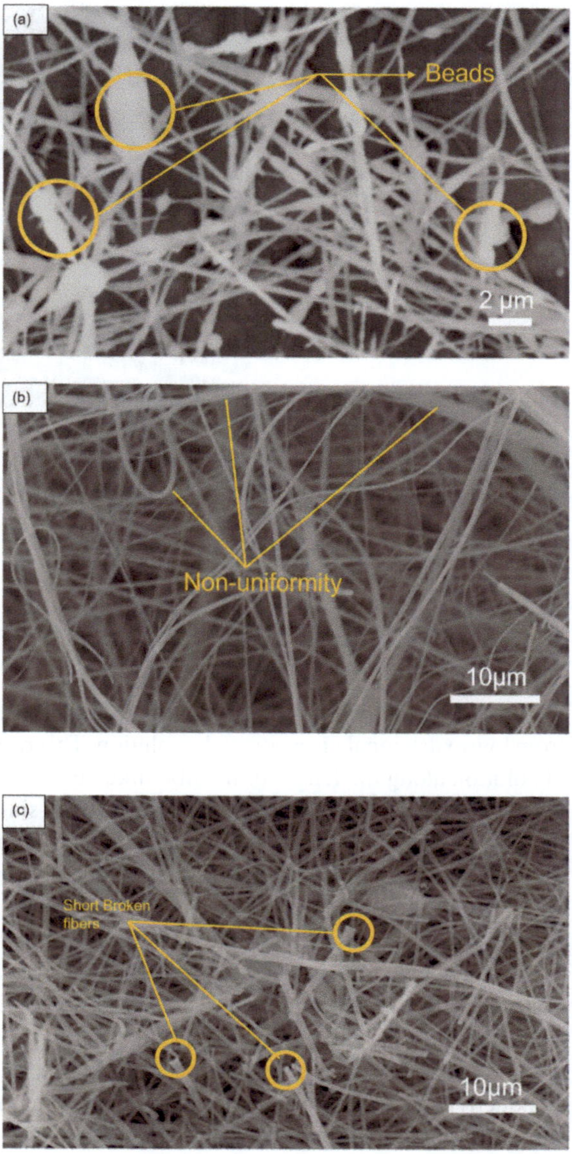

Fig. 7 WSEM images of **a** sample with highly beaded fibres, **b** sample with low to no beads and continuous fibres, **c** sample with short broken fibres

hydrophobic to hydrophilic character through the hydrolysis of Si–O-Si bridges into silanol (Si–OH) groups. This surface chemistry transition has important implications for the material's interfacial properties and requires further investigation to fully understand its effects on phenomena such as protein adsorption, cellular interactions, and environmental stability. The drying-induced modifications in silica nanofibres highlight the need to carefully control storage conditions and account for atmospheric

Fig. 8 Fibreous mat being dried in an electric oven

exposure when characterizing these materials for biomedical or filtration applications (Fig. 8).

17 Physics Behind Beaded Fibres

The phenomenon behind bead formation is called Rayleigh instability which result from low extensional viscosity of the solution. This results in the formation of clusters of solution along the length of the fibre like beads on a necklace. It is considered as a defect when working with skin tissue scaffolds, since, it creates non uniformity in the fibrous mat and reduces the surface area to volume ratio which affects cellular infiltration and might also negatively affect its mechanical properties (Fig. 9).

The formation of defect-free nanofibres depends critically on optimizing solution properties such as viscosity, surface tension, and conductivity. While studies suggest an ideal viscosity range (typically 100–200 mPa·s for polymeric and sol–gel systems) [20], the optimal value varies with material composition and processing conditions. For instance, in silica sols, the viscosity window coincides with a distinct volume reduction during heating, marking the point of optimal spinnability [25]. Conductivity also plays a pivotal role in electrospinning,while some systems achieve sufficient jet polarization without additives, this raises broader questions: How does salt-mediated conductivity enhancement affect spinnability across different materials? What post-treatment trade-offs emerge (e.g., salt residues, thermal stability? Equally critical are solvent evaporation dynamics and precursor stoichiometry—incomplete drying or deviations from optimal reactant ratios (e.g., acid/TEOS < 0.05 consistently lead to bead defects [20, 25].

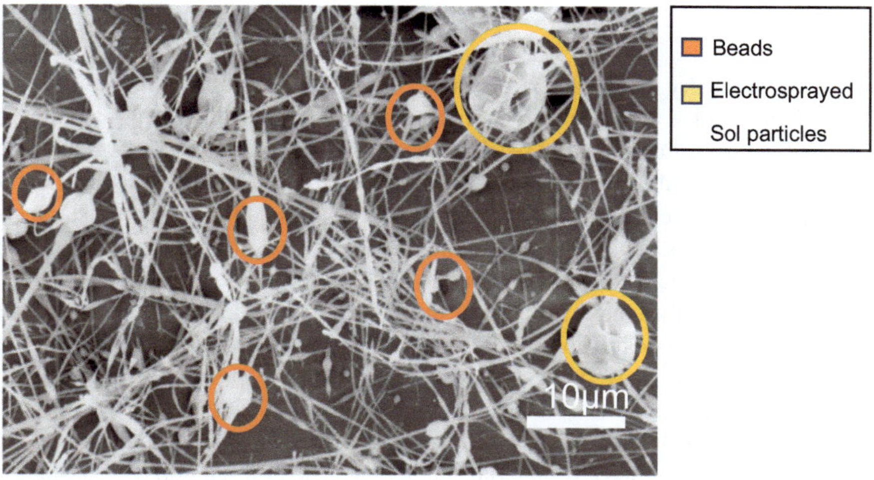

Fig. 9 Highly beaded fibres of sample r-0.05-SiO_2c under WSEM

18 Contact Angle Goniometer

Silica ceramic nanofibres demonstrate humidity-responsive wettability transitions that are particularly advantageous for wound healing applications. Their surfaces undergo a controlled shift from hydrophobic to super-hydrophilic states under ambient conditions, mediated by the conversion of surface siloxane (Si–O–Si) to silanol (Si–OH). This dynamic amphiphilic behavior, which is consistent with established literature, enables optimal fluid management throughout the healing process—initially limiting excessive exudate absorption before gradually promoting cellular interactions as hydrophilicity increases. The material's environmental sensitivity allows autonomous adaptation to wound conditions while maintaining structural integrity. Combined with the inherent biocompatibility of silica and the biomimetic architecture of nanofibrous scaffolds, these tunable wettability properties create an advanced platform for wound regeneration that responds naturally to the changing requirements of healing tissue.

19 Swelling Test

Swelling tests are essential for evaluating the fluid uptake capacity of nanofibrous ceramic materials intended for wound healing applications. The test is typically conducted by immersing dried samples in phosphate-buffered saline (PBS) or simulated physiological fluids to assess their hydration behavior. The swelling percentage is calculated using the standard formula:

$$\text{Swelling\%} = [(W_d - W_f)/W_f] \times 100\%$$

where W_d is the dry weight and W_f is the wet weight after immersion.

For hydrophobic silica-based nanofibrous mats, a gradual swelling response is expected, with a saturation point reached after prolonged exposure. However, due to silica's inherent solubility in aqueous environments, some degree of structural degradation may occur over time, potentially limiting long-term stability. In contrast, superhydrophilic variants may exhibit rapid initial swelling but are prone to accelerated dissolution before achieving full saturation.

To optimize performance, material modifications such as crosslinking, composite formation with biodegradable polymers, or surface functionalization can be employed to enhance swelling capacity while mitigating premature degradation. Controlling pore structure, fibre density, and silica composition further influences fluid retention and mechanical integrity during hydration. These adjustments are critical for ensuring that the material maintains structural support and sustained functionality in wound healing applications.

20 Fourier Transform Infrared Spectroscopy (FTIR)

Fourier-transform infrared spectroscopy (FTIR) is a widely used analytical method for characterizing the chemical structure and bonding interactions in silica-based nanofibrous materials. The technique identifies functional groups and molecular vibrations through distinct absorption peaks in the infrared spectrum. For silica-based systems, the spectrum typically reveals Si–O–Si asymmetric stretching vibrations in the range of 1050–1100 cm^{-1}, along with Si–O bending vibrations near 780–800 cm^{-1}. A peak around 950 cm^{-1} corresponds to Si–OH stretching vibrations, while a broad absorption band near 3350–3400 cm^{-1} arises from O–H stretching, indicating the presence of surface silanol groups or adsorbed water.

The absence of organic residues, such as C–H stretching vibrations (2800–3000 cm^{-1}) or C = O/C–H bending modes (1400–1800 cm^{-1}), confirms the complete hydrolysis of alkoxide precursors like tetraethyl orthosilicate (TEOS) into silanol groups. Environmental factors, particularly humidity, strongly influence the spectral features: hydrophilic silica samples often exhibit enhanced O–H and H–O–H stretching bands due to water absorption, whereas hydrophobic variants display more intense Si–O–Si network vibrations, reflecting higher crosslinking density and reduced hygroscopicity. By correlating these spectral features with material properties, FTIR analysis provides critical insights into compositional integrity, hydrolysis efficiency, and environmental stability, making it indispensable for developing optimized silica-based nanofibres for biomedical applications.

21 Biodegradability

In vitro biodegradation testing is a critical evaluation method used to assess the stability and degradation behavior of biomaterials under physiologically relevant conditions. This standardized test involves incubating material samples in simulated body fluid (SBF), a solution with ionic composition similar to human blood plasma, at 37 °C to mimic physiological temperature. The protocol typically requires preparing multiple samples from different regions of the scaffold to account for material heterogeneity.

During testing, samples are periodically removed from the SBF environment, carefully dried, and weighed at predetermined intervals (e.g., 1, 3, 5, and 7 days). The fractional weight change is calculated using the formula:

$$\text{Fractional weight change} = [(W_i - W_f)/W_i]$$

where W_i represents the initial dry weight and W_f denotes the weight after immersion. For highly porous silica-based scaffolds, the degradation profile typically follows a characteristic pattern: an initial weight gain phase due to fluid absorption into the porous network, followed by a gradual weight loss as material dissolution occurs. The degradation rate can be quantified by measuring the slope of weight loss over time, with evaluation periods typically spanning one week to one month depending on the expected material lifetime.

22 Mechanical Characterization

The evaluation of mechanical properties is essential for assessing the suitability of silica-based nanofibrous mats as wound healing biomaterials, as they must withstand physiological stresses while maintaining structural integrity. Tensile testing is the primary method for determining key mechanical parameters, including tensile strength, elongation at break, and Young's modulus, which collectively define the material's response to applied loads. For nanofibrous ceramic mats, uniaxial tensile tests are performed on standardized specimens using a universal testing machine, where stress–strain curves are generated to quantify stiffness, strength, and ductility. Young's modulus, a measure of elastic deformation resistance, is derived from the linear elastic region of the curve, while tensile strength indicates the maximum stress the material can endure before failure.

In addition to tensile testing, nanoindentation is employed to evaluate localized mechanical behavior, providing insights into hardness and elastic modulus at the nanoscale. This technique is particularly useful for assessing heterogeneous fibre networks, where variations in density and porosity influence mechanical performance. Micro-indentation methods may also be applied to measure compressive strength, which is critical for understanding how the material behaves under pressure in a wound environment. Furthermore, dynamic mechanical analysis (DMA) can be utilized to study viscoelastic properties, such as storage and loss moduli, under cyclic loading conditions that simulate physiological stresses.

Since ceramic nanofibres can exhibit brittleness, optimizing mechanical properties often involves reinforcing the silica matrix with polymeric or bioactive additives to enhance flexibility without compromising structural stability. These characterization techniques collectively ensure that the material possesses the necessary mechanical robustness for wound healing applications, balancing strength, elasticity, and durability to support tissue regeneration while resisting premature degradation.

23 Conclusion and Outlook on Electrospun Ceramic Mats

Nanofibrous ceramics based on silica have emerged as a compelling class of biomaterials for advanced wound care applications, demonstrating the successful convergence of materials science and biomedical engineering. The development of these materials through sol–gel electrospinning techniques represents a significant advancement in fabrication methodologies, offering precise control over fibre morphology and composition at the nanoscale. This approach has opened new possibilities for creating biomimetic scaffolds that combine the structural advantages of ceramic matrices with the functional requirements of wound healing environments.

The unique properties of silica nanomats—including their tunable porosity, high surface area, and tailorable degradation kinetics—position them as versatile platforms for wound management. Their ability to mimic the extracellular matrix architecture while providing bioactive surfaces for cell interaction addresses critical needs in both acute and chronic wound treatment. The inherent versatility of the silica matrix allows for incorporation of therapeutic agents, growth factors, and antimicrobial components, enabling multifunctional wound dressing capabilities.

Looking forward, several promising directions emerge for this technology. Future research should focus on three key areas: (1) optimization of mechanical properties to balance flexibility and durability for clinical handling, (2) development of standardized protocols for assessing long-term biocompatibility and degradation profiles, and (3) exploration of hybrid material systems that combine silica with natural polymers or other bioactive ceramics to enhance biological performance. The translation of these materials from laboratory research to clinical applications will require concerted efforts in scalable manufacturing processes and rigorous preclinical validation.

As the field progresses, silica-based nanofibrous ceramics are poised to make significant contributions to regenerative medicine. Their potential extends beyond wound healing to applications in drug delivery, tissue engineering, and biosensing. The continued evolution of these materials will benefit from interdisciplinary collaborations that bridge materials science, biology, and clinical medicine, ultimately leading to innovative solutions for complex healthcare challenges. The journey from fundamental research to clinical implementation will undoubtedly present challenges, but the unique advantages of these nanofibrous ceramic systems offer compelling reasons to pursue their development as next-generation wound care technologies.

References

1. Hench LL (1991) Bioceramics: from concept to clinic. J Am Ceram Soc 74(7):1487–1510. https://doi.org/10.1111/j.1151-2916.1991.tb07132.x
2. Macha IJ, Sufi S (2020) Novel slow drug release bio ceramic composite materials for wound dressing applications: potential of natural materials. SN Appl Sci 2(2):176. https://doi.org/10.1007/s42452-020-1980-4
3. Esfahani H, Jose R, Ramakrishna S (2017) Electrospun ceramic nanofiber mats today: synthesis, properties, and applications. Materials 10(11):1238. https://doi.org/10.3390/ma10111238
4. Memic A, Abdullah T, Mohammed HS, Joshi Navare K, Colombani T, Bencherif SA (2019) Latest progress in electrospun nanofibers for wound healing applications. ACS Appl Bio Mater 2(3):952–969. https://doi.org/10.1021/acsabm.8b00637
5. Gupta KC, Haider A, Choi Y-R, Kang I-K (2014) Nanofibrous scaffolds in biomedical applications. Biomater Res 18(1):5. https://doi.org/10.1186/2055-7124-18-5
6. Xue J, Wu T, Dai Y, Xia Y (2019) Electrospinning and electrospun nanofibers: methods, materials, and applications. Chem Rev 119(8):5298–5415. https://doi.org/10.1021/acs.chemrev.8b00593
7. Brinker CJ, Assink RA (1989) Spinnability of silica sols structural and rheological criteria. J Non-Crystall Solids 111(1):48–54. https://doi.org/10.1016/0022-3093(89)90263-4
8. Tang Y, Chen Z, Sun X, Chong C, Yan X, Li M, Xu J (2022) Electrospun nanofiber-based membranes for water treatment. Polymers 14(10):2004. https://doi.org/10.3390/polym14102004
9. Geltmeyer J, Van Acker T, Lava K, De Buysser K, De Clerck K (2016) The influence of tetraethoxysilane sol preparation on the electrospinning of silica nanofibres. J Sol-Gel Sci Technol 77(2):453–462. https://doi.org/10.1007/s10971-015-3868-0
10. Zhou Y, Liu Y, Zhang M, Feng Z, Yu D-G, Wang K (2022) Electrospun nanofiber membranes for air filtration: a review. Nanomaterials 12:1077. https://doi.org/10.3390/nano12071077
11. Haddadi SA, Ramazani A, Farshid B (2018) Mechanical and piezoelectric characterizations of electrospun PVDF-nanosilica fibrous scaffolds for biomedical applications. Mater Today: Proceed 5(7):15710–15716. https://doi.org/10.1016/j.matpr.2018.04.157
12. Mir M, Ali MN, Barakullah A, Gulzar A, Arshad M, Fatima S, Asad M (2018) Synthetic polymeric biomaterials for wound healing: a review. Prog Biomater 7(1):1–21. https://doi.org/10.1007/s40204-018-0083-4
13. Ravichandran R, Venugopal JR, Sundarrajan S, Mukherjee S, Sridhar R, Ramakrishna S (2013) Hierarchical mesoporous silica nanofibers as multifunctional scaffolds for bone tissue regeneration. J Biomater Sci Polym Ed 24(17):1988–2005. https://doi.org/10.1080/09205063.2013.814100
14. Ma Z, He J, Tan D, Teng Y, Dong G, Zhou J, Qiu J, Zhang M (2011) Silver nanoparticles decorated, flexible SiO_2 nanofibers with long-term antibacterial effect as reusable wound cover. Colloids Surf, A 387(1–3):57–64. https://doi.org/10.1016/j.colsurfa.2011.07.019
15. Ji L, Yao Y, Toprakci O, Lin Z, Liang Y, Shi Q, Medford AJ, Millns CR, Zhang X (2010) Fabrication of carbon nanofiber-driven electrodes from electrospun polyacrylonitrile/polypyrrole bicomponents for high-performance rechargeable lithium-ion batteries. J Power Sourc 195(7):2050–2056. https://doi.org/10.1016/j.jpowsour.2009.10.021

16. Waldrop K, Pintauro PN (2020) Application of electrospinning for the fabrication of proton-exchange membrane fuel cell electrodes. Curr Opin Electrochem 21:257–264. https://doi.org/10.1016/j.coelec.2020.03.007
17. Mogoşanu GD, Grumezescu AM (2014) Natural and synthetic polymers for wounds and burns dressing. Int J Pharm 463(2):127–136. https://doi.org/10.1016/j.ijpharm.2013.12.015
18. Lee MH, Ducheyne P, Lynch L, Boettiger D, Composto RJ (2006) Effect of biomaterial surface properties on fibronectin-α5β1 integrin interaction and cellular attachment. Biomaterials 27(9):1907–1916. https://doi.org/10.1016/j.biomaterials.2005.11.003
19. Veverková I, Lovětinská-Šlamborová I (2016) Modified silica nanofibers with antibacterial activity. J Nanomater 2016:2837197. https://doi.org/10.1155/2016/2837197
20. Choi S-S, Lee SG, Joo CW, Im SS, Kim SH (2003) Silica nanofibers from electrospinning/sol-gel process. J Mater Sci Lett 22(12):891–893. https://doi.org/10.1023/A:1023794114561
21. Loccufier E, De Geyter N, Neyts E, De Clerck K (2019) Electrospinning of silica nanofibers without carrier polymer for advanced engineering applications. In: 19th World Textile Conference (AUTEX 2019): textiles at the crossroads. AUTEX, Ghent, Belgium
22. Han D, Gouma PI (2006) Electrospun bioscaffolds that mimic the topology of extracellular matrix. Nanomed Nanotechnol Biol Med 2(1):37–41. https://doi.org/10.1016/j.nano.2005.12.003
23. Sakka S, Kozuka H (1988) Rheology of sols and fiber drawing. J Non-Crystall Solids 100(1–3):142–153. https://doi.org/10.1016/0022-3093(88)90015-5
24. Zhou W, Xie Y, Zhang L, Sun X, Ma J, She S (1992) Spinnability of $Si(OC_2H_5)_4$-C_2H_5OH-H_2O-HCl solutions. Mater Lett 13(2–3):166–168. https://doi.org/10.1016/0167-577X(92)90062-3
25. Geltmeyer J, De Roo J, Van den Broeck F, Martins JC, De Buysser K (2013) Optimum sol viscosity for stable electrospinning of silica nanofibres. J Sol-Gel Sci Technol 67(1):188–195. https://doi.org/10.1007/s10971-013-3068-8

Underpinning the Therapeutic Potential of Biopolymer-Based Nano Medications in Wound Healing

Amit Ghosh and Amlan Das

Abstract Wound healing is a complex, tightly controlled biological process that involves four main stages: hemostasis, inflammation, proliferation, and remodelling. When these stages are disrupted by microbial infection or external stressors, healing can be delayed, potentially leading to chronic wounds. Conventional wound dressings have limited ability to fight infection and may not adequately support the intricate process of tissue regrowth. The development of biopolymer-based nanotechnology offers a novel solution, combining antimicrobial action with improved compatibility and degradability within the body. Biopolymers, including polysaccharides, proteins, and synthetic polymers, serve as adaptable frameworks for crafting sophisticated wound dressings. When integrated into nanoparticle-mediated delivery systems, they facilitate regulated drug dissemination, focused antimicrobial action, and the adjustment of cellular reactions vital for tissue restoration. Polymeric nanoparticles, hydrogels, and nanofiber-based dressings, when impregnated with bioactive compounds like antimicrobial peptides, metallic nanoparticles, and growth factors, have exhibited enhanced wound healing properties. These dressings effectively hinder bacterial settlement and biofilm development while promoting fibroblast multiplication, collagen accumulation, and angiogenesis, all of which are essential for successful tissue mending. Biopolymer wound dressings, enhanced by nanotechnology, provide numerous benefits such as moisture retention, oxygen permeability, and exudate absorption, thereby establishing an ideal environment for healing. Additionally, they lessen the reliance on traditional antibiotics, thus reducing the possibility of antimicrobial resistance. Cutting-edge progress in biomaterial engineering underscores the capacity of biopolymer-derived nanomedicines to transform wound management through expedited healing, decreased susceptibility to infection, and facilitated

A. Ghosh
Department of Biological Sciences, School of Life Sciences and Biotechnology, Adamas University, Kolkata, India

A. Das (✉)
Department of Microbiology, Royal School of Biosciences, The Assam Royal Global University, Guwahati, India
e-mail: adas5@rgu.ac

Department of Biochemistry, Royal School of Biosciences, The Assam Royal Global University, Guwahati, India

A. K. Mandal et al. (eds.), *Functionalized Nanoparticles Hydrogels for Wound Healing*, Smart Nanomaterials Technology, https://doi.org/10.1007/978-981-95-0720-7_7

tissue regeneration. Future investigations should prioritize clinical implementation, guaranteeing that these novel therapies gain broad acceptance in medical practice.

Keywords Wound healing · Hemostasis · Inflammation · Proliferation · Epithelialization · Extracellular matrix

1 Introduction

Wound healing is a meticulously orchestrated biological process that repairs damaged tissue through four overlapping stages: hemostasis, Inflammation, proliferation, and remodeling [1–3]. Disruptions caused by external or internal factors can delay these stages, resulting in incomplete healing and potentially leading to chronic wound conditions. One of the most prevalent obstacles in wound recovery is the infiltration and colonization of pathogens at the injury site, which can significantly impede natural healing mechanisms [4].

As the body's largest organ, the skin serves as a primary barrier against environmental damage and microbial invasion while also playing vital roles in fluid regulation and nutrient retention [5, 6]. Structurally, the skin comprises two main layers: the epidermis and the dermis [7]. The epidermis, the outermost layer, contains sub-layers that vary by body region and primarily consist of keratinocytes, melanocytes, Langerhans cells, and Merkel cells. Beneath it, the dermis provides support through extracellular matrix proteins like collagen, elastin, and glycosaminoglycans, all produced by fibroblasts [8, 9]. When injured, the skin activates a multi-phase healing process that includes hemostasis, Inflammation, proliferation, epithelialization, and maturation, each involving complex cellular and biochemical interactions aimed at restoring tissue integrity and function [10, 11].

The skin's microbiota naturally guards against harmful bacteria. However, when pathogenic microorganisms surpass critical thresholds and form biofilms, they can significantly hinder wound healing. For example, Staphylococcus aureus frequently disrupts the early healing stages, while Pseudomonas aeruginosa and Escherichia coli are more common in chronic wounds, where they invade deeper tissues [12]. These infections exacerbate Inflammation, prolong wound recovery, and may necessitate advanced antimicrobial interventions. Effective wound management requires antimicrobial dressings that not only prevent infections but also foster a conducive healing environment. Nevertheless, the rise of antibiotic-resistant bacteria underscores the urgent need for innovative, non-toxic wound care materials [13–15].

An ideal wound dressing should address barriers to healing by integrating antibacterial properties, biocompatibility, and biodegradability. It must maintain an optimal moisture level, facilitate oxygen exchange, absorb exudates, minimize irritation, and promote cellular activity to enhance tissue regeneration [11, 16, 17]. Antibacterial agents—including antibiotics, nanoparticles (NPs), and cationic organic compounds—play a critical role in infection prevention and are commonly incorporated into advanced wound dressings such as films, hydrogels, sponges, scaffolds, and foams. These materials often combine natural or synthetic polymers with bioactive agents to regulate bacterial growth and stimulate extracellular matrix formation, thereby accelerating cellular activity and healing [18–25].

Recent advancements in nanotechnology have revolutionized wound care by introducing polymeric nanoparticles, which offer unique benefits such as targeted delivery, enhanced antibacterial activity, and the ability to mimic natural tissue environments. Polysaccharide-based biopolymers and synthetic nanoparticles have emerged as effective tools in developing next-generation wound dressings. By incorporating these advanced materials, researchers aim to overcome existing challenges, offering new hope for efficient and expedited wound healing solution [26–29].

2 Different Stages of Wound Healing

Wound healing consists of five distinct phases: hemostasis, Inflammation, proliferation, epithelialization, and maturation. These phases involve intricate cellular and mediator interactions. These phases enhance wound healing and restore the skin's normal structure and function [30–32] (Fig. 1).

a. **Hemostasis**

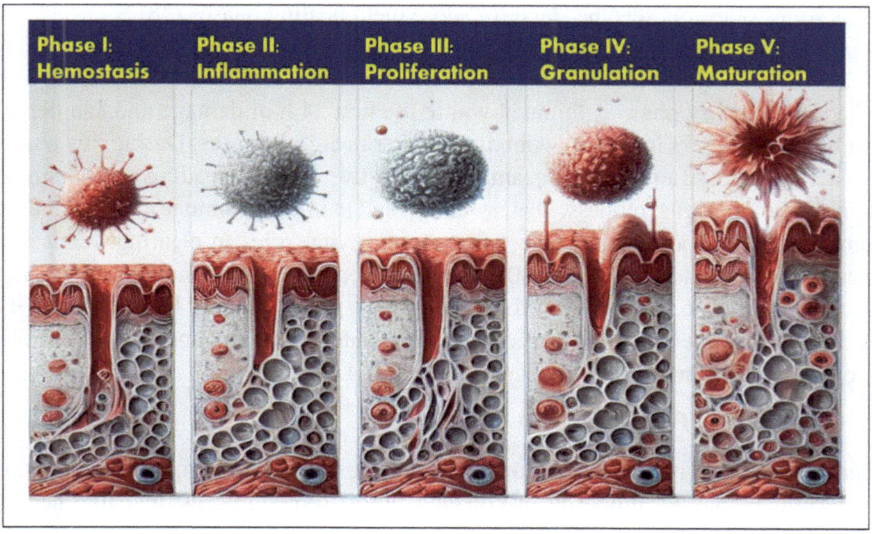

Fig. 1 A diagram showing the steps of wound healing. Phase I: Hemostasis—the body's quick reaction to stop bleeding by narrowing blood vessels and forming a clot. Phase II: Inflammation—immune cells move in to clean the wound, remove damaged tissue, and release signals for healing. Phase III: Proliferation—cells called fibroblasts help repair the tissue, build collagen, and create new blood vessels. Phase IV: Granulation—new tissue forms with tiny blood vessels and a supportive structure for healing. Phase V: Maturation—the wound gets stronger as collagen reorganizes, and the tissue returns to normal function

Hemostasis begins immediately after an injury to prevent excessive blood loss. Vasoconstriction occurs at the injury site, triggering platelet activation, adhesion, and aggregation [33]. Platelets detect exposed extravascular collagen, such as type I collagen, via specific integrin receptors on their surface [34, 35]. Upon binding to collagen, platelets release soluble mediators, including growth factors and cyclic AMP, as well as adhesive glycoproteins, which promote their aggregation. Key glycoproteins, such as fibrinogen, fibronectin, thrombospondin, and von Willebrand factor, are secreted from platelet alpha granules. These proteins facilitate the formation of a fibrin clot, which acts as a temporary matrix for wound repair. Platelets become trapped within the fibrin network, forming the bulk of the clot, while their membranes provide a surface for activating clotting enzyme proteases that drive the coagulation cascade [36].

In addition to clot formation, platelet alpha granules release important cytokines, including platelet-derived growth factor (PDGF), transforming growth factors beta and alpha (TGF-β and TGF-α), insulin-like growth factor-1 (IGF-1), basic fibroblast growth factor (bFGF) and vascular endothelial growth factor (VEGF) [34, 37]. These factors recruit neutrophils and monocytes to the site, initiating the inflammatory response. Other chemotactic agents, such as complement fragment C5a and bacterial byproducts like f-Met-Leu-Phe, also draw neutrophils to the wound. VEGF, TGF-α, and bFGF stimulate endothelial cells to promote new blood vessel formation, while PDGF attracts fibroblasts to produce extracellular matrix components like collagen and glycosaminoglycans, which are crucial for cellular migration and interaction. This series of events sets the stage for subsequent healing phases [38].

b. **Inflammation**

The inflammatory phase is initiated within the first 24 h of damage and can persist for up to two weeks in typical wounds or longer in chronic cases. During this phase, mast cells release enzymes, histamine, and active amines, leading to Inflammation's hallmark signs: redness, heat, swelling, and pain. Key immune cells, including neutrophils, monocytes, and macrophages, play vital roles in clearing debris and pathogens from the wound [39, 40]. They also secrete pro-inflammatory cytokines such as IL-1, IL-6, IL-8, and TNF-α, along with growth factors like PDGF, TGF-β, and IGF-1. These mediators recruit and activate fibroblasts and epithelial cells, preparing the wound for tissue regeneration [41, 42].

c. **Proliferative Phase**

The proliferative phase focuses on replacing the initial fibrin matrix with a new extracellular matrix composed of collagen fibres, proteoglycans, and fibronectin. This stage restores tissue structure and functionality. Fibroblasts play a pivotal role during this phase, producing extracellular matrix components and promoting tissue regeneration [43, 44]. This phase includes the following key events:

i. **Fibroblast Migration:** Fibroblasts migrate to the wound site in response to soluble mediators released by platelets and macrophages. Their movement is guided by interactions with extracellular matrix components like fibronectin, vitronectin, and fibrin. Fibroblasts bind to these molecules via integrin receptors and secrete proteases to clear a path through the matrix. Migration is influenced by chemotactic gradients of growth factors and the alignment of matrix fibrils, with fibroblasts typically moving along these structures. Once at the wound site, fibroblasts produce collagen, elastin, and proteoglycans, forming the foundation of granulation tissue [45, 46].

ii. **Collagen and Extracellular Matrix Production:** Fibroblasts synthesize granulation tissue components, primarily collagen and proteoglycans. PDGF and TGF-β are essential regulators of fibroblast activity, promoting their proliferation, migration, and collagen production [47, 48]. TGF-β also suppresses protease activity and enhances tissue integrity by increasing tissue inhibitors of metalloproteinases (TIMPs). Newly formed type III collagen is gradually replaced by type I collagen, providing greater tensile strength to the tissue [49]. Proper collagen production requires hydroxylation of proline and lysine, a process dependent on sufficient vitamin C levels. Collagen cross-linking by lysyl oxidase strengthens the matrix, ensuring stable wound repair [50].

iii. **Angiogenesis:** New blood vessel formation is critical for maintaining tissue viability in the wound. Low oxygen levels, low pH, and high lactate concentrations in the wound microenvironment stimulate angiogenesis [46, 51]. Endothelial cells respond to factors like VEGF and bFGF, which promote migration and proliferation. Hypoxia-inducible factors (HIF) regulate VEGF expression, driving capillary growth under low oxygen conditions. Endothelial cells extend sprouts into the matrix, forming new vascular loops to restore blood flow. These capillary networks give granulation tissue its characteristic granular appearance [51, 52].

iv. **Granulation:** Granulation tissue temporarily replaces the normal dermis during healing and consists of a dense network of capillaries, fibroblasts, and macrophages. This tissue supports cell migration and protein synthesis, enabling further wound healing. Over time, granulation tissue matures into scar tissue, characterized by a lower cellular density and collagen remodeling [46, 53].

d. **Epithelialization**

Epithelialization involves the migration of epithelial cells from wound edges and skin appendages. Growth factors such as epidermal growth factor (EGF) and keratinocyte growth factor (KGF) stimulate epithelial cells to detach, migrate, and proliferate. These cells form a monolayer over the wound and produce proteolytic enzymes to penetrate scabs or necrotic tissue. Once the wound is covered, the epidermis stratifies and matures, restoring the skin barrier [54].

e. **Remodeling**

In the final phase of healing, granulation tissue is replaced by a scar. Type III collagen is replaced by type I collagen, improving tensile strength. Collagen fibers undergo reorganization and cross-linking, increasing stability. Although remodelled tissue regains up to 80% of its original strength, it remains weaker than uninjured skin. Scar tissue is less organized and lacks the resilience of a normal dermis [43, 55, 56].

3 Application of Nanoparticles in Wound Healing

The application of nanoparticles (NPs) in wound healing has expanded significantly, with clinical research actively investigating the potential of nanotechnology-based therapies for improving wound care. Nanoparticles, ranging in size from 1 to 1000 nm, are at the forefront of this innovation. Their ability to revolutionize modern healthcare stems from their unique physicochemical properties. One of the standout advantages of NPs in wound management is their capacity for controlled drug delivery. Encapsulating therapeutic agents within NPs ensures localized and sustained release directly at the wound site, reducing systemic side effects while enhancing efficacy. NPs can be engineered to target specific cell types, such as immune cells, fibroblasts, or endothelial cells, thereby maximizing their therapeutic effects. They can also mimic the extracellular matrix (ECM), creating an environment conducive to cell adhesion, migration, and proliferation. This biomimicry facilitates tissue regeneration and accelerates wound closure. Furthermore, NPs can modulate inflammatory responses at the wound site, promoting a transition from a pro-inflammatory to a pro-healing state. For instance, NPs loaded with anti-inflammatory agents can reduce excessive Inflammation, while those delivering growth factors or cytokines stimulate tissue repair.

Inherent properties of certain nanoparticles, such as antimicrobial activity, is important in preventing or treating infections—a common barrier to effective wound healing. Additionally, NPs with optical, magnetic, or mechanical properties are being used for diagnostics and real-time monitoring of wound healing progression. Their customizable size, shape, and surface charge allow precise tailoring for different wound types and healing stages. Smaller NPs, in particular, penetrate tissues more effectively, reaching deeper wound layers and ensuring uniform drug distribution. After penetrating the wound site, NPs undergo biodistribution and can reach various organs depending on their size and surface characteristics. Smaller particles are often cleared by the kidneys, while others may accumulate in organs like the liver or spleen. The surface charge also influences their behaviour, such as positively charged NPs interacting favourably with negatively charged cell membranes, enhancing cellular absorption and aiding targeted drug delivery to cells such as fibroblasts and immune cells. Conversely, negatively charged NPs may face reduced internalization but contribute to tissue regeneration through other mechanisms.

The dimensions and morphology of NPs further enhance their utility in wound healing. Their ability to release drugs, penetrate tissues, and exhibit antimicrobial activity supports various cellular processes critical for repair. Studies have shown that silver nanoparticles (AgNPs) and other biomaterial composites are effective in preventing bacterial infections caused by E. coli, Staphylococcus aureus, and methicillin-resistant S. aureus (MRSA). For example, silver nanoparticles synthesized under specific conditions demonstrated toxicity against multiple bacterial strains, validating their antimicrobial efficacy. Two main categories of nanomaterials are prevalent in wound care: those with intrinsic properties that promote healing and those that serve as carriers for therapeutic agents. Polymeric nanoparticles (PNPs), with their dynamic structures and morphologies, exemplify the versatility of nanotechnology. These materials, which can range from 1 to 1000 nm in size, hold significant promise for advancing drug delivery systems and enhancing wound healing outcomes.

4 Wound Healing Application of the Biopolymers

Biopolymers, often referred to as natural polymers, are organic compounds produced by living organisms. Structurally, these polymers consist of repeating units called monomers, which include amino acids, monosaccharides, nucleotides, or esters. These monomers are connected via covalent bonds, forming larger compounds such as peptides, polysaccharides, polyphenols, and polyesters [57]. Biopolymers are derived from a variety of natural sources, including plants (e.g., starch, cellulose, natural rubber), animals (e.g., collagen, hyaluronic acid, chitosan), fungi (e.g., chitin), bacteria (e.g., bacterial cellulose, exopolysaccharides), and algae (e.g., alginate). Due to their biocompatibility, biodegradability, and reduced allergenic potential, biopolymers are preferred over synthetic materials and are widely used in wound healing applications [58]. Common biopolymers include collagen, cellulose, chitosan, alginate, hyaluronan, fucoidan, and carrageenan, which exhibit properties such as antibacterial, anti-inflammatory, and proliferative effects, promoting targeted cellular activity. Polymeric nanoparticles (PNPs, made using biopolymers like poly(lactide (PLA, poly(lactide-co-glycolic acid (PLGA, polycaprolactone (PCL, and natural polymers such as alginate, chitosan, gelatin, and albumin, offer significant promise for wound care. These nanoparticles enable controlled and sustained drug release, facilitating localized delivery of therapeutic agents directly to the wound site. This precise targeting enhances the efficacy of drugs, growth factors, and bioactive molecules, optimizing their impact on wound healing [59, 60] (Fig. 2).

4.1 Polylactic Acid (PLA)

Poly (lactic acid) (PLA) is a versatile thermoplastic polymer that is widely acknowledged for its high biodegradability, compatibility, and robust mechanical properties. These attributes make PLA particularly suitable for wound care applications, especially in the development of advanced wound dressings [61–63]. PLA-based dressings provide a protective barrier that shields the wound from environmental contaminants while permitting the exchange of oxygen and moisture. This creates an optimal microenvironment conducive to tissue regeneration and healing. PLA undergoes degradation through hydrolysis, breaking down into lactic acid. This sustainable production framework makes PLA an environmentally friendly material. Despite its advantages, the synthesis of PLA involves complex polymerization processes that

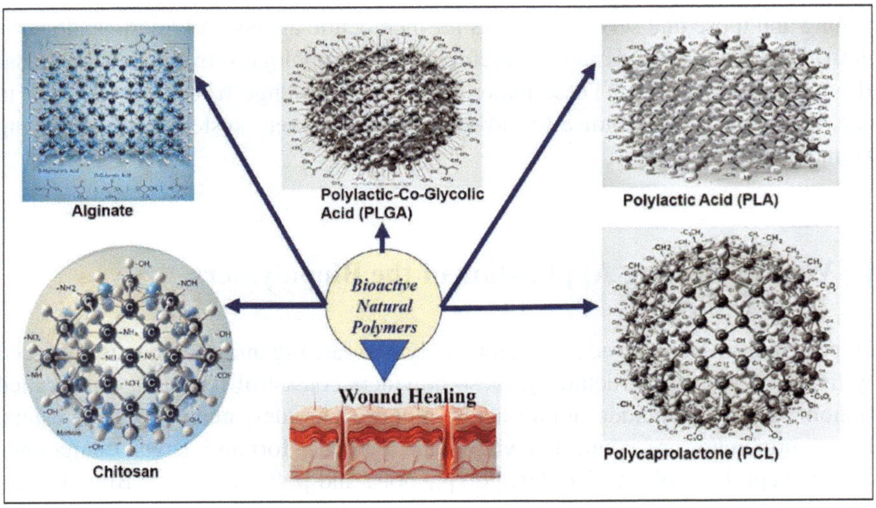

Fig. 2 Biopolymers help in wound healing. Alginate: Stops bleeding and keeps the wound moist to help it heal. Polylactic Co-Glycolic Acid (PLGA): Releases medicine slowly and helps new tissue grow. Polylactic Acid (PLA): Helps cells stick to it and acts as a framework for tissue repair. Chitosan: Fights bacteria, speeds up wound healing, and helps new skin form. Polycaprolactone (PCL): Provides strong, long-lasting support and helps cells grow and tissues rebuild

require precise parameter control, extended reaction durations, and significant energy inputs. Nevertheless, its biodegradability, moisture retention properties, and ability to promote collagen deposition and angiogenesis make PLA an excellent candidate for wound healing [62].

In one study, Munteanu et al. developed PLA-based silver nanoparticles (AgNPs) combined with Vitamin E, which displayed strong antibacterial and antioxidant activities. These nanofibers were highly effective against pathogens such as *Salmonella typhimurium*, *Escherichia coli* and *Listeria monocytogenes.* Hasan et al. created clindamycin-loaded PLA nanoparticles specifically for treating MRSA-infected wounds. The nanoparticles demonstrated excellent adhesion to bacterial cells and significantly enhanced wound healing and re-epithelialization in mouse models of MRSA infections, with no adverse effects on healthy fibroblast cells. Silva et al. formulated PLA nanoparticles encapsulating red propolis (RP). These nanoparticles, averaging 194.1 nm in size with an encapsulation efficiency of 85.4%, exhibited high stability and demonstrated potent antioxidant activity, contributing to improved wound healing outcomes [64–66].

4.2 Polylactic-Co-glycolic Acid (PLGA)

Polylactic-co-glycolic acid (PLGA) is a copolymer synthesized by combining PLA and polyglycolic acid (PGA). As a widely utilized biodegradable and biocompatible material, PLGA has gained prominence in drug delivery and wound healing applications. The asymmetric α-carbon in PLA can exist as D or L enantiomers, with a mixture of both forming poly D, L-lactic-co-glycolic acid. PLGA's degradation produces lactate, which promotes angiogenesis and activates pro-collagen factors, further enhancing wound repair processes. Several studies have highlighted PLGA's efficacy in wound healing [67–69]. Nurhasni et al. developed PLGA nanoparticles conjugated with polyethyleneimine for prolonged nitric oxide delivery. These nanoparticles displayed substantial bactericidal activity against MRSA and Pseudomonas aeruginosa, adhering effectively to bacterial surfaces and releasing nitric oxide for up to six hours. Similarly, You et al. incorporated graphene oxide (GO) into PLGA composites, which exhibited improved wound-specific mechanical properties, biocompatibility, and antibacterial activity. The addition of electrical stimulation enhanced cell survival and expedited healing, particularly in full-thickness wounds. Similarly, Ebrahimi et al. fabricated PLGA nanoparticles loaded with an ethanolic extract of propolis. These nanoparticles promoted early wound contraction and stimulated angiogenesis, fibroblast proliferation, and granulation tissue formation. Similarly, Lee et al. developed PLGA nanoparticles encapsulating S-nitrosoglutathione (GSNO) for MRSA-infected wounds. The nanoparticles demonstrated improved nitric oxide release, facilitating enhanced wound healing in mouse models with infected cutaneous lesions [70].

4.3 Collagen

Collagen is the most abundant protein in the human body, playing a fundamental role in wound healing by providing structural support and promoting tissue repair. This protein is instrumental in attracting fibroblasts to the wound site and stimulating the synthesis of new collagen within the wound bed [71–73]. Fibroblasts, along with other cells, synthesize collagen and transform it into complex structures to facilitate tissue regeneration [74]. The tensile strength of repaired skin depends on the type, quantity, and organization of collagen, which undergoes dynamic changes during the healing process. In the early stages of wound repair, type III collagen is predominantly synthesized, providing initial support. As healing progresses, type III collagen is gradually replaced by type I collagen, which is the primary form found in mature skin. The initial random deposition of collagen fibers during granulation tissue formation is strengthened through covalent cross-linking mediated by lysyl oxidase. This cross-linking helps collagen mature into highly organized structures that restore tensile strength. Collagen remodeling continues for several months after wound closure, ultimately enhancing tissue strength to approximately 80–85% of its original condition under normal healing processes [75, 76]. Recent studies have demonstrated the effectiveness of collagen-based formulations in wound healing. For example, Tanha et al. developed collagen-coated nanoparticles (NPs) loaded with granulocyte colony-stimulating factor (G-CSF) for wound healing applications. Results from in vivo experiments and histopathological analyses indicated that these nanoparticles significantly accelerated skin regeneration compared to conventional treatments. The collagen coating enhanced fibroblast maturation, improved collagen deposition, and reduced inflammatory cell infiltration, making it a superior alternative to standard wound dressings [77]. In another study, Gopinath et al. fabricated collagen-based films containing curcumin (CFC). Biochemical and histological evaluations showed that these films enhanced wound healing, promoted cell proliferation, and exhibited strong free radical scavenging activity. The increased shrinkage temperature of the CFC films suggested improved hydrothermal stability compared to conventional collagen films. Furthermore, spectroscopic analysis revealed that curcumin bound effectively to collagen without disrupting its triple-helical structure, ensuring its stability and functionality [78].

Malathi et al. explored the application of collagen-based zinc oxide (ZnO) nanoparticles in wound healing. Their findings highlighted the ability of these bionanocomposites to expedite the healing process. These composites exhibited strong antibacterial activity, excellent biocompatibility, and scalability, making them a cost-effective and efficient solution for wound management. Their rapid production and flexible design further underscore their potential for widespread application in biomedical contexts, particularly in wound care [79].

4.4 Polycaprolactone (PCL)

Polycaprolactone (PCL) is a synthetic aliphatic polyester composed of repeating hexanoate units known for its biodegradability and compatibility with other materials [80]. Its unique mechanical properties, low melting temperature, and excellent miscibility with various polymers make PCL an attractive material for biomedical applications, particularly in wound healing and bone tissue regeneration [81]. The semi-crystalline nature of PCL, coupled with its hydrophobic characteristics and solubility at ambient temperatures, allows for easy processing and versatility. It's crystallinity and molecular weight are key determinants of its mechanical, thermal, and physical behaviour, influencing its suitability for various medical uses [82, 83]. Studies have consistently shown that PCL-based composites can significantly enhance wound healing. For instance, Augustine et al. incorporated zinc oxide (ZnO) nanoparticles into PCL membranes, which were subsequently implanted subcutaneously in guinea pigs. The study demonstrated improved cell adhesion and migration in membranes containing ZnO nanoparticles [84]. Histological, immunological, and macroscopic analyses confirmed that these membranes facilitated wound healing without causing irritation or scarring. Nawaz et al. developed acriflavine-loaded PCL nanoparticles using the double-emulsion solvent evaporation method.

These nanoparticles exhibited a biphasic drug release profile, beginning with an initial burst followed by sustained release over 24 h [85]. Morphologically, the nanoparticles were spherical with smooth surfaces, and the acriflavine-loaded formulations demonstrated significant wound healing potential, evidenced by enhanced re-epithelialization compared to untreated controls. Nhi et al. prepared PCL-coated gelatin-silver nanoparticles (Gel-Ag NPs) to evaluate their antibacterial efficacy in wound healing applications. The study found that increasing the number of Gel-Ag coatings enhanced the antibacterial properties of the membranes, as confirmed by both in vitro and in vivo results [86]. Moreover, the membranes exhibited non-adherent properties, reducing the risk of wound damage during dressing removal, thus making them ideal for wound dressing applications. In another investigation, Ghorbani et al. evaluated the wound healing potential of PCL nanoparticles loaded with Matricaria chamomilla extracts. These nanoparticles demonstrated robust antibacterial activity, against Staphylococcus aureus and Escherichia coli, respectively. These findings suggest that PCL-based nanoparticles have significant potential as bioactive wound dressings due to their antibacterial efficacy and ability to support tissue regeneration [87].

4.5 Alginate

Alginate, an anionic biopolymer derived from brown algae, is widely recognized for its biocompatibility, non-toxicity, and ability to create an optimal wound-healing environment [17, 88]. Structurally, alginate consists of β-D-mannuronate (M) and α-L-guluronate (G) residues arranged in blocks or alternating sequences, influencing its molecular weight and physical properties. Alginate dressings are known for maintaining a moist wound environment, absorbing excess exudates, and facilitating granulation tissue formation. These dressings also assist in hemostasis by forming a gel-like layer that adheres to the wound bed, controlling bleeding while providing a scaffold for cell migration and tissue formation. Recent studies highlight the versatility of alginate-based formulations in wound healing. For instance, Oliveira et al. developed polymyxin-alginate nanoparticles coated with lidocaine to control bleeding and prevent infection [89]. The formulation demonstrated significant antibacterial activity against Staphylococcus aureus and Pseudomonas aeruginosa, as confirmed by microbiological testing. Additionally, in vivo studies revealed that these nanoparticles significantly enhanced wound healing, as evidenced by the wound retraction index and histological analysis. Zhu et al. explored Esculentoside A (EsA)-loaded alginate/chitosan nanoparticles for burn wound healing [90]. These hybrid nanoparticles exhibited high encapsulation efficiency and prolonged EsA release, which modulated macrophage activity by reducing Inflammation and enhancing anti-inflammatory cytokine production. This led to faster and higher-quality healing of burn wounds, demonstrating the potential of EsA-alginate nanoparticles for treating burn-related injuries. Manuja et al. investigated alginate-based ZnO nanoparticles for wound healing. These nanoparticles, when applied topically, accelerated healing in rabbit models with full-thickness excision wounds. The enhanced healing was attributed to the synergistic effects of alginate and ZnO, which promoted cell adhesion, proliferation, and granulation tissue formation. Techniques such as transmission electron microscopy, collagen content analysis, and histological observations provided detailed insights into the healing mechanism [91].

4.6 Chitosan

Chitosan is a linear amino-polysaccharide composed of glucosamine and N-acetyl glucosamine units linked via β-(1–4) glycosidic bonds. This biopolymer is derived from the N-deacetylation of chitin, a naturally occurring polymer found abundantly in the exoskeleton strucuture of crustaceans and fungal cell walls. As the second most prevalent biopolymer after cellulose, chitin serves as the precursor for producing chitosan, which is widely utilized in various biomedical applications due to its unique properties. Structurally, chitosan is a polycationic polymer with free amine groups and hydroxyl functionalities attached to glucopyranose rings. These functional groups allow for chemical modifications, enabling a broad spectrum of functionalization to suit specific applications [92, 93].

Chitosan nanoparticles and biopolymers have been extensively explored for their potential in wound healing due to their ability to encapsulate and deliver bioactive compounds. For instance, Hou et al. developed chitosan NPs loaded with vaccarin (VAC) for wound healing applications, which enhanced the proliferation and migration of human umbilical vein endothelial cells (HUVECs), promoted angiogenesis by upregulating IL-1β and PDGF-BB expression, facilitating new blood vessel formation or angiogenesis, and accelerating wound healing [94, 95]. In another study, Reddy et al. formulated curcumin-loaded chitosan NPs for diabetic wound healing. The nanohybrid scaffold significantly enhanced wound contraction, epithelialization, and granulation tissue formation in diabetic wound models compared to placebo scaffolds. Curcumin's anti-inflammatory and antioxidant properties, combined with chitosan's drug delivery capabilities and collagen's scaffold attributes, created a synergistic effect that effectively addressed diabetic wound healing challenges. Similarly, Fang et al. developed curcumin-loaded chitosan nanoparticles (Cur-CS-NPs) for treating diabetic wounds. These nanoparticles exhibited sustained drug release and high cellular absorption, significantly reducing macrophage-induced Inflammation while promoting angiogenesis [96].

In another innovative approach, Zhengguo et al. developed chitosan-based silver nanoparticles (AgNPs) for wound healing. These nanoparticles exhibited strong antibacterial activity against Staphylococcus aureus, Escherichia coli, and Pseudomonas aeruginosa. The AgNPs, incorporated into a sponge matrix, demonstrated excellent hemostatic properties, achieving rapid blood clotting and effective wound healing in both in vitro and in vivo experiments. This enhanced performance was attributed to the combined effects of high blood absorption capacity and interactions with amino and sulfhydryl groups, effectively preventing wound infection and promoting tissue repair [97, 98].

5 Conclusion and Future Perspectives

The application of biopolymer-based nano-medications in wound healing offers a promising avenue for advancing modern therapeutics. These materials, with their intrinsic biocompatibility, biodegradability, and bioactivity, address critical challenges in wound care, such as infection control, inflammation modulation, and tissue regeneration. Various biopolymers like chitosan, alginate, collagen, and hyaluronic acid have been extensively studied, demonstrating their ability to serve as scaffolds, drug carriers, and agents that enhance cellular responses. When incorporated with nanoparticles, these biopolymers not only provide sustained drug release but also mimic the extracellular matrix, thus promoting cell proliferation and angiogenesis.

Despite significant progress, challenges such as scalability, cost, and regulatory compliance continue to hinder widespread adoption. The long-term safety, efficacy and sustainability of these advanced materials require further investigation, particularly concerning their interactions within complex biological systems. Future research should focus on developing smart biomaterials with responsive drug delivery systems, integrating diagnostic capabilities, and enhancing antimicrobial efficacy to combat the growing threat of drug-resistant pathogens. Collaborative efforts across material science, biotechnology, and clinical research are crucial for translating laboratory findings into commercially viable solutions. With continued innovation, biopolymer-based nanomedications hold the potential to revolutionize wound care, offering hope for more effective, patient-friendly, and sustainable therapeutic interventions.

References

1. Eming SA, Martin P, Tomic-Canic M (2014) Wound repair and regeneration: mechanisms, signaling, and translation. Sci Transl Med 6(265). https://doi.org/10.1126/scitranslmed.3009337
2. Guo S, DiPietro LA (2010) Factors affecting wound healing. J Dent Res 89(3):219–229. https://doi.org/10.1177/0022034509359125
3. Gurtner GC, Werner S, Barrandon Y, Longaker MT (2008) Wound repair and regeneration. Nature 453(7193):314–321. https://www.nature.com/articles/nature07039
4. Edwards R, Harding KG (2004) Bacteria and wound healing. Curr Opin Infect Dis 17(2):91–96. https://journals.lww.com/co-infectiousdiseases/fulltext/2004/04000/bacteria_and_wound_healing.4.aspx
5. Kumar M, Mandal UK, Mahmood S (2024) Dermatological formulations. In: Dosage forms, formulation developments and regulations. Elsevier, pp 613–642. https://www.sciencedirect.com/science/article/pii/B9780323918176000218

6. Litany RJ, Praseetha PK (2022) Tiny tots for a big-league in wound repair: tools for tissue regeneration by nanotechniques of today. J Controll Release 349:443–459. https://www.sciencedirect.com/science/article/pii/S016836592200414X
7. Gushiken LFS, Beserra FP, Bastos JK, Jackson CJ, Pellizzon CH (2021) Cutaneous wound healing: an update from physiopathology to current therapies. Life 11(7):665. https://www.mdpi.com/2075-1729/11/7/665
8. Aitcheson SM, Frentiu FD, Hurn SE, Edwards K, Murray RZ (2021) Skin wound healing: Normal macrophage function and macrophage dysfunction in diabetic wounds. Molecules 26(16):4917. https://www.mdpi.com/1420-3049/26/16/4917
9. Kumar M, Hilles AR, Ge Y, Bhatia A, Mahmood S (2023) A review on polysaccharides mediated electrospun nanofibers for diabetic wound healing: their current status with regulatory perspective. Int J Biol Macromol 234:123696. https://www.sciencedirect.com/science/article/pii/S0141813023005895
10. GU P, BS, U, J S, MG A, Mohan D, Pillai KR, TT S (2022) Electrospun polysaccharide scaffolds: wound healing and stem cell differentiation. J Biomater Sci Poly Edn 33(7):858–877. https://doi.org/10.1080/09205063.2021.2024053
11. Kumar M, Keshwania P, Chopra S, Mahmood S, Bhatia A (2023) Therapeutic potential of nanocarrier-mediated delivery of phytoconstituents for wound healing: their current status and future perspective. AAPS PharmSciTech 24(6):155. https://doi.org/10.1208/s12249-023-02616-6
12. Kim MH (2016) Nanoparticle-based therapies for wound biofilm infection: opportunities and challenges. IEEE Trans Nanobiosci 15(3):294–304. https://ieeexplore.ieee.org/abstract/document/7423795/
13. Abdollahi M, Baharian A, Mohamadhoseini M, Hassanpour M, Makvandi P, Habibizadeh M, Jafari B, Nouri R, Mohamadnia Z, Nikfarjam N (2024) Advances in ionic liquid-based antimicrobial wound healing platforms. J Mater Chem B. https://pubs.rsc.org/en/content/articlehtml/2024/tb/d4tb00841c
14. Babuskin S, Babu PAS, Sasikala M, Sabina K, Archana G, Sivarajan M, Sukumar M (2014) Antimicrobial and antioxidant effects of spice extracts on the shelf life extension of raw chicken meat. Int J Food Microbiol 171:32–40. https://www.sciencedirect.com/science/article/pii/S0168160513005321
15. Yousefian F, Hesari R, Jensen T, Obagi S, Rgeai A, Damiani G, Bunick CG, Grada A (2023) Antimicrobial wound dressings: a concise review for clinicians. Antibiotics 12(9):1434. https://www.mdpi.com/2079-6382/12/9/1434
16. Shi C, Wang C, Liu H, Li Q, Li R, Zhang Y, Liu Y, Shao Y, Wang J (2020) Selection of appropriate wound dressing for various wounds. Front Bioeng Biotechnol 8:182. https://doi.org/10.3389/fbioe.2020.00182/full
17. Wang L, Zhou M, Xu T, Zhang X (2022) Multifunctional hydrogel as wound dressing for intelligent wound monitoring. Chem Eng J 433:134625. https://www.sciencedirect.com/science/article/pii/S1385894722001334
18. Bal-Öztürk A, Özkahraman B, Özbaş Z, Yaşayan G, Tamahkar E, Alarçin E (2021) Advancements and future directions in the antibacterial wound dressings—a review. J Biomed Mater Res B Appl Biomater 109(5):703–716. https://doi.org/10.1002/jbm.b.34736
19. Homaeigohar S, Boccaccini AR (2020) Antibacterial biohybrid nanofibers for wound dressings. Acta Biomaterialia 107:25–49. https://www.sciencedirect.com/science/article/pii/S174270612030101X
20. Jia B, Li G, Cao E, Luo J, Zhao X, Huang H (2023) Recent progress of antibacterial hydrogels in wound dressings. Mater Today Bio 19:100582. https://www.sciencedirect.com/science/article/pii/S259000642300042X
21. Khubiev OM, Egorov AR, Kirichuk AA, Khrustalev VN, Tskhovrebov AG, Kritchenkov AS (2023) Chitosan-based antibacterial films for biomedical and food applications. Int J Mol Sci 24(13):10738. https://www.mdpi.com/1422-0067/24/13/10738

22. Kumar M, Kumar D, Singh S, Chopra S, Mahmood S, Bhatia A (2024) Quality by design perspective for designing foam-based formulation: current state of art. Curr Pharmaceutic Design 30(6):410–419. https://www.ingentaconnect.com/content/ben/cpd/2024/00000030/00000006/art00002
23. Lanno G-M, Ramos C, Preem L, Putrinš M, Laidmäe I, Tenson T, Kogermann K (2020) Antibacterial porous electrospun fibers as skin scaffolds for wound healing applications. ACS Omega 5(46):30011–30022. https://doi.org/10.1021/acsomega.0c04402
24. Li S, Zhang Y, Ma X, Qiu S, Chen J, Lu G, Jia Z, Zhu J, Yang Q, Chen J, Wei Y (2022) Antimicrobial lignin-based polyurethane/ag composite foams for improving wound healing. Biomacromol 23(4):1622–1632. https://doi.org/10.1021/acs.biomac.1c01465
25. Tamahkar E, Özkahraman B, Özbaş Z, İzbudak B, Yarimcan F, Boran F, Öztürk AB (2021) Aloe vera-based antibacterial porous sponges for wound dressing applications. J Porous Mater 28(3):741–750. https://doi.org/10.1007/s10934-020-01029-1
26. Alavi M, Rai M (2020) Topical delivery of growth factors and metal/metal oxide nanoparticles to infected wounds by polymeric nanoparticles: an overview. Expert Rev Anti Infect Ther 18(10):1021–1032. https://doi.org/10.1080/14787210.2020.1782740
27. Kamaly N, Xiao Z, Valencia PM, Radovic-Moreno AF, Farokhzad OC (2012) Targeted polymeric therapeutic nanoparticles: design, development and clinical translation. Chem Soc Rev 41(7):2971–3010. https://pubs.rsc.org/en/content/articlehtml/2012/cs/c2cs15344k
28. Kumar M, Mandal UK, Mahmood S (2023) Novel drug delivery system. In: Advanced and modern approaches for drug delivery. Elsevier, pp 1–32. https://www.sciencedirect.com/science/article/pii/B9780323916684000125
29. Ribeiro MC, Correa VLR, da Silva FKL, Casas AA, das Chagas A, de L, de Oliveira LP, Miguel MP, Diniz DGA, Amaral AC, de Menezes LB (2020) Wound healing treatment using insulin within polymeric nanoparticles in the diabetes animal model. Euro J Pharmaceutic Sci 150:105330. https://www.sciencedirect.com/science/article/pii/S0928098720301196
30. Enoch S, Leaper DJ (2008) Basic science of wound healing. Surgery (Oxford) 26(2):31–37. https://www.sciencedirect.com/science/article/pii/S026393190700316X
31. Strodtbeck F (2001) Physiology of wound healing. Newborn Infant Nurs Rev 1(1):43–52. https://www.sciencedirect.com/science/article/pii/S1527336901700542
32. Stroncek JD, Reichert WM (2008) Overview of wound healing in different tissue types. Indwelling Neural Implants: Strateg Contend Vivo Environ 1:3–41. https://books.google.com/books?hl=en&lr=&id=ghLLo6ZzAwYC&oi=fnd&pg=PA3&dq=Different+stages+of+wound+healing&ots=KFeNyXIGpi&sig=cHCHRVXDPuDHA1oRfWpxbyTvOqA
33. Dorgalaleh A, Daneshi M, Rashidpanah J, Roshani Yasaghi E (2018) An overview of hemostasis. In: Dorgalaleh A (ed) Congenital bleeding disorders. Springer International Publishing, pp 3–26. https://doi.org/10.1007/978-3-319-76723-9_1
34. Golebiewska EM, Poole AW (2015) Platelet secretion: from haemostasis to wound healing and beyond. Blood Rev 29(3):153–162. https://www.sciencedirect.com/science/article/pii/S0268960X1400085X
35. Smyth SS, McEver RP, Weyrich AS., Morrell CN, Hoffman MR, Arepally GM, French PA, Dauerman HL, Becker RC, THE F (2009) Platelet functions beyond hemostasis. J Thromb Haemost 7(11):1759–1766. https://www.sciencedirect.com/science/article/pii/S1538783622106070
36. Naik B, Karunakar P, Jayadev M, Marshal VR (2013) Role of platelet rich fibrin in wound healing: a critical review. J Conserv Dentistr Endod 16(4):284–293. https://journals.lww.com/jcde/fulltext/2013/16040/role_of_platelet_rich_fibrin_in_wound_healing__a.3.aspx
37. Pierce GF, Mustoe TA, Altrock BW, Deuel TF, Thomason A (1991) Role of platelet-derived growth factor in wound healing. J Cell Biochem 45(4):319–326. https://doi.org/10.1002/jcb.240450403
38. Nurden AT, Nurden P, Sanchez M, Andia I, Anitua E (2008) Platelets and wound healing. Front Biosci 13(9):3532–3548. https://article.imrpress.com/bri/Landmark/articles/pdf/Landmark2947.pdf

39. Koh TJ, DiPietro LA (2011) Inflammation and wound healing: the role of the macrophage. Expert Rev Mol Med 13:e23. https://www.cambridge.org/core/journals/expert-reviews-in-molecular-medicine/article/inflammation-and-wound-healing-the-role-of-the-macrophage/9B0BF84E8A239160D32F2EE8214633D3
40. Turabelidze A, Dipietro LA (2011) Inflammation and wound healing. Endod Top 24(1):26–38. https://doi.org/10.1111/etp.12012
41. Henry G, Garner WL (2003) Inflammatory mediators in wound healing. Surgic Clinic 83(3):483–507. https://www.surgical.theclinics.com/article/S0039-6109(02)00200-1/abstract
42. Nosenko MA, Ambaryan SG, Drutskaya MS (2019) Proinflammatory cytokines and skin wound healing in mice. Mol Biol 53(5):653–664. https://doi.org/10.1134/S0026893319050121
43. Landén NX, Li D, Ståhle M (2016) Transition from inflammation to proliferation: a critical step during wound healing. Cell Mol Life Sci 73(20):3861–3885. https://doi.org/10.1007/s00018-016-2268-0
44. Wallace HA, Basehore BM, Zito PM (2017) Wound healing phases. https://europepmc.org/books/nbk470443
45. Addis R, Cruciani S, Santaniello S, Bellu E, Sarais G, Ventura C, Maioli M, Pintore G (2020) Fibroblast proliferation and migration in wound healing by phytochemicals: evidence for a novel synergic outcome. Int J Med Sci 17(8):1030. https://www.ncbi.nlm.nih.gov/pmc/articles/PMC7211158/
46. Hosgood G (2006) Stages of wound healing and their clinical relevance. Vet Clinic Small Anim Pract 36(4):667–685. https://www.vetsmall.theclinics.com/article/S0195-5616(06)00023-4/abstract
47. Carlson MA, Longaker MT (2004) The fibroblast-populated collagen matrix as a model of wound healing: a review of the evidence. Wound Repair Regenerat 12(2):134–147. https://doi.org/10.1111/j.1067-1927.2004.012208.x
48. Diller RB, Tabor AJ (2022) The role of the extracellular matrix (ECM) in wound healing: a review. Biomimetics 7(3):87. https://www.mdpi.com/2313-7673/7/3/87
49. Cook H, Stephens P, Davies KJ, Thomas DW, Harding KG (2000) Defective extracellular matrix reorganization by chronic wound fibroblasts is associated with alterations in TIMP-1, TIMP-2, and MMP-2 activity. J Investigat Dermatol 115(2):225–233. https://www.sciencedirect.com/science/article/pii/S0022202X15409509
50. Kumari S, Panda TK, Pradhan T (2017) Lysyl oxidase: its diversity in health and diseases. Indian J Clin Biochem 32(2):134–141. https://doi.org/10.1007/s12291-016-0576-7
51. Nissen NN, Polverini P, Koch AE, Volin MV, Gamelli RL, DiPietro LA (1998) Vascular endothelial growth factor mediates angiogenic activity during the proliferative phase of wound healing. Am J Pathol 152(6):1445. https://www.ncbi.nlm.nih.gov/pmc/articles/PMC1858442/
52. Tonnesen MG, Feng X, Clark RA (2000) Angiogenesis in wound healing. J Investigat Dermatol Sympos Proceed 5(1):40–46. https://www.sciencedirect.com/science/article/pii/S0022202X15528571
53. Alhajj M, Goyal A (2020) Physiology, granulation tissue. https://europepmc.org/article/nbk/nbk554402
54. Pastar I, Stojadinovic O, Yin NC, Ramirez H, Nusbaum AG, Sawaya A, Patel SB, Khalid L, Isseroff RR, Tomic-Canic M (2014) Epithelialization in wound healing: a comprehensive review. Adv Wound Care 3(7):445–464. https://doi.org/10.1089/wound.2013.0473
55. Gonzalez AC, de O, Costa TF, Andrade Z, de A, Medrado ARAP (2016) Wound healing-A literature review. Anais Brasileiros de Dermatologia 91(5):614–620. https://www.scielo.br/j/abd/a/tqnxHTLMnj4pfrhrRdfLG6K/
56. Reinke JM, Sorg H (2012) Wound repair and regeneration. Euro Surgic Res 49(1):35–43. https://karger.com/esr/article-abstract/49/1/35/128119
57. Smith AM, Moxon S, Morris GA (2016) Biopolymers as wound healing materials. In: Wound healing biomaterials. Elsevier, pp 261–287. https://www.sciencedirect.com/science/article/pii/B9781782424567000131
58. Sahana TG, Rekha PD (2018) Biopolymers: applications in wound healing and skin tissue engineering. Mol Biol Rep 45(6):2857–2867. https://doi.org/10.1007/s11033-018-4296-3

59. Kumar M, Mahmood S, Chopra S, Bhatia A (2024) Biopolymer based nanoparticles and their therapeutic potential in wound healing–a review. Int J Biologic Macromol 131335. https://www.sciencedirect.com/science/article/pii/S0141813024021408
60. Xu R, Fang Y, Zhang Z, Cao Y, Yan Y, Gan L, Xu J, Zhou G (2023) Recent advances in biodegradable and biocompatible synthetic polymers used in skin wound healing. Mater 16(15):5459. https://www.mdpi.com/1996-1944/16/15/5459
61. Ali IH, Khalil IA, El-Sherbiny IM (2023) Design, development, in-vitro and in-vivo evaluation of polylactic acid-based multifunctional nanofibrous patches for efficient healing of diabetic wounds. Sci Rep 13(1):3215. https://www.nature.com/articles/s41598-023-29032-x
62. Chen HL, Chung JWY, Yan VCM, Wong TKS (2023) Polylactic acid-based biomaterials in wound healing: a systematic review. Adv Skin Wound Care 36(9):1–8. https://journals.lww.com/aswcjournal/fulltext/2023/09000/polylactic_acid_based_biomaterials_in_wound.13.aspx
63. Pawar PR, Tekale US, Shisodia US, Totre TJ, Domb JA (2014) Biomedical Applications of Poly(Lactic Acid). Rec Pat Regenerat Med 4(1):40–51. https://doi.org/10.2174/2210296504666140402235024
64. da Costa Silva V, do Nascimento TG, Mergulhão NL, Freitas JD, Duarte IFB, de Bulhões LCG, Dornelas CB, de Araújo JX, Dos Santos J, Silva AC (2022) Development of a polymeric membrane impregnated with poly-lactic acid (pla) nanoparticles loaded with red propolis (RP). Molecules 27(20):6959. https://www.mdpi.com/1420-3049/27/20/6959
65. Hasan N, Cao J, Lee J, Hlaing SP, Oshi MA, Naeem M, Ki MH, Lee BL, Jung Y, Yoo JW (2019) Bacteria-targeted clindamycin loaded polymeric nanoparticles: effect of surface charge on nanoparticle adhesion to MRSA, antibacterial activity, and wound healing. Pharmaceutics 11(5):236. https://www.mdpi.com/1999-4923/11/5/236
66. Munteanu BS, Aytac Z, Pricope GM, Uyar T, Vasile C (2014) Polylactic acid (PLA)/Silver-NP/VitaminE bionanocomposite electrospun nanofibers with antibacterial and antioxidant activity. J Nanopart Res 16(10):2643. https://doi.org/10.1007/s11051-014-2643-4
67. Blasi P (2019) Poly(lactic acid)/poly(lactic-co-glycolic acid)-based microparticles: an overview. J Pharm Investig 49(4):337–346. https://doi.org/10.1007/s40005-019-00453-z
68. Gentile P, Chiono V, Carmagnola I, Hatton PV (2014) An overview of poly (lactic-co-glycolic) acid (PLGA)-based biomaterials for bone tissue engineering. Int J Mol Sci 15(3):3640–3659. https://www.mdpi.com/1422-0067/15/3/3640
69. Makadia HK, Siegel SJ (2011) Poly lactic-co-glycolic acid (PLGA) as biodegradable controlled drug delivery carrier. Polymers 3(3):1377–1397. https://www.mdpi.com/2073-4360/3/3/1377
70. Yoo JW, Nurhasni H, Cao J, Choi M, Kim I, Lee BL, Jung Y (2015) Nitric oxide-releasing poly(lactic-co-glycolic acid)-polyethylenimine nanoparticles for prolonged nitric oxide release, antibacterial efficacy, and in vivo wound healing activity. Int J Nanomed 3065. https://doi.org/10.2147/IJN.S82199
71. Chen K, Sivaraj D, Davitt MF, Leeolou MC, Henn D, Steele SR, Huskins SL, Trotsyuk AA, Kussie HC, Greco AH, Padmanabhan J, Perrault DP, Zamaleeva AI, Longaker MT, Gurtner GC (2022) Pullulan-Collagen hydrogel wound dressing promotes dermal remodelling and wound healing compared to commercially available collagen dressings. Wound Repair Regen 30(3):397–408. https://doi.org/10.1111/wrr.13012
72. Harsha L, Brundha MP (2020) Role of collagen in wound healing. Drug Invent Today 13(1). https://search.ebscohost.com/login.aspx?direct=true&profile=ehost&scope=site&authtype=crawler&jrnl=09757619&AN=141552664&h=%2BaIrRGzecw4aTuEWaDjpLkOMYwuS2dlEXdRfWf7eeTCgUmf7j48U1RRvgLFuRd88ZaXkbeZgtzmiaLgL8sp%2Bcw%3D%3D&crl=c
73. Mahmoud AA, Salama AH (2016) Norfloxacin-loaded collagen/chitosan scaffolds for skin reconstruction: preparation, evaluation and in-vivo wound healing assessment. Euro J Pharmaceutic Sci 83:155–165. https://www.sciencedirect.com/science/article/pii/S0928098715301007
74. Reilly DM, Lozano J (2021) Skin collagen through the lifestages: importance for skin health and beauty. Undefined 8:N-A. https://www.oaepublish.com/articles/3863

75. Mathew-Steiner SS, Roy S, Sen CK (2021) Collagen in wound healing. Bioeng 8(5):63. https://www.mdpi.com/2306-5354/8/5/63
76. Schultz GS, Chin GA, Moldawer L, Diegelmann RF (2011) Principles of wound healing. In: Mechanisms of vascular disease: a reference book for vascular specialists [Internet]. https://www.ncbi.nlm.nih.gov/books/NBK534261/
77. Tanha S, Rafiee-Tehrani M, Abdollahi M, Vakilian S, Esmaili Z, Naraghi ZS, Seyedjafari E, Javar HA (2017) G-CSF loaded nanofiber/nanoparticle composite coated with collagen promotes wound healing *in vivo*. J Biomed Mater Res, Part A 105(10):2830–2842. https://doi.org/10.1002/jbm.a.36135
78. Gopinath D, Ahmed MR, Gomathi K, Chitra K, Sehgal PK, Jayakumar R (2004) Dermal wound healing processes with curcumin incorporated collagen films. Biomaterials 25(10):1911–1917. https://www.sciencedirect.com/science/article/pii/S0142961203006252
79. Zayed HS, Saleh S, Omar AE, Saleh AK, Salama A, Tolba E (2024) Development of collagen–chitosan dressing gel functionalized with propolis–zinc oxide nanoarchitectonics to accelerate wound healing. Int J Biologic Macromol 261:129665. https://www.sciencedirect.com/science/article/pii/S0141813024004689
80. Mohamed RM, Yusoh K (2016) A review on the recent research of polycaprolactone (PCL). Adv Mater Res 1134:249–255. https://www.scientific.net/AMR.1134.249
81. Katsogiannis KAG, Vladisavljević GT, Georgiadou S (2015) Porous electrospun polycaprolactone (PCL) fibres by phase separation. Euro Polym J 69:284–295. https://www.sciencedirect.com/science/article/pii/S0014305715000403
82. Augustine R, Saha A, Jayachandran VP, Thomas S, Kalarikkal N (2015) Dose-dependent effects of gamma irradiation on the materials properties and cell proliferation of electrospun polycaprolactone tissue engineering scaffolds. Int J Polym Mater Polym Biomater 64(10):526–533. https://doi.org/10.1080/00914037.2014.977900
83. Trakoolwannachai V, Kheolamai P, Ummartyotin S (2019) Characterization of hydroxyapatite from eggshell waste and polycaprolactone (PCL) composite for scaffold material. Compos Part B: Eng 173:106974. https://www.sciencedirect.com/science/article/pii/S1359836819314222
84. Augustine R, Dominic EA, Reju I, Kaimal B, Kalarikkal N, Thomas S (2014) Electrospun polycaprolactone membranes incorporated with ZnO nanoparticles as skin substitutes with enhanced fibroblast proliferation and wound healing. Rsc Adv 4(47):24777–24785. https://pubs.rsc.org/en/content/articlehtml/2014/ra/c4ra02450h
85. Nawaz T, Iqbal M, Khan BA, Nawaz A, Hussain T, Hosny KM, Abualsunun WA, Rizg WY (2021). Development and optimization of acriflavine-loaded polycaprolactone nanoparticles using box–behnken design for burn wound healing applications. Polymers 14(1):101. https://www.mdpi.com/2073-4360/14/1/101
86. Thanh NT, Hieu MH, Phuong NTM, Thuan TDB, Thu HNT, Do Minh T, Dai HN, Thi HN (2018) Optimization and characterization of electrospun polycaprolactone coated with gelatin-silver nanoparticles for wound healing application. Mater Sci Eng C 91:318–329. https://www.sciencedirect.com/science/article/pii/S092849311733388X
87. Nezhad-Mokhtari P, Kazeminava F, Abdollahi B, Gholizadeh P, Heydari A, Elmi F, Abbaszadeh M, Kafil HS (2023) Matricaria chamomilla essential oil-loaded hybrid electrospun nanofibers based on polycaprolactone/sulfonated chitosan/ZIF-8 nanoparticles for wound healing acceleration. Int J Biologic Macromol 247:125718. https://www.sciencedirect.com/science/article/pii/S0141813023026120
88. Lee KY, Mooney DJ (2012) Alginate: properties and biomedical applications. Progr Polym Sci 37(1):106–126. https://www.sciencedirect.com/science/article/pii/S0079670011000918
89. Oliveira DML, Rezende PS, Barbosa TC, Andrade LN, Bani C, Tavares DS, da Silva CF, Chaud MV, Padilha F, Cano A (2020) Double membrane based on lidocaine-coated polymyxin-alginate nanoparticles for wound healing: In vitro characterization and in vivo tissue repair. Int J Pharmaceut 591:120001. https://www.sciencedirect.com/science/article/pii/S0378517320309868
90. Zhu Z, He F, Shao H, Shao J, Li Q, Wang X, Ren H, You C, Zhang Z, Han C (2023) Chitosan/Alginate nanoparticles with sustained release of esculentoside a for burn wound healing. ACS Appl Nano Mater 6(1):573–587. https://doi.org/10.1021/acsanm.2c04714

91. Manuja A, Raguvaran R, Kumar B, Kalia A, Tripathi BN (2020) Accelerated healing of full thickness excised skin wound in rabbits using single application of alginate/acacia based nanocomposites of ZnO nanoparticles. Int J Biologic Macromol 155:823–833. https://www.sciencedirect.com/science/article/pii/S0141813020328087
92. Berger J, Reist M, Chenite A, Felt-Baeyens O, Mayer JM, Gurny R (2005) Pseudo-thermosetting chitosan hydrogels for biomedical application. Int J Pharmaceut 288(1):17–25. https://www.sciencedirect.com/science/article/pii/S0378517304005538
93. Hassan SS, Abd El-Aziz ME, Fayez AES, Kamel AH, Youssef AM (2024) Synthesis and characterization of bio-nanocomposite based on chitosan and CaCO3 nanoparticles for heavy metals removal. Int J Biologic Macromol 255:128007. https://www.sciencedirect.com/science/article/pii/S0141813023049061
94. Hou B, Qi M, Sun J, Ai M, Ma X, Cai W, Zhou Y, Ni L, Hu J, Xu F (2020) Preparation, characterization and wound healing effect of vaccarin-chitosan nanoparticles. Int J Biologic Macromol 165:3169–3179. https://www.sciencedirect.com/science/article/pii/S0141813020348340
95. Loo HL, Goh BH, Lee LH, Chuah LH (2022) Application of chitosan-based nanoparticles in skin wound healing. Asian J Pharmaceutic Sci 17(3):299–332. https://www.sciencedirect.com/science/article/pii/S1818087622000411
96. Li F, Shi Y, Liang J, Zhao L (2019) Curcumin-loaded chitosan nanoparticles promote diabetic wound healing via attenuating Inflammation in a diabetic rat model. J Biomater Appl 34(4):476–486. https://doi.org/10.1177/0885328219860929
97. Pansara C, Mishra R, Mehta T, Parikh A, Garg S (2020) Formulation of chitosan stabilized silver nanoparticle-containing wound healing film: In vitro and in vivo characterization. J Pharmaceutic Sci 109(7):2196–2205. https://www.sciencedirect.com/science/article/pii/S0022354920301878
98. Zhao F, Liu Y, Song T, Zhang B, Li D, Xiao Y, Zhang X (2022) A chitosan-based multifunctional hydrogel containing in situ rapidly bioreduced silver nanoparticles for accelerating infected wound healing. J Mater Chemistr B 10(13):2135–2147. https://pubs.rsc.org/en/content/articlehtml/2022/tb/d1tb02850b

Surfactant Based Gels for Topical Applications

Priyanka Raul, Maitreyee Mitra, Sahadeb Panja, Tuhin Manna, Subhadip Mandal, Chandradipa Ghosh, and Amiya Kumar Panda

Abstract Surfactant, an acronym of "SURFace ACTive AgeNT", typically comprises hydrophilic head and hydrophobic tail that help to form a variety of aggregates beyond a threshold concentration. Surfactant gels are formed when surfactant molecules self-assemble into a three-dimensional network structure in a solvent, resulting in the entrapment of solvent molecules within the gel matrices. Formation of surfactant gels are influenced by different factors, viz., surfactant concentration, temperature, pH, and the presence of additives (like polymers, co-solvents, electrolytes, etc.). Formulation of surfactant gels involve the preparation of surfactant solutions followed by inducing gelation through methods, viz., temperature variation, pH adjustment, salinity or addition of co-solvents. Commonly used surfactants for gel formation include anionic, cationic, non-ionic, and zwitterionic surfactants, each imparting unique property to the resulting gel. Characterization of surfactant gel is crucial for understanding their structure–property relationships and optimizing their performances. Rheology, microscopy, UV–visible absorption and emission spectroscopy, and thermal analysis are usually employed to evaluate parameters like gel strength, microstructure, viscoelastic behaviour, and thermal stability. Application of surfactant gels span a wide range that include pharmaceuticals, cosmetics, personal care products, agriculture, and environmental remediation. In cosmetics and personal care products, they serve as thickeners, emulsifiers, and stabilizers in formulations

P. Raul
Centre for Life Sciences, Vidyasagar University, Midnapore, India

M. Mitra · S. Panja · S. Mandal · A. K. Panda (✉)
Department of Chemistry and Chemical Technology, Vidyasagar University, Midnapore, India
e-mail: akpanda@mail.vidyasagar.ac.in

T. Manna
Department of Human Physiology, Vidyasagar University, Midnapore, India

Present Address:
C. Ghosh
Sadhu Ramchand Murmu University of Jhargram, Jhargram 721514, West Bengal, India

A. K. Panda
Department of Chemistry, Rani Rashmoni Green University, Hooghly, Tarakeswar, West Bengal, India

A. K. Mandal et al. (eds.), *Functionalized Nanoparticles Hydrogels for Wound Healing*, Smart Nanomaterials Technology, https://doi.org/10.1007/978-981-95-0720-7_8

like creams, lotions, and gels. In agriculture, surfactant gels are employed for soil stabilization, water retention, and pesticide delivery. Additionally, surfactant gels find use in environmental applications, viz., oil spill clean-up and groundwater remediation, owing to their ability to encapsulate and remove contaminants. The present chapter emphasizes the importance of surfactant-based gels in the topical delivery of drugs. These surfactant based gel formulations are useful in skin debridement as hydrophobic regions of the preformed micelles can trap the necrotic parts of tissues infected with microbes and the hydrophilic regions allow it to be washed away. The main use for such combination of surfactants include antimicrobial and antibiotic carriers as well as cleaning solutions and wound washes. This chapter explores the multifaceted role of surfactants in wound care, emphasizing their dual function as cleansing agents and carriers for antimicrobial agents through comprehensive examinations of the applications and advancements in surfactant-based wound care. Further research and development efforts in surfactant gel technology hold great potential for addressing current challenges and driving innovation in various industries.

Keywords Surfactant gel · Topical applications · Antimicrobials · Burn treatment

1 Introduction

SURFace ACTive AgeNTs, or surfactants, are substances that change the physicochemical properties of the interface. They are composed of a hydrophilic (water-soluble) and a lipophilic (fat-soluble) part. The lipophilic portion adheres to dirt and oil, while the hydrophilic portion allows it to be removed by washing. They find uses in a wide range of applications due to their amphiphilic nature [1]. Interfacial tension between two immiscible layers can be lowered by using surfactants. In this case, the surface tension continues to decrease as the surfactant concentration rises, especially at the interface. Surfactant molecules, in water, eventually form micelles when they reach a threshold concentration called critical micelle concentration (CMC). Surface tension does not change beyond the CMC value. Micelles are highly effective cleaning agents because they can trap hydrophobic molecules at their hydrophobic core and act as a wetting agent.

Surfactants can be categorised as cationic, anionic, non-ionic, amphoteric, or zwitterionic depending on their charge carried by the hydrophilic group [2]. Anionic surfactants, which have a negative charge, are detergents and soaps, like sodium lauryl sulphate or sodium sulfosuccinate. Quaternary ammonium compounds (QAC) are a class of cationic surfactants. Nonionic surfactants include alkyl polyglycosides and polyethylene glycols. Strong detergents, anionic surfactants are not thought to be particularly antimicrobial, however it has been observed that they can lyse Gram-negative bacteria [3]. Cationic surfactants are beneficial in cleaning and deodorising [2] and they an be employed as disinfectants in the food industry and in many medical facilities [4]. QACs can get electrostatically attached to negatively charged regions of microorganisms that causes lysis, stress on the cell wall, and ultimately death.

Additionally, these surfactants have the ability to denaturize proteins that alter the permeability of cell walls and decreases nutrient intake. Nonionic surfactants have minimal capacity to irritate cells because they do not ionize when exposed to water. Because they are thought to be non-cytotoxic, nonionic surfactants are found to be helpful in wound treatment [5]. Solvent molecules are trapped inside the matrix of surfactant gels when surfactant molecules self-assemble into a three-dimensional network structure. Several factors like temperature, pH, surfactant concentration, and the presence of additives (such as polymers, co-solvents, electrolytes, etc.) can affect the formation of surfactant gels.

Surfactant gels are widely used in different industries as dispersants, emulsifiers, wetting/foaming agents, and detergents [6]. Those are found in several wound cleaners that promote healing because of their cleaning qualities and capacity to eliminate bacteria. Surfactant molecules can increase the slippery nature of molecules, which decreases their tendency to adhere together. For this reason, gels have been used mostly as antimicrobials and antibiotic transporters [7], as well as wound washes and cleaning solutions [8]. Several studies suggest potential functions of the combination of antimicrobials and surfactants, indicating that such combination may boost the antimicrobial effectiveness [9]. Both synthetic and natural surfactants play important roles in the prevention and control of biofilm, that can provide protection to bacteria against the environmental constraints to it [10]. Biosurfactants can also perform different functions in biofilm fighting approach. Biofilm can be defined as a cluster of microorganisms that are commonly associated with either living or non-living surfaces [11]. Surfactants exhibit the capacity to induce the dispersion of biofilms produced by *Pseudomonas aeruginosa* [12]. Surfactant gels have the ability to break down exopolysaccharide (EPS) that represents crucial component of the biofilm, and managing the biofilm by controlling it [13]. These surfactants exert their antimicrobial activity by increased reactive oxygen species (ROS) generation and cell membrane disruption of bacterial cells [14]. Additionally, in molecular biology techniques nonionic surfactants are frequently used as essential agents to solubilize and disassemble proteins, obstruct surfaces that promote protein adhesion, and stop suspended cells in culture from adhering together. Uses of surfactants to bone, muscle, and orthopaedic implants have demonstrated that the removal of short-term biofilms can be enhanced by the use of any one of numerous surfactants in a fluid jet-based cleaning device [15]. Additionally, surfactant gels are utilised in cosmetics such as shampoos, perfumes, toothpastes, and skin and hair care products [16] (Figs. 1 and 2).

2 Characteristics of Gel

- Surfactant gels need to be safe, and should not react with any other ingredients in the formulation.
- An appropriate antimicrobial property should be present.
- The surfactant gel should not have any stickiness.

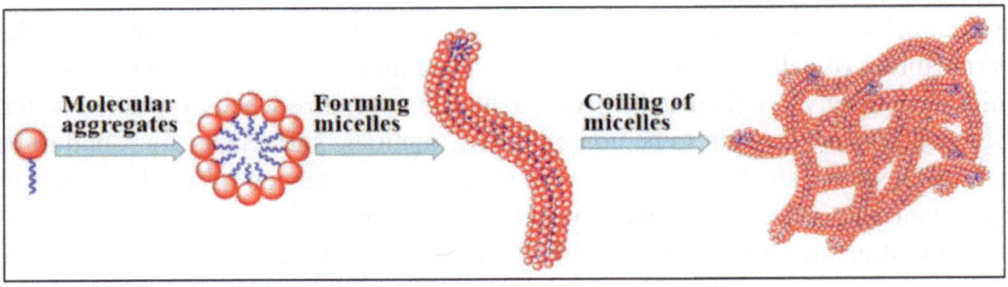

Fig. 1 Schematic diagram of surfactant gel formation

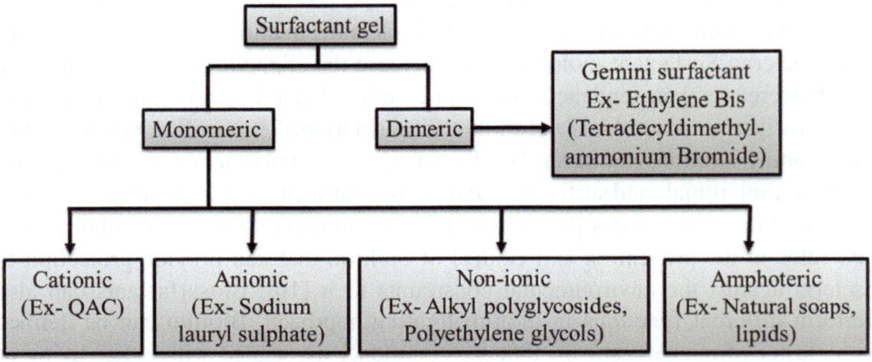

Fig. 2 Classification of surfactant gel on the basis of charge

- The surfactant gel should qualify for sterilization.
- As the effective crosslink density of the gel grows, the apparent viscosity or gel strength also rises. However, depending on the molecular interactions between the solvent and polymer, a temperature increase may cause the apparent viscosity to change.
- The gels should show the solid state's mechanical properties.
- The gels' concentration should be less than 10%, ranging from 0.5 to 2% with few exceptions.
- Gels should be easily spreadable, greaseless, thixotropic, non-staining and water-soluble.

3 Factors Affecting Gel Formation: pH, Temperature, and Concentration

3.1 pH

pH level of a solution can significantly impact gel formation. Many gelling agents and surfactants (like hyaluronic acid-CTAB nano hydrogel) exhibit pH-dependent behavior, meaning their ability to form gels may vary with changes in pH. For example, some gelling agents require specific pH conditions to form stable gel structures. Similarly, certain surfactants may experience changes in solubility or aggregation behavior at different pH levels, affecting their capacity to stabilize the gel. Additionally, pH can influence the ionization state of the functional groups in the formulation, thereby altering interactions between molecules and impacting gel network formation.

As an example, the light scattering studies of a system named hyaluronic acid (HyA)-CTAB nano hydrogel depends on pH within the range of 3.6 to 5.5. Formation of HyA-CTAB gels in the acetate buffer medium occurred later compared to their formation in the previously examined highly purified water medium. By examining the particle size values in the initial quasi-constant sections, it was observed that at pH 3.6 and 4.0, particles were approximately 125 nm in size. However, at pH 4.5, the hydrodynamic diameter of the particles increased to about 173 nm, with a further increase to approximately 265 nm observed at pH 5.5. This systematic change in particle size can be attributed to variations in the molar ratio values at the aggregation points, illustrating the effect of pH on the process [17, 18].

3.2 Temperature

Effect of temperature on surfactant gels are significantly different between crystalline and amorphous assemblies. In crystalline gels, increasing temperature can lead to a gel-to-sol transition as the crystalline structure breaks down, accompanied by a notable decrease in viscosity and elasticity due to the melting of crystals. Upon cooling, these gels may revert to their original structure. Conversely, amorphous gels exhibit a more gradual gel-to-sol transition with temperature increases, leading to smoother changes in viscosity and structural integrity without a defined melting point. Elasticity and modulus of both the types of gels decrease with rising temperature, but the changes are more pronounced in crystalline gels. Understanding these temperature effects are crucial for optimizing the use of surfactant gels in various applications [19] (Figs. 3 and 4).

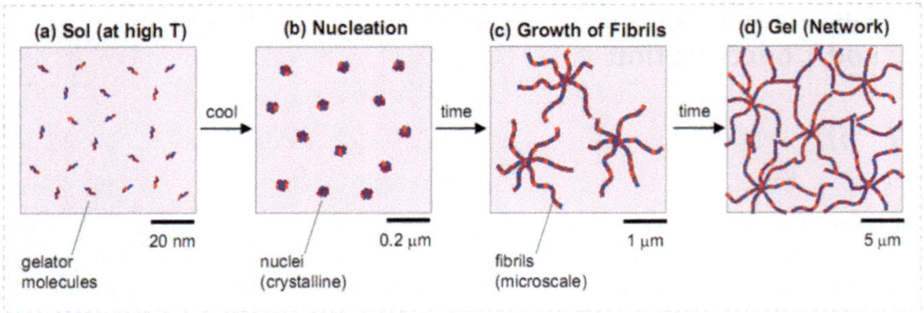

Fig. 3 Schematic representation of crystalline gel formulation

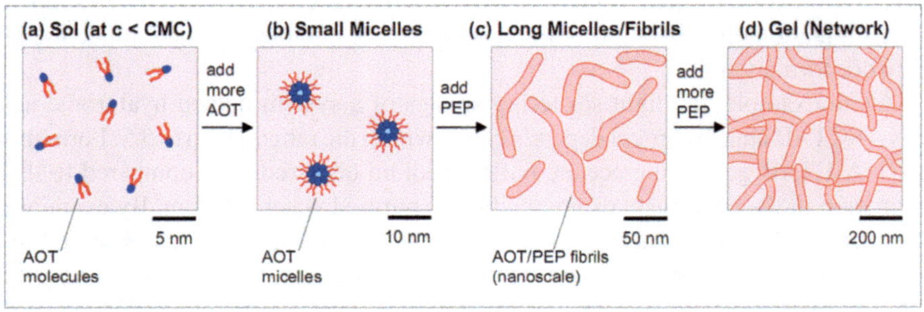

Fig. 4 Schematic representation of amorphous gel formulation

3.3 Concentration and Mole Ratio in Surfactant Gels

Concentration of gelling agents, mixed surfactants, and other additives in a formulation significantly influence the formation and properties of gels. This section explores how varying the concentration and mole ratio of these components impacts the viscosity, gel network density, and overall stability of surfactant gels.

3.3.1 Effect of Concentration on Gel Formation

Increasing the concentration of gelling agents and surfactants typically enhances the viscosity of the solution, leading to a denser gel network. This increase in viscosity is primarily due to enhanced interactions between polymer chains in the case of gelling agents or between surfactant molecules in the case of micelle formation. Higher surfactant concentration facilitates the formation of micelles at lower critical micelle concentrations (CMCs) improve interfacial adsorption which, in turn, stabilizes the gel. However, excessively high concentration of certain ingredients can lead to undesirable effects such as phase separation, reduced solubility, and altered rheological behavior. These changes can compromise the stability and performance of the gel. Therefore, finding the optimal concentration of each component is crucial for achieving the desired gel characteristics and functionality.

3.3.2 Influence of Mole Ratio on Surfactant Mixtures

Mole ratio of mixed surfactants, particularly when combining cationic and anionic surfactants, play crucial role in determining the properties of the gel. An optimal mole ratio can promote micelle formation and enhance gel stability. Conversely, an inappropriate mole ratio may lead to poor micelle formation and reduced gel quality.

3.3.3 Case Study

In a study by Barai et al. [20], viscosity of surfactant mixtures containing 20–40 mol% amino acid-based surfactants (AAS) in combination with the cationic surfactant hexadecyltrimethylammonium bromide (HTAB/CTAB) was investigated. The results demonstrated that these mixtures exhibited high viscosity, which was consistent with the larger micelle sizes observed in dynamic light scattering (DLS) measurements. These mixed surfactant systems also displayed unusual rheological properties, such as shear thickening (where viscosity increases with shear rate) and increased viscosity with rising temperature. These behaviors indicate complex interactions within the micellar structures, likely due to the formation of larger and more intricate aggregates at these specific mole ratios. [20].

3.4 *Gelling Agents*

Gelling agents are substances that create a three-dimensional network within a gel matrix, thereby providing structural integrity. These agents include a range of materials, from synthetic polymers such as hyaluronic acid, carbomers, hydroxyethyl cellulose, and polyacrylamide to inorganic materials like colloidal silica and bentonite [18]. When oppositely charged surfactants, such as amino acid-based surfactants and cationic surfactants are combined in specific molecular ratio, interact to form a viscous gel. This interaction results in a gel with specific rheological properties, allowing it to maintain its shape and consistency. Studies by Lee et al. (2015) and Barai et al. [20] have shown that these surfactant interactions are crucial for the formation of gels with desired characteristics (Figs. 5 and 6).

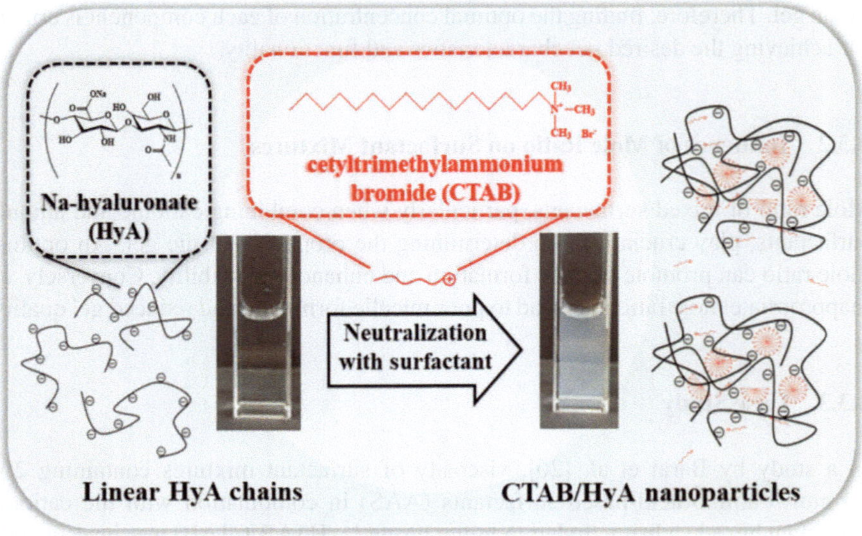

Fig. 5 Formation of surfactant-based nano hydrogel in the presence of hyaluronic acid as a gelling agent

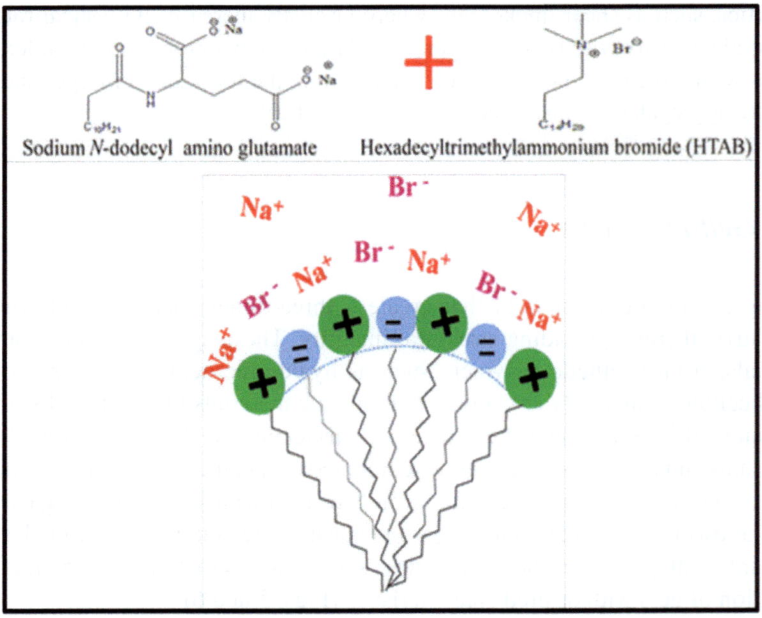

Fig. 6 Formation of surfactant-based gel in the presence of amino acid based surfactants

4 Characterization of Surfactant-Based Gels

- Physical characterization techniques: rheology, dynamic light scattering (DLS), turbidity, and viscosity.
- Microscopic techniques: scanning electron microscopy (SEM), transmission electron microscopy (TEM) and polarized optical microscopy (POM).

4.1 Rheology, Turbidity and Zeta Potential Studies

The turbidity and rheology studies on the interaction between hyaluronic acid (HyA), a hydrophilic polymer, and HTAB, a cationic surfactant, reveal significant structural changes. The experiments showed that adding HTAB to HyA solutions reduces viscosity due to the collapse of HyA's gel-like structure. Beyond a certain CTAB concentration, viscosity no longer decreases, indicating complete structural collapse. Increasing HyA concentration resulted in lower surfactant/polymer, monomer molar ratios and higher viscosity at the characteristic linear sections (Fig. 7).

This behaviour was consistent across light scattering and zeta-potential measurements. The break points of viscosity coincided with aggregation and neutralization points, showing temperature dependence with ratios ranging from n_{CTAB}/n_{HyA}, monomer ~ 0.85 at 15 °C to ~ 0.97 at 35 °C. Higher pH increased the surfactant/polymer ratios at breaking points due to variations in ionic strength. These findings highlight that both temperature and ionic strength significantly influence the HyA-CTAB interaction, impacting structural changes and aggregation processes. [18].

4.2 Viscosity Studies

According to Barai et al. viscosity behavior of mixed surfactants comprising dicarboxylic amino acids (AAS) and hexadecyltrimethylammonium bromide (HTAB/CTAB) at different ratio, demonstrating thixotropic properties where viscosity decreases with increasing shear rate. [20, 21] Surfactant mixtures with 20–40 mol% AAS are particularly viscous, showing a viscosity sequence of disodium N-dodecyl aminomalonate ($C_{12}MalNa_2$) + HTAB > disodium N-dodecyl aspartate ($C_{12}AspNa_2$) + HTAB > disodium N-dodecyl glutamate ($C_{12}GluNa_2$) + HTAB. This sequence correlates with the increasing distance between the two carboxylate groups in the AASs, enhancing hydrophobic interactions (Fig. 8).

Initially, viscosity rises sharply with the shear rate until 97 s^{-1}, then declines due to shear thinning. For $C_{12}GluNa_2$-HTAB mixtures at 40 mol% AAS, viscosity increases with temperature due to the formation of wormlike micelles, which enhances viscoelasticity. The complex aggregates form disordered structures up to a point, where further ordering is kinetically hindered, increasing fluidity and viscosity. The

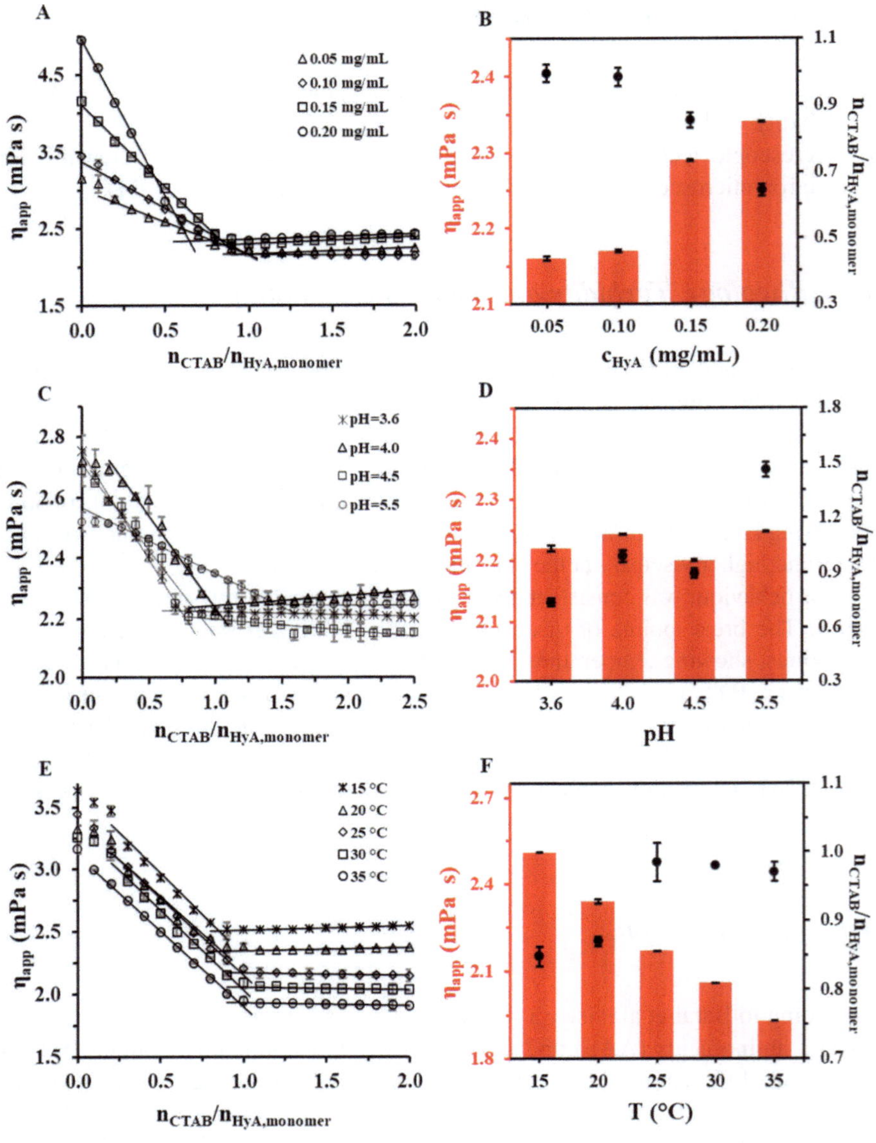

Fig. 7 Rheological measurements were conducted on hyaluronic acid (HyA) titrated with 25 mM CTAB under varying conditions: **a** different polymer concentrations ([HyA] = 0.05–0.20 mg/mL; Vol. of HyA = 19 mL; highly purified water), **c** different pH values ([HyA] = 0.10 mg/mL; Vol. of HyA = 19 mL; pH = 3.6–5.5 acetate buffer), and **e** different temperatures ([HyA] = 0.10 mg/mL; Vol. of HyA = 19 mL; T = 15–35 °C; highly purified water). Diagrams **b, d, f** illustrate the apparent viscosity and the n_{CTAB}/n_{HyA}, monomer molar ratios at the points where the curves break. Adopted from [17, 18]

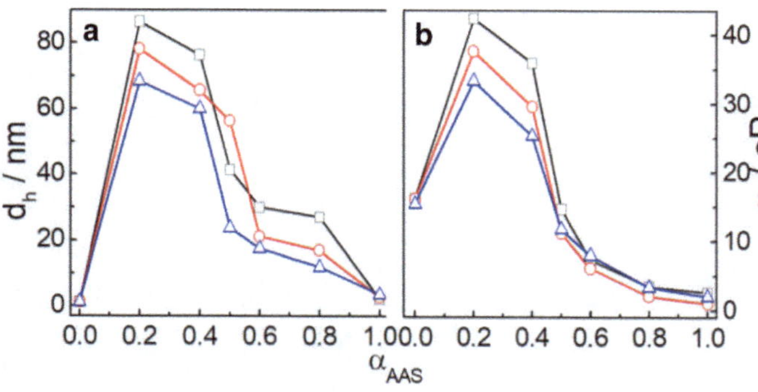

Fig. 8 Variation of hydrodynamic diameter and viscosity with the mole fraction of amino acid-based surfactants (AAS). Adopted from [21]

higher viscosity at 40 mol% AAS is attributed to a higher aggregation number of HTAB, as correlated with DLS studies. The η_0 *vs.* α_{AAS} profiles for different mixtures show a polynomial decrease, fitting a two-degree polynomial with a high regression coefficient, indicating a strong correlation.[20].

4.3 SEM and POM Studies

The structural study of $C_{12}MalNa_2$ and HTAB, along with other AASs ($C_{12}AspNa_2$ and $C_{12}GluNa_2$) in combination with HTAB, reveals significant insights into the gel formation and microstructural variations observed through various techniques. The ternary phase diagram demonstrates that increasing the proportion of HTAB in surfactant mixtures leads to the formation of gels, with the prevalence of viscous and gel states increasing in the order of $C_{12}MalNa_2$ + HTAB > $C_{12}AspNa_2$ + HTAB > $C_{12}GluNa_2$ + HTAB. The primary driving force for the formation of different aggregate types is the hydrophobic interaction between AASs and HTAB, supplemented by electrostatic attraction. At a 1:2 molar ratio, AASs and HTAB form gels, particularly at equimolar regions due to the dominance of HTAB molecules (Fig. 9).

Polarizing optical microscopy (POM) and field emission scanning electron microscopy (FE-SEM) were used to investigate the microstructures of the surfactant mixtures at various concentrations and compositions. POM studies revealed liquid crystal textures in gels, which included nematic, smectic, spherulite, cholesteric, calamitic, and flower-like textures, particularly in the $C_{12}MalNa_2$ + HTAB system. These textures are indicative of the alignment and aggregation behaviors of the surfactant molecules (Fig. 10).

FE-SEM studies provided further insights, showing interconnected morphologies such as flower-like, coral-like, porous, flake, leaf, wrinkled, sheet-like, fibrous, granular, and cuboid structures. $C_{12}MalNa_2$ + HTAB gels exhibited a variety of these structures due to the interaction between protrusions and larger channels,

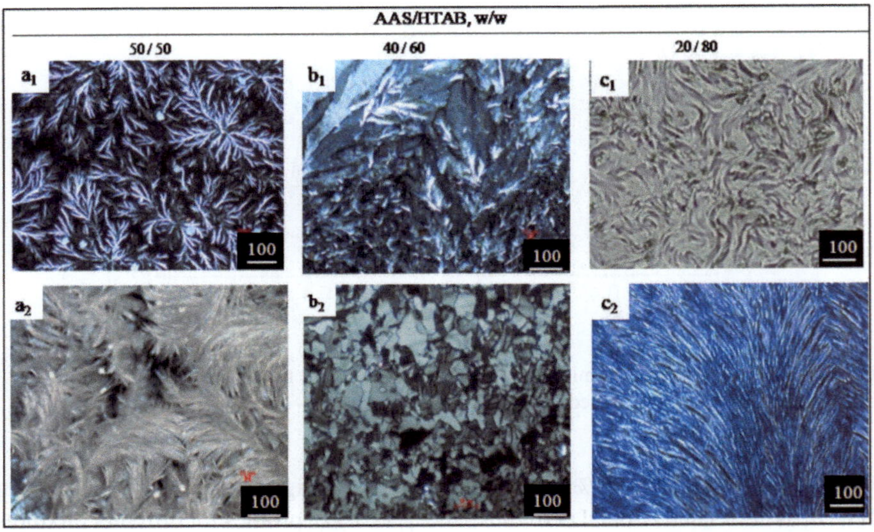

Fig. 9 Polarizing optical microscopic (POM) study of surfactant mixture at various compositions. Adopted from [21]

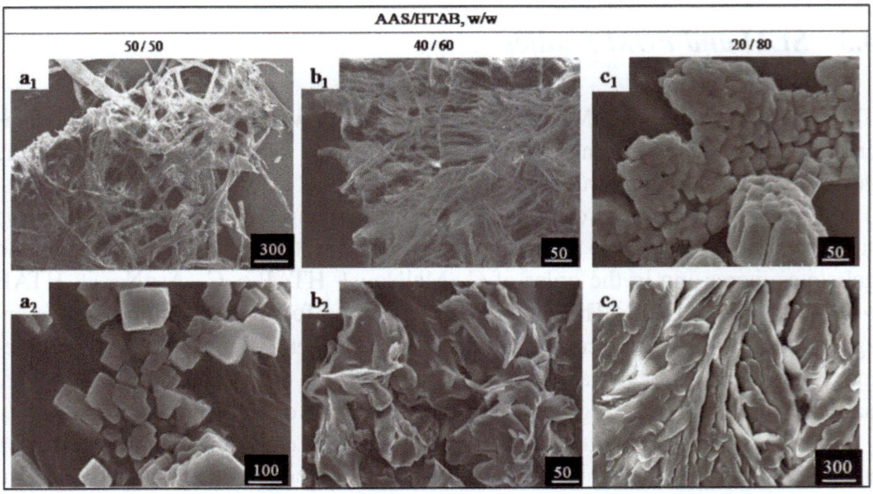

Fig. 10 Field emission-scanning electron microscopic study of surfactant mixture at various compositions. Adopted from [21]

leading to extended flat features and larger bundled-fiber networks. On the other hand, $C_{12}AspNa_2$ + HTAB and $C_{12}GluNa_2$ + HTAB gels showed fibrous and cuboid structures, indicating the formation of dense networks capable of holding water molecules. Overall, the combination of $C_{12}MalNa_2$ and HTAB formed more complex and prominent structures compared to $C_{12}AspNa_2$ and $C_{12}GluNa_2$. The study highlights the role of HTAB in enhancing aggregation and forming porous-like morphologies, corroborated by phase manifestation and POM studies.

4.4 TEM Studies

For a more detailed characterization of the system, TEM images were taken of the particles at an n_{CTAB}/n_{HyA}, monomer ratio of 0.8. The images show that as the concentration of the gelling agent, hyaluronic acid (HyA), increases, the size of the gel particles also increases gradually. This indicates a direct correlation between the amount of HyA and the dimensions of the resulting gel structures. As HyA concentration rises, more extensive cross-linking and interaction among the particles occur, leading to the formation of larger gel aggregates (Fig. 11).

5 Safety and Toxicity Concern

The use of surfactant gels for debridement and wound healing has gained significant attention in recent years. They are compatible with the body tissue and facilitate the healing process by creating an optimal microenvironment for cell proliferation and differentiation [22]. These gels offer several advantages, such as enhanced wound cleansing and debridement properties, improved tissue penetration, and a controlled release of active agents. Furthermore, surfactant gels have shown excellent biocompatibility, as they do not cause any adverse reactions or toxicity in the wound environment [23]. Some biocompatible surfactant have significant cleansing properties and capability to take out microorganisms. Now the question is: how do they work? Simply, surfactants make the molecules more slippery and thus lessen their ability to stick with each other. With this ability they prevent the adherence of microbes to

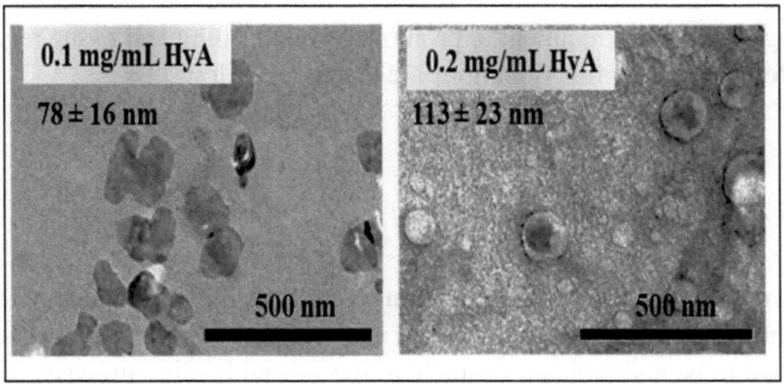

Fig. 11 Transmission electron microscopic study of CTAB/HyA nano hydrogel. Adopted from [18]

surfaces and together as well as accumulation within the wound [24]. This biocompatibility is particularly crucial in ensuring minimal tissue irritation or adverse reactions when in contact with the wound bed, thereby promoting a favourable healing environment.

A highly concentrated Poloxamer-188 based and water-soluble hydrogel, Plurogel micelle matrix (PMM) illustrates a considerable biocompatibility [25]. Poloxamer-188, as the key constituent of PMM, can control the blood flow in the arterioles and venules to forbid the build-up blood clots and debris, thus enhances oxygenation of wounded tissue [26]. Controlled entrapment and release of DNA can be used for chronic wound-healing process by preparing novel biocompatible tool with surfactants having the cationic functionality based on amino acid skeleton. Arginine-N-lauroyl amide dihydrochloride and N^{α}-lauroyl-arginine-methyl ester hydrochloride, two different single-chain amino acid based surfactants, are mixed with either single-strained or double-stranded DNA to form DNA gel particles [27]. Moreover, low toxicity of surfactant gels add to their attractiveness as platforms for wound healing and debridement. Ability to function effectively under extreme conditions further extends the potential applications of surfactant gels in addressing a wide range of wounds, including chronic and difficult-to-heal wounds. As research into surfactant gels for wound healing and debridement continues, understanding their interaction with biological systems and the mechanisms underlying their therapeutic effects will be pivotal in harnessing their full potential. This deeper comprehension will pave the way for the development of advanced surfactant gel formulations tailored for specific wound types and patient needs.

6 Applications of Surfactant-Based Gels in Wound Care

Surfactant based gels have gained attention of the scientific community lately due to their widespread utilities as antimicrobial agents as well as antimicrobial drug delivery systems for sustained drug release in the target areas to treat skin wounds and infections. Interestingly, these gels have also exhibited satisfactory results as curative agents in burn treatment to promote wound healing and tissue regeneration. Their different types of functionalities are discussed below.

6.1 As Anti-microbial Agents

Chronic wounds due to microbial infection pose a considerable challenge for patients and healthcare providers. The underlying cause of this burden is microbial biofilms which prove to be a significant obstacle to the wound healing process. Biofilm can be defined as a cluster of microorganisms that are commonly associated with either living or non-living surfaces [11]. This cluster is held together by exo-polymeric subtances (EPS), secreted by the microbes themselves. The production of this matrix facilitates attachment of the microbes to one another, while acting as a protective shield against the infiltration of antimicrobial agents [28]. It is believed that 60% of chronic wound infections are due to virtue of microbial biofilms [29]. Several studies reported antimicrobial properties of surfactant-based gels. Furthermore, surfactant gels containing antibacterial properties have proven to be successful in decreasing the density of bacterial cells within wound biofilms, encompassing both aerobic and anaerobic species. Gels, loaded with cationic surfactants like benzalkonium chloride (BAK) and di-dodecyl dimethyl ammonium bromide (DDAB), have demonstrated effectiveness against bacteria such as *Staphylococcus aureus* and *Escherichia coli*, with DDAB showing superior antimicrobial effects compared to BAK [30]. A concentrated surfactant gel formulated with antibacterial preservative agents, have been found to be effective in disrupting wound-relevant single species (*Pseudomonas aeruginosa*) as well as polymicrobial biofilms (*Acinetobacter baumannii, Staphylococcus aureus and Staphylococcus epidermidis*) by reducing bacterial cell density [31]. Studies revealed that, a gel derived from a combination of at least one polyethylene glycol and an aqueous solution with a pH ranging from 2 to 4 with a total solute concentration between 1.8 to 4.0 gm/L, along with a quantity of one cationic surfactant falling within the range of 0.9 to 1.7 g/L was beneficial for diminishing bacterial colonization in the vicinity of a wound. Effectiveness of the afore-mentioned gel was found to be independent of additional antimicrobial agents like antibiotics or antifungals [32]. Dicarboxylic amino acid-based surfactants when combined with HTAB exhibited potent antibacterial activity against most common skin pathogen *S. aureus* without causing considerable toxicity to human lymphocytes as well as mice skin suggesting their compelling dermal applications [21]. A surfactant based wound dressing SPD had shown promising anti-biofilm activity solitary or synergistically with antibiotics against pathogens like *S. aureus* and *P. aeruginosa* [33]. It is evident from the above studies that these types of gels are being developed as prospective therapeutic agents to treat cuts and wounds and thereby prevent septicemia associated risk factors.

6.2 As Drug Delivery Agent

The topical route for pharmaceutical drug delivery is considered to be more widely accepted and easily accessible by patients, as it has been enhanced and optimized through the integration of cutting-edge technologies aimed at improving the overall effectiveness of drug delivery mechanisms. Topical dosage forms have crucial impacts in the field of medicine, especially in the domain of localized therapy where they are primarily used and appreciated for their capacity to focus on specific areas of the body for treatment, thus maximizing therapeutic results and reducing potential side effects. [34]. Surfactants play a crucial role as wetting agents within the realm of formulating highly effective drug delivery systems such as gels, owing to their extensive utilization in pharmaceutical formulations aimed at enhancing the processes of dissolution and absorption of inadequately soluble medications [35]. It is reported that lipophilic non-ionic surfactants are superior to ionic surfactants in competence to dissolve poorly soluble drug molecules. Tween 80, Tween 20 and Brij 97 are the most commonly used non-ionic surfactants in drug delivery systems [35]. In recent times, surfactant-based gels are used in delivery of natural products to the skin for treating skin inflammation caused due to eczema, psoriasis, and dermatitis, etc. Ursolic acid, a beneficial natural compound used in the treatment of skin conditions, demonstrates limited solubility in water, inadequate absorption and reduced bioavailability. Oleic acid and PPG-5-CETETH-20 surfactant mixed with water were used to form gel to enhance the sustained release and bioavailability of urosolic acid [36]. Olive oil, Tween 20 and PEG 400 microemulsion based topical gel containing sulconazole, has been found to be efficient in the treatment of dermatophytes and other fungi to treat skin infections [37]. Micro-emulsion gel formed using karanj oil, surfactants (Tween 80 and diamethyl carbinol) with water shows better release profile of the loaded drug itraconazole than marketed [38]. Another interesting concept of emugels are being considered now-a-days especially for delivery of bioactive phytocompounds to the target sites. Most natural occurring phytochemicals are hydrophobic and to enhance their absorption to skin during topical applications mixture of emulsions and gels are prepared, which are termed as emugels [39]. These are thermodynamically stable formulation with low interfacial tension surfactants, viz., sorbitan esters, polysorbates, polyethylene glycols, etc., are considered very important in formation of emugels [39]. It is expected that these surfactant-gel based drug delivery systems would be significantly efficient in reducing drug toxicity as well as increased bioavailability of drugs at site specific targets in delivering synthetic drugs as well as novel phytochemicals to treat different types of complicated skin diseases.

6.3 In Burn Care Treatment

In burn care, wound debridement remains to be one of the most important issues. For the proper recovery of burnt tissues, wounds are required to be free from necrotic tissues. Surfactants are assessed to be promising that could be used in wound debridement as they contain hydrophilic and hydrophobic regions that results in micelle formation in aqueous systems. The core hydrophobic part can trap the necrotic tissues and external hydrophilic parts. Concentrated surfactant gels (CSGs) comprising higher proportion of Poloxamer provides moisture around the wound while softening necrosis with time. It was observed that CSGs significantly contributed to complete re-epithelialization or wound healing, even in deeper burns [40]. Another study conducted in 2019 revealed that CSGs helped burnt patients to relieve pain up to 90% than other topical agents available in the market. Post twenty eight days of treatment with CSGs, patients did not require any surgery [41]. Another report reveals that water-soluble surfactant dressing (WSD) used twice a day for treatment on patients with 70% flame injuries. It was found that such patients could recover from burn injury with a median of 7 days. Administration of WSD was well received by patients, with no instances of wound infection occurring in the treated area. No deaths

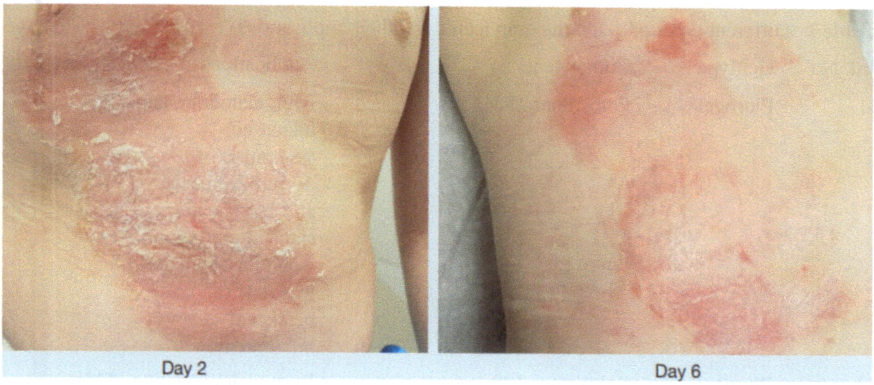

Fig. 12 A two-year-old girl who got burns from hot liquid contact. After six days, PluroGel micelle matrix (PMM) was shown to have debrided non-viable tissue [24]

were reported though 22% of patients were diabetic [42]. In another study, surfactant gels with encapsulated fluoxetine demonstrated potential for diabetic wound healing by enhancing drug delivery and promoting tissue repair [43]. Thus, surfactant-based gels could be considered as promising formulations in diabetic wound management and healing as well burn care treatments (Fig. 12, Table 1).

7 Future Prospects

Surfactant-based gels are considered to have prospects in topical applications due to their diverse functionalities. Studies have highlighted the formation of liquid crystal phases, gels with antibacterial effects, and minimal toxicity to skin cells [21, 47]. Additionally, the development of microemulsion-based gels (MBGs) has resulted in enhanced drug delivery capabilities, such as improved drug release and permeation compared to traditional formulations [48]. These gels offer advantages like controlled release through skin barriers, high stability, and biocompatibility, making them attractive for topical pharmaceutical applications. The incorporation of nano-emulsion into gels enhances drug penetration and therapeutic activity, overcoming limitations associated with conventional dosage forms [49]. Overall, surfactant-based gels present a promising avenue for the development of advanced topical drug delivery systems. However, further studies are needed to investigate the incorporation of different drugs into surfactant-based gels to enhance drug penetration and therapeutic activity. Continued research is warranted to optimize the formulations of surfactant-based gels for enhanced drug release, permeation, and stability, addressing current limitations in conventional dosage forms. Mostly, the antibacterial and antibiofilm experiments are conducted in laboratory environments which might not be the exact representatives of the microbial biofilm in the wounds. Fruther in vivo studies on animal

Table 1 Different types of gel with components and their applications

Sl. No.	Gel type	Components	Applications	References
1	Plurogel	Poloxamer-188	Pain, malodour, and increased generation of devitalized tissue and exudate are signs of chronic wounds, which are in a state of persistent inflammation	[25]
2	Plurogel	Poloxamer-188	Burn wound healing	[24]
3	Zwitterionic	Amino acid-based surfactants + hexadecyltrimethyl-ammonium bromide (AAS + HTAB)	Against causative agent *Staphylococcus aureus*, which results in chronic infections of the skin and soft tissues	[21]
4	Cationic polymer	Chitosan	Pharmaceutical and cosmetic purposes	[44]
5	Plurogel	F-127	Pharmaceutical	[45]
6	Anionic surfactant	Carbopol 934 + SDS	Drug solubilization	[46]
7	Non ionic surfactant	Carbopol 934 + Tween 80	Drug solubilization	[45]

models are necessary to assess the efficacy and toxicological aspects of the developed formulations to provide valuable insights into the potential of these gels for clinical applications.

References:

1. Singh P, Cameotra SS (2004) Potential applications of microbial surfactants in biomedical sciences. Trends Biotechnol 22:142–146
2. McDonnell G, Russell AD (1999) Antiseptics and disinfectants: activity, action, and resistance Clin Microbiol Rev 12:147–179
3. Glover RE, Smith RR, Jones MV et al (1999) An EPR investigation of surfactant action on bacterial membranes. FEMS Microbiol Lett 177:57–62
4. Mereghetti L, Quentin R, Marquet-Van Der Mee N, Audurier A (2000) Low sensitivity of Listeria monocytogenes to quaternary ammonium compounds. Appl Environ Microbiol 66:5083–5086
5. Zölß C, Cech JD (2016) Efficacy of a new multifunctional surfactant-based biomaterial dressing with 1% silver sulphadiazine in chronic wounds. Int Wound J 13:738–743

6. Percival SL, Mayer D, Malone M et al (2017) Surfactants and their role in wound cleansing and biofilm management. J Wound Care 26:680–690
7. Faulkner DM, Sutton ST, Hesford JD et al (1997) A new stable pluronic® F68 gel carrier for antibiotics in contaminated wound treatment. Am J Emerg Med 15:20–24
8. Yang Q, Larose C, Della Porta AC et al (2017) A surfactant-based wound dressing can reduce bacterial biofilms in a porcine skin explant model. Int Wound J 14:408–413
9. Belfield K, Bayston R, Hajduk N et al (2017) Evaluation of combinations of putative anti-biofilm agents and antibiotics to eradicate biofilms of Staphylococcus aureus and Pseudomonas aeruginosa. J Antimicrob Chemother 72:2531–2538
10. Satputea KS, Banpurkar GA, Banat MI et al (2016) Multiple roles of biosurfactants in biofilms. Curr Pharm Des 22:1429–1448
11. Flemming H-C, Wingender J (2010) The biofilm matrix. Nat Rev Microbiol 8:623–633
12. Díaz De Rienzo MA, Stevenson PS, Marchant R, Banat IM (2016) *Pseudomonas aeruginosa* biofilm disruption using microbial surfactants. J Appl Microb 120:868–876
13. Patel D, Patel B, Thakkar H (2021) Lipid based nanocarriers: promising drug delivery system for topical application. Eur J Lipid Sci Tech 123:2000264
14. Nakata K, Tsuchido T, Matsumura Y (2011) Antimicrobial cationic surfactant, cetyltrimethylammonium bromide, induces superoxide stress in *Escherichia coli* cells. J Appl Microbiol 110:568–579
15. Anglen J, Gainor B, Simpson W, Christensen G (2003) The use of detergent irrigation for musculoskeletal wounds. Int Orthop 27:40–46
16. Rathod H, Mehta D (2015) A review on pharmaceutical gel. Int J Pharm Sci 1:33–47
17. Larrañeta E, Henry M, Irwin NJ et al (2018) Synthesis and characterization of hyaluronic acid hydrogels crosslinked using a solvent-free process for potential biomedical applications. Carbohydr Polym 181:1194–1205
18. Seres L, Csapó E, Varga N, Juhász Á (2023) The effect of concentration, temperature, and ph on the formation of hyaluronic acid–surfactant nanohydrogels. Gels 9:529
19. Raghavan SR (2009) Distinct character of surfactant gels: a smooth progression from micelles to fibrillar networks. Langmuir 25:8382–8385
20. Barai M, Mandal MK, Karak A et al (2019) Interfacial and aggregation behavior of dicarboxylic amino acid-based surfactants in combination with a cationic surfactant. Langmuir 35:15306–15314
21. Barai M, Manna E, Sultana H et al (2021) Micro-structural investigations on oppositely charged mixed surfactant gels with potential dermal applications. Sci Rep 11:15527
22. Mir M, Ali MN, Barakullah A et al (2018) Synthetic polymeric biomaterials for wound healing: a review. Prog Biomater 7:1–21
23. Garcia-Orue I, Pedraz JL, Hernandez RM, Igartua M (2017) Nanotechnology-based delivery systems to release growth factors and other endogenous molecules for chronic wound healing. J Drug Deliv Sci Technol 42:2–17
24. Kirsner RS, Amaya R, Bass K et al (2019) Effects of a surfactant-based gel on acute and chronic paediatric wounds: a panel discussion and case series. J Wound Care 28:398–408
25. Woo K, Hill R, LeBlanc K et al (2018) Effect of a surfactant-based gel on patient quality of life. J Wound Care 27:664–678
26. Birchenough SA, Rodeheaver GT, Morgan RF et al (2008) Topical poloxamer-188 improves blood flow following thermal injury in rat mesenteric microvasculature. Ann Plast Surg 60:584–588
27. Morán MC, Infante MR, Miguel MG et al (2010) Novel biocompatible DNA gel particles. Langmuir 26:10606–10613
28. Vert M, Doi Y, Hellwich K-H et al. (2012) Terminology for biorelated polymers and applications (IUPAC Recommendations 2012). Pure Appl Chem 84:377–410
29. Kim D, Namen Ii W, Moore J et al (2018) Clinical assessment of a biofilm-disrupting agent for the management of chronic wounds compared with standard of care: a therapeutic approach. Wounds 30:120–130

30. Teng X, Clerkin, Caoimhe G, McCoy, Colin P (2022) Antimicrobial cationic surfactant-loaded hydrogel coatings in preventing medical device-associated infections. Br J Pharm 7
31. Salisbury AM, Mullin M, Foulkes L et al (2022) Controlled-release iodine foam dressings demonstrate broad-spectrum biofilm management in several in vitro models. Int Wound J 19:1717–1728
32. Myntti M (2019) Translation of antibiofilm technologies to wounds and other clinical care. In: Williams DL (ed) targeting biofilms in translational research, device development, and industrial sectors. Springer International Publishing, Cham, pp 85–95
33. Das Ghatak P, Mathew-Steiner SS, Pandey P et al (2018) A surfactant polymer dressing potentiates antimicrobial efficacy in biofilm disruption. Sci Rep 8:873
34. Roberts MS, Mohammed Y, Pastore MN et al (2017) Topical and cutaneous delivery using nanosystems. J Control Release 247:86–105
35. Kaur P, Garg T, Rath G et al (2016) Surfactant-based drug delivery systems for treating drug-resistant lung cancer. Drug Deliv 23:727–738
36. Fonseca-Santos B, Araujo GA, Ferreira PS et al (2023) Design and characterization of lipid-surfactant-based systems for enhancing topical anti-inflammatory activity of ursolic acid. Pharm 15:366
37. Payyal SP, NC R, Sathyanarayana SD et al. (2020) Microemulsion based gel of sulconazole nitrate for topical application. Turk J Pharm Sci 17(3):259–264
38. Md Hamid A, Ashish J, Mahadev K, Akhlesh Kumar S (2024) Formulation and evaluation of microemulsion based gel of Calendula officinalis extract and its anti-microbial activity. Int J Multidiscip Res Stud 7:40–51
39. Olayemi O, Chukwuemeka A (2023) Emulgel: a promising technology for topical delivery of herbal extracts. Br J Pharm 8
40. Pittinger T, Curran D, Hermans M (2020) Treatment of burns in adult patients with a concentrated surfactant gel: a real-life retrospective evaluation. Wounds 32:339–344
41. Coffey R, Penny R (2021) 556 The use of a concentrated surfactant based gel to promote burn wound healing. J Burn Care Res 42:S126–S127
42. Hill DM, Mahmood AO, Mitchell MB et al (2021) 555 Early application of a water-soluble surfactant dressing for partial-thickness burn wounds to reduce wound conversion. J Burn Care Res 42:S126–S126
43. Fatima F, Aleemuddin M, Ahmed MM et al (2023) Design and evaluation of solid lipid nanoparticles loaded topical gels: repurpose of fluoxetine in diabetic wound healing. Gels 9:21
44. van der Lubben IM, Verhoef JC, Borchard G, Junginger HE (2001) Chitosan and its derivatives in mucosal drug and vaccine delivery. Eur J Pharm Sci 14:201–207
45. Alvarez-Lorenzo C, Concheiro A (2003) Effects of surfactants on gel behavior. Am J Drug Deliv 1:77–101
46. Ashton P, Hadgraft J, Brain KR et al (1988) Surfactant effects in topical drug availability. Int J Pharm 41:189–195
47. Roy R, Majumder J, Datta HK et al (2022) Supramolecular hydrogels developed from mafenide and indomethacin as a plausible multidrug self-delivery system as antibacterial and anti-inflammatory topical gels. ACS Appl Bio Mater 5:610–621
48. Yasir Siddique M, Nazar MF, Mahmood M, Saleem MA, Alwadai N, Almuslem AS, Alshammari FH, Haider S, Akhtar MS, Hussain SZ, Safdar M, Akhlaq M (2021) Microemulsified gel formulations for topical delivery of clotrimazole structural and invitro evaluation. Langmuir 37:13767–13777
49. Daryab M, Faizi M, Mahboubi A, Aboofazeli R (2022) Preparation and characterization of lidocaine-loaded, microemulsion-based topical gels. Iran J Pharm Res 21:e123787

Hydrogel Dressing for Wound Repair

Rumeysa Sarigol, Seyfure Adiguzel, Rabia Betul Aygun, Miray Karamese, and Sevde Altuntas

Abstract Wound healing is a complex biological process consisting of hemostasis, inflammation, proliferation, and remodeling phases. Throughout this process, key biological events such as angiogenesis, phagocytosis, and collagen remodeling are regulated and supported by growth factors and cellular activities. The design of wound dressings plays a critical role in promoting healing, preventing infection, and reducing pain by maintaining an optimal moist environment. In this context, research focused on developing cost-effective, biocompatible, and efficient materials has gained increasing importance. Hydrogels stand out as prominent materials in wound care and healing processes due to their high water retention capacity, biocompatibility, and antibacterial properties. The high water content of hydrogels helps maintain moisture balance in the wound area, preventing tissue desiccation and accelerating the healing process, while simultaneously alleviating pain and pressure. By mimicking the extracellular matrix, hydrogels promote cell proliferation and granulation tissue formation, while minimizing scar formation and optimizing healing. Their biocompatibility makes hydrogels an ideal choice for wound healing. Additionally, the incorporation of growth factors (e.g., FGF, VEGF) and nanoparticles into hydrogel polymer matrices enhances their therapeutic effects, further improving the healing process. Through controlled drug release, antimicrobial agents, and strategies targeting biofilm, hydrogels play a crucial role in wound healing by combining tissue regeneration and infection control.

R. Sarigol · R. B. Aygun
Department of Molecular Biology and Genetics, Faculty of Arts and Sciences, Yildiz Technical University, Istanbul, Turkey

R. Sarigol · S. Adiguzel · M. Karamese · S. Altuntas
Experimental Medicine Research and Application Center, Validebağ Research Park, University of Health Sciences Turkey, Istanbul, Turkey

S. Adiguzel
Department of Molecular Biology and Genetics, Institution of Health Sciences, University of Health Sciences, Istanbul, Turkey

M. Karamese · S. Altuntas (✉)
Department of Tissue Engineering, Institution of Health Sciences, University of Health Sciences Turkey, Istanbul, Turkey
e-mail: sevde.altuntas@sbu.edu.tr

A. K. Mandal et al. (eds.), *Functionalized Nanoparticles Hydrogels for Wound Healing*, Smart Nanomaterials Technology, https://doi.org/10.1007/978-981-95-0720-7_9

Keywords Hydrogel dressings · Tissue regeneration · Biocompatibility · Extracellular matrix

1 Introduction: Hydrogel Dressings, Composition and Properties of Hydrogel Dressings

Skin wounds arise from the disruption of skin integrity caused by internal factors, including infection, metabolic disorders, and neoplastic illnesses, or exterior factors like trauma, burns, or surgery [1, 2]. Throughout history, various medical interventions and materials have been used to optimize the wound healing process [3]. One of these materials, wound dressings, is essential to the healing process because it helps create granulation tissue and isolates the injured tissue from the surrounding environment [4]. Historical records show that the use of bandages for skin wound care dates back to 2500 BC [5]. In Mesopotamian tablets, there are findings on methods such as washing wounds with water or milk, applying plasters, and treating them with natural products such as honey, resin [6]. According to Hippocrates, the father of medicine, wounds were cleansed with oils, honey, and wine and covered with woolen cloths boiled in wine or water [7]. After Hippocrates used woolen towels to make basic bandages, wound care materials have evolved over time. As sterilizing methods advanced in the late 19th and early twentieth centuries, gauze and other absorbent, sterile materials became widely employed [8]. For many years, gauze has been a common material for wound dressings because of its benefits, which include its availability, non-toxicity, and affinity for the skin [9]. Gauze dressings only offer passive protection, thus they are not a sustainable treatment option because they need to be changed frequently, dry up the wound bed, do not react to changing wound conditions, and harm tissue when removed [3, 10]. In contrast to a dry wound environment, a moist wound environment was shown to accelerate wound healing by 50% in a 1995 study by George Winter on porcine tissue [11, 12]. Modern wound care products that are founded on the principles of moist wound healing lessen pain, promote keratinocyte migration to speed up re-epithelialization, prevent tissue dehydration, speed up angiogenesis, and actively aid in the healing process by lowering infection rates, dressing changes, and pain [13–16]. As a result, it is critical to produce low-cost wound dressing materials that are designed to the specifics of the wound, do not harm the body, limit the risk of infection, offer a moist environment, and promote skin regeneration [17]. In this context, the first moist dressings designed following Winter's findings were transparent polyurethane films [18]. These films are an alternative dressing type to gauze because of their transparency, which allows the wound to be seen, reduces the risk of infection by separating the wound from the external environment, does not cause pain during dressing changes because they do not stick to the wound surface, provides a moist wound environment, is permeable to air and water vapor, and retains moisture. However, because these films lack a fluid absorption characteristic, they are unsuitable for wounds that produce exudate [10,

19]. Foam dressings are a versatile treatment that is widely used in wound care [20]. Because of their non-sticky surfaces, high absorption capacity, and deformability, they effectively manage wound exudate. In addition to these characteristics, they protect the wound bed by providing thermal insulation and forming a barrier against mechanical stress. They also aid in epithelialization by maintaining a moist wound environment. By using various components, foam dressings can suit a wide range of clinical needs [21, 22]. For example, silver ion-containing foams minimize the risk of infection due to their antibacterial properties [23]. Similarly, foams containing anti-inflammatory medicines like ibuprofen help to alleviate wound discomfort. Clinical trials have demonstrated that ibuprofen-containing foams dramatically reduce discomfort in the donor area, particularly in patients undergoing skin transplants. Finally, foam dressings are a versatile tool for wound treatment [24]. They can be used successfully on a wide range of wound types due to their high absorption capacity, malleability, and ability to be enhanced with additional components. Hydrogels are three-dimensional (3D) network structures made of polymer-based materials that are cross-linked physically, chemically, or biologically [25]. The extracellular matrix structure found in live tissues is reflected in the architecture, viscoelasticity, and porousness of hydrogels, which allow them to retain large volumes of liquid [26]. Hydrogels can be used in a variety of applications, including tissue engineering as cell scaffolds, drug delivery systems as biomolecule carriers, biosensors for diagnosis and treatment, and wound dressings as moist environments due to their biocompatibility and naturally tissue-like, tunable physical characteristics [27]. Depending on the polymers that make up the majority of these materials, they are categorized as natural, synthetic, or hybrid. Synthetic polymers like alginate and chitosan are more robust and long-lasting, whereas natural polymers like collagen, hyaluronic acid, and elastin are better for biocompatibility. Hydrogels have a variety of qualities due to the cross-links in their structure [28]. The production source of natural polymers can affect their susceptibility to enzymatic breakdown and pathogen contamination [29, 30]. Therefore, these obstacles limit the use of hydrogels formed from natural polymers. However, hydrogel dressings made of synthetic polymers are not subject to the hazards and drawbacks of animal origin, and they offer the benefit of possibly being able to be specially made to serve particular purposes [19]. However, the inability of synthetic polymers to actively engage in the healing process limits their efficacy. Therefore, bioartificial hydrogel-based dressings that combine natural and synthetic materials have garnered a lot of interest in the fields of wound healing, medication delivery, and tissue engineering [31]. The features that demonstrate the properties of hydrogels in wound healing are briefly shown in Fig. 1.

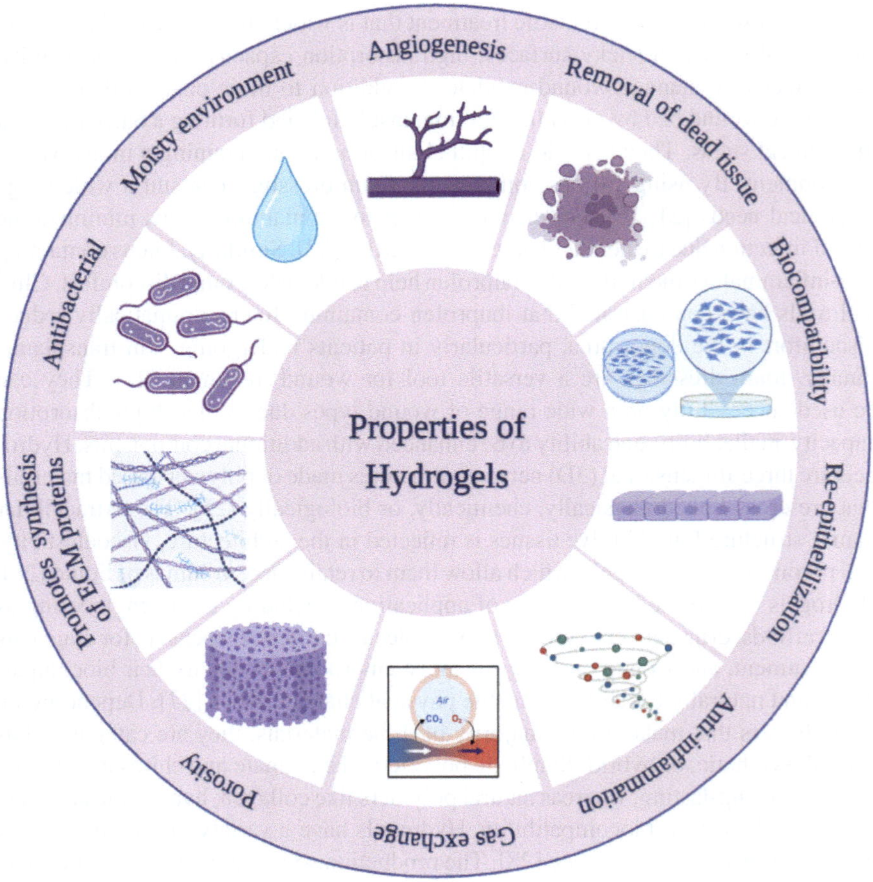

Fig. 1 Properties that demonstrate the properties of using hydrogels as wound dressings

2 Wound Healing Processes and the Role of Hydrogel Dressings in Wound Healing

Wound healing consists of a series of biological processes aimed at repairing damaged tissue and occurs in four main stages. Hemostasis occurs immediately after the injury to reduce the size of the wound; inflammation occurs within the first few days after the injury to clean the wound of dead tissue and bacteria; proliferation can last for weeks after the injury and involves the formation of new tissue; remodeling can take up to two years after the injury and involves the replacement of new tissue [32]. In these stages, complex biological events such as angiogenesis, proliferation, chemotaxis, phagocytosis, and the breakdown and reconstruction of collagen occur. Many cell types and growth factors work together and in harmony during these stages [33].

In the first stage of wound healing, known as the hemostasis phase, the initial response to prevent excessive blood loss is vasoconstriction, meaning the blood vessels constrict to minimize bleeding. Then, platelets come together to form blood clots [34]. Blood clots not only stop bleeding but also serve as a temporary support structure for inflammatory cells, fibroblasts, and growth factors. Additionally, the release of factors such as vascular endothelial growth factor (VEGF), platelet-derived growth factor (PDGF), transforming growth factor-β (TGF-β), and a group of cytokines like IL-1, IL-6, IL-8, and IL-10 occurs at this stage, regulating the immune response in the wound area and supporting the healing process [35, 36]. In the inflammation phase, which begins with the release of VEGF, the wound starts to be cleaned of dead tissue and bacteria approximately 24 h after the injury and continues for up to two weeks [37]. In the inflammation phase, neutrophils are the first to respond to the injured area, followed by M1 and M2 macrophages. Neutrophils are the first cells to reach the injured area, and their function is to clean up debris such as bacteria, necrotic tissues, and dead cells, and to release reactive oxygen species into the environment to initiate the phagocytic activity of macrophages. In response to inflammatory signals, chemokines and cytokines rapidly accumulate at the wound site and, along with peroxidase enzymes, help eliminate pathogens, dead cells, and tissue debris [38]. Matrix metalloproteinases (MMPs), which are involved in both the reconstruction and degradation of extracellular matrix components secreted by neutrophils, can prolong the wound healing process and lead to chronic wounds if they remain in the wound area for an extended period. Neutrophils typically die by apoptosis or are phagocytosed by macrophages about three to seven days after injury [39]. Macrophages are the second inflammatory cells to reach the injured area and are the key cells used in the wound healing phase. In the wound area, M1 macrophages are first formed from monocytes, cleaning the wound of dead tissue and bacteria. M2 macrophages, on the other hand, form when the wound begins to heal and have anti-inflammatory effects. Additionally, M2 macrophages support the migration and proliferation of fibroblasts, keratinocytes, and endothelial cells for the repair of the skin layers (dermis, epidermis) and blood vessels. Immune system cells involved in inflammation (Langerhans cells, dendritic cells, and T cells) become active to ht foreign antigens. Macrophages reach their highest concentration approximately 48 h after injury, secreting epidermal growth factor (EGF), VEGF, and fibroblast growth factor (FGF), thereby initiating the processes of angiogenesis and granulation [40]. When growth factors and cytokines are released to the wound area, they support the reformation of the extracellular matrix and collagens, and the wound closes afterward [41, 42].

Proliferation is the stage at which new tissue forms and can continue for weeks following an injury. In the proliferation phase, fibroblasts and myofibroblasts play a key role in the reconstruction of granulation tissue [43]. A significant increase in the production of keratinocytes, fibroblasts, and endothelial cells is observed; this significantly contributes to wound healing and ECM accumulation. Especially fibroblasts play a critical role in transforming the fibronection-rich provisional matrix into a more resilient and mature granulation tissue. The proliferation phase represents a dynamic process that continues until functional wound closure is achieved [44, 45].Remodeling can take up to two years after the injury and alters the integrity and functionality of the new tissue. The composition of the extracellular matrix evolves over time and can lead to the reorganization of collagen fibers temporarily formed in the injured area, with soft and gelatinous type III collagen being replaced by well-structured type I collagen. This leads to a decrease in the amount of granulation tissue in the wound bed [38].

Many internal and external local and systemic factors affect wound healing. This healing phase can last for months or even years. Scarring is a normal part of the healing process. The functional restoration of damaged tissue and the replacement of normally non-regenerating tissue are mediated by an integrated set of cellular processes, including inflammation, angiogenesis, matrix accumulation, and tissue remodeling [46].

Due to a disorder that develops during wound healing, which can lead to reduced tissue function, fibrotic mismatch, and further complications, optimal wound repair requires the proper resolution of damaged tissue. Unfortunately, non-healing chronic wounds are an increasingly common health issue affecting millions of people [47]. As a result, these issues make it beneficial to develop new therapeutic strategies and materials to intervene in or improve specific stages of this complex process [48].

Hydrogels are a versatile class of biomaterials that can hold water up to thousands of times their weight. Water-soluble hydrophilic macromolecules and a simple regular three-dimensional structure are two prerequisites for a material to be used as a hydrogel [49]. Cross-linked hydrophilic polymer networks, commonly found in both natural and synthetic polymers, are characterized by unique properties such as high hydrophilicity, biocompatibility, and porous structures (Yao [50]). Due to these properties, hydrogels play an important role in fields such as biomedical applications, tissue engineering, drug delivery systems, and wound healing. The use of hydrogels in these areas is highly valuable due to their biocompatibility and the mechanical properties they offer [51]. Hydrogels are primarily divided into two categories: natural and synthetic types, which may differ in their applications in medicine. Firstly, hydrogels composed of natural polysaccharides such as alginate, chitosan, and hyaluronic acid primarily function as extracellular matrices in organisms, providing good tissue compatibility and adjustable surface modification [52]. Secondly, hydrogels composed of synthetic polymers such as poly-ethylene glycol (PEG), poly-vinyl alcohol (PVA), and poly-(acrylic acid (PAA), are widely used clinical medical materials that are easy to modify during the process and have a good drug loading effect [53]. With scientific and technological innovations, the continuous improvement of hydrogel technology enables hydrogels to possess high performance, good flexibility, ease of use, long-lasting effects, simple sterilization, low toxicity, and mass production capability.

Various dressings of critical importance in skin repair have been developed. These dressings support wound healing by providing a moist environment on the wound surface while also protecting the wound from all external harmful agents. Among these, hydrogel dressings are considered one of the best options for healing and other medical applications due to their plant origins, adjustable hydrophilicity, swelling capacities, biocompatibility, non-adhesiveness to wounds, and transparency. Hydrogel dressings can promote cell growth by maintaining a moist environment for the cells [54].

The primary purpose of hydrogels used in wound healing is to provide an optimal environment from the initial stage of hemostasis. The ideal hydrogel has high tensile strength and elasticity, high bioactivity, can be easily shaped to the defect area, has the ability to control the release rate, and supports good vascularization as well as the potential for stem cell, cell migration, proliferation, and growth to replace the extracellular matrix [55].

Another characteristic of hydrogel dressings is that they promote antibacterial activity. Hydrogels can be beneficial to the wound area by exhibiting antibacterial properties in small injuries such as cuts, abrasions, burns, and ulcers, as well as diabetic foot and surgical wounds [56]. Like other dressings, hydrogels can be used to create an effective healing environment that accelerates the healing of various wounds, thereby preventing the need for strong medical treatments. Hydrogels love and can maintain a moist environment. The moist environment supported by hydrogels can promote the natural migration of leukocytes and defense cells to the wound [45]. While inflammation, harmful bacteria, and immune cells can be removed from the wound, endogenous growth factors and additional cytokines can be simultaneously increased by the immune cells and leukocytes surrounding the wound, thereby accelerating tissue repair [57]. Hydrogels, especially in burn wounds, function as coolants due to the water they contain around the wound. Hydrogels alleviate pain in the wound area with their cooling effect while also protecting the wound from infection [58].

The biocompatibility of hydrogels, due to their structural similarity to the natural extracellular matrix, makes them good candidates for wound dressings. Additionally, the composition, mechanical properties, and functionalization of hydrogels can be manipulated to support the overall wound healing process and reduce the risk of scar formation [55]. Hydrogels composed of proteins, peptides, and polysaccharides can help regulate wound healing, resist infections, and promote tissue regeneration. Due to their availability, biodegradability, biocompatibility, and similarity to the extracellular matrix, hydrogels are gaining increasing interest as a wound healing platform [59]. Since hydrogels contain a significant amount of water, organizations such as the extracellular matrix and cells easily adhere to their surfaces. Therefore, they can contribute to the proliferation phase and similarly stimulate the healing tissues. With the active components and growth factors to be used, hydrogels can be further developed to actively control cell proliferation. These can be used in pathological conditions such as slow or rapid healing, while maintaining the healing rates of natural tissue, and in cases where intensive healing is required, such as severe burns [60, 61].

Nanoparticles (NPs) combined with hydrogels can be used in different stages of wound healing, directly affect growth factors, and can be used with growth factors in NP-hydrogel composites. In the inflammation stage, NPs are used to reduce inflammation by encapsulating anti-inflammatory agents. In the angiogenic process associated with platelet activation, NPs can simulate the accumulation of macrophages and fibroblasts in hydrogels [62, 63]. The combination of hydrogels with nanoparticles enhances applicability by providing greater control over response and function. The combined applications of hydrogels with drug delivery and antimicrobial properties significantly enhance wound healing capabilities [64, 65].

3 Applications of Hydrogel Dressings in Wound Care

Hydrogels, due to their high water content, help maintain the moisture balance over damaged tissue, while providing a cooling effect that alleviates pain and reduces pressure on the wound [66]. Thanks to their high water-holding capacity, they prevent the desiccation of the damaged tissue, thereby accelerating the healing process. The moist environment created by the hydrogel mimics the extracellular matrix, promoting cell proliferation, supporting granulation tissue formation, and preventing scar tissue development [67, 68]. Additionally, hydrogels serve as a protective barrier, reducing the risk of infection in the damaged tissue. Recent research in the field of hydrogels has focused on integrating growth factor release [69], metallic nanoparticles [70], controlled drug delivery [71], and sensors [72], offering potential solutions for chronic wounds (Fig. 2).

Hydrogels, while functioning as a protective barrier in wound healing, also provide an environment that prevents enzymatic degradation of growth factors [73]. The biocompatibility of the hydrogel and its high water content create a suitable environment for healing in the wound area, while the integrated growth factors, particularly fibroblast growth factors (FGF), vascular endothelial growth factor (VEGF), and acidic fibroblast growth factor (aFGF), promote cell proliferation and the formation

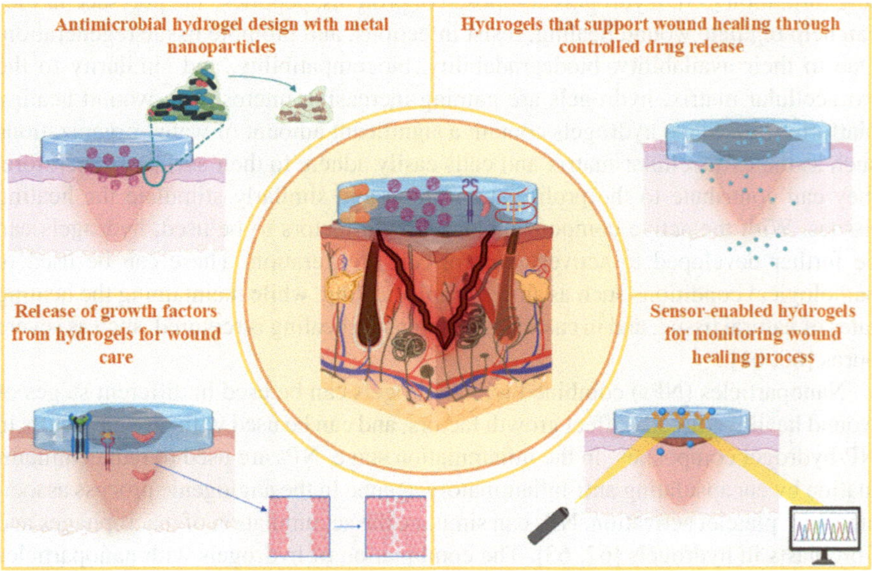

Fig. 2 Applications of hydrogels in wound care

of granulation tissue in the wound area [69, 74]. These factors also enhance angiogenesis, facilitating the formation of new blood vessels in the wound area, which helps transport oxygen and nutrients [67]. Hydrogels protect the biological activity of these growth factors by preventing their enzymatic degradation, ensuring their controlled release [75].

Inflammatory and hypoxic wound tissues have been observed to have acidic pH levels due to increased glycolysis, elevated lactic acid production, and carbon dioxide accumulation. This acidic environment makes it an ideal option for early intervention in wound healing, particularly due to the acidic isoelectric point of aFGF. Gao and colleagues have leveraged this advantage of aFGF by integrating nanoparticles with a biocompatible hydrogel to develop a new therapeutic platform. In this study, a complex hydrogel was designed to accelerate diabetic wound healing, effectively control inflammation and oxidative stress, promote the transition of the wound into the proliferative phase, and enhance angiogenesis. The skeleton of this hydrogel was produced using biocompatible and biodegradable hyaluronic acid. In the HAMA-mSAM@aFGF hydrogel, which integrates aFGF, a structural mesophase with gold nanoparticles (Au) exhibiting nanozyme properties was used. This nanomaterial contributed to wound healing by exhibiting dual functions, such as neutralizing reactive oxygen species (ROS), alleviating inflammation, promoting cell proliferation, and stimulating angiogenesis. Experimental studies revealed that this multifunctional wound dressing was effective in treating diabetic wounds. In vivo studies with mSAM and aFGF examined the therapeutic effects of the hydrogel on wound healing. mSAM neutralized ROS, protecting cells from oxidative stress, and promoted cell proliferation in response to induced oxidative stress. Furthermore, in the presence of aFGF, mSAM enhanced the migration of fibroblasts and endothelial cells, contributing to the healing process. In vivo data confirmed the effectiveness of HAMA-mSAM@aFGF and HAMA hydrogels in a full-thickness diabetic wound model in db/db mice. After 10 days, the HAMA-mSAM@aFGF group showed 88.46% healing, while the control group achieved only 53.66%. When combined with mSAM@aFGF, this hydrogel exhibited multifunctional wound dressing properties, such as anti-inflammatory effects, modulation of macrophage polarization, enhanced angiogenesis, and promotion of collagen deposition. This innovative system was designed to alleviate excessive inflammatory responses, reverse chronic inflammation, and facilitate the transition of wounds into the proliferative phase, thereby accelerating the healing of diabetic ulcers. This multifunctional nanozyme-based therapeutic strategy offers a promising solution to advance diabetic wound care [67].

Chronic skin wounds are often difficult to heal due to severe bacterial infections, prolonged inflammation, and damaged vascular structures. The inflammation response begins when macrophages differentiate from monocytes and migrate to the affected area, where they rapidly exert their effects through phagocytosis. Macrophages also contribute to tissue remodeling by modulating the immune response and accelerating the healing process through the secretion of anti-inflammatory cytokines and growth factors. As a result, macrophage-related research is gaining increasing interest in developing therapeutic strategies for treating skin wounds. However, the presence of classical macrophages, known as the M1 phenotype, is associated with pro-inflammatory cytokines and excessive inflammation, which can negatively affect tissue remodeling processes. Therefore, promoting effective M2 macrophage polarization has become a popular research topic in wound healing. For instance, findings suggest that miRNAs enhance M2 macrophage polarization during osteoarthritis progression. In the design of wound dressing materials, it is important to address both macrophage modulation and antibacterial activity together. Even against antibiotics, a wide variety of antimicrobial peptides (AMPs) provide effective defense mechanisms in organisms. Wu and colleagues developed a biomass hydrogel incorporating modified antimicrobial peptides (mAMPs) and miR-301a. The positively charged mAMP and the negatively charged miR-301a-3p mimic molecule self-assemble in solution to form spherical nanoparticles (mAMP/miR-301a). These nanoparticles are encapsulated within the hydrogel to provide physical stability and allow for controlled cellular uptake. This approach takes advantage of the controlled release properties of hydrogels, preventing systemic drug administration and enabling localized treatment. Furthermore, the hydrogel absorbs inflammatory exudate, regulates the wound microenvironment, reduces inflammation, and prevents infection. In vitro studies have demonstrated that the mAMP/miR-301a@CMCs system is an effective biomaterial with low toxicity, promoting cell proliferation and migration, biocompatibility, and controlled release properties. This system activates the PI3Kγ/mTOR signaling pathway by inhibiting PTEN expression, and regulates macrophage polarization from M1 to M2, increasing the expression of anti-inflammatory genes (Arg-1, TGF-β, IL-10). This interaction provides an anti-inflammatory effect and is considered a promising biomaterial for chronic wound treatment. In vivo studies in a diabetic wound model, supported by in vitro results, have shown that mAMP/miR-301a@CMCs reduces inflammation by regulating M2 macrophage polarization, promotes angiogenesis, and ensures dynamic renewal of the dermis and epidermis. Additionally, by increasing collagen deposition and promoting rapid tissue regeneration with a thin keratin layer, it presents a promising therapeutic method for accelerating healing in diabetic infected wounds [76].

It is possible to integrate a variety of molecules into the polymer network structure of hydrogels using encapsulation or impregnation methods [77, 78]. Hydrogels serve as protective barriers, shielding damaged tissue from inflammation, however, they are insufficient in the face of inflammation that arises during the injury. To address this limitation, metallic nanoparticles are integrated into hydrogels to provide antimicrobial effects and enhance the healing process [79]. Due to the mechanisms of bacterial resistance to antibiotics, non-antibiotic metallic nanoparticles offer significant advantages with their antimicrobial properties in infection treatments [80].

Bacterial cellulose (BC) is a natural biosynthetic cellulose hydrogel that is frequently used in wound dressing applications. However, the lack of intrinsic antimicrobial properties of BC has led to efforts to enhance this material with antimicrobials, anti-inflammatories, antioxidants, and other healing agents. In this context, silver nanoparticles (AgNPs) stand out due to their remarkable antimicrobial properties. The antibacterial effect of AgNPs is based on mechanisms such as increasing cell membrane permeability, inhibiting DNA replication, denaturing bacterial proteins, and creating an effective antibacterial environment by releasing silver ions inside the cell. Additionally, the antifungal effect of AgNPs is reported to cause cell death by increasing reactive oxygen species in fungal cells and enhancing membrane permeability. Gupta and colleagues characterized curcumin (CUR)-derived AgNPs (cAgNPs) produced through a green synthesis method and loaded them into BC hydrogels. In this study, the curcumin plant, known for its antimicrobial, antioxidant, and anti-inflammatory properties, was used in the green synthesis process. BC produced using Gluconacetobacter xylinus bacteria was functionalized via a physical impregnation method with cAgNP nanoparticles. The evaluations confirmed that free-form cAgNPs exhibited lower cytotoxicity when loaded into BC, while providing broad-spectrum antimicrobial activity. Notably, cAgNPs demonstrated a stronger antibacterial effect on P. aeruginosa compared to S. aureus. Furthermore, cAgNP prepared with CUR:HPβCD exhibited $76.65 \pm 3.21\%$ antioxidant activity against DPPH radicals, and when loaded into BC, it was shown to significantly contribute to the wound healing process. These data suggest that cAgNP integrated into BC offers potential as an antimicrobial wound dressing material and holds promise in supporting wound healing [81].

Hydrogels have the ability to encapsulate various drugs within three-dimensional network structures due to their adjustable pore structures [78]. Additionally, their adjustable physical properties enable long-term and controlled drug release. Hydrogels are classified into two groups based on the stimuli they respond to: endogenous stimuli (e.g., pH, glutathione, redox gradients, and enzyme concentrations) and exogenous stimuli (e.g., light, magnetic fields, and temperature) [82]. Thanks to these characteristics, hydrogels can precisely control drug distribution and release to maximize therapeutic effects. Controlled release prevents the rapid loss of the drug in the biological environment and enhances therapeutic efficacy through localized treatment [83].

In diabetic foot ulcers (DFU), hyperglycemia induces oxidative stress, increasing the production of proinflammatory cytokines and leading to complex bacterial biofilm infections, which result in prolonged inflammation and chronic wounds. Current DFU treatments are often inadequate. As an alternative, nanoenzymes, with their enzyme-like properties, offer a promising approach for DFU treatment. Shang and colleagues designed and developed an effective material for DFU by encapsulating nanozymes in hydrogels and providing controlled release. Nanozymes prevent ROS-induced inflammation, regulate redox homeostasis, and support wound healing by integrating nitric oxide synthase (NOS) mimics. They also regulate cytokine levels and break down biofilms to reduce bacterial adhesion. Nanoenzymes encapsulated in hydrogels create a moist microenvironment that fills irregular wounds and accelerates healing. Spray hydrogel applications significantly inhibited the growth of pathogens such as S. aureus, MRSA, E. coli, and P. aeruginosa. Studies conducted on diabetic rats showed that ACPCAH and ACPCAH + US treatments reduced inflammation and accelerated wound healing by 98.3%. Both in vitro and in vivo tests demonstrated that ACPCAH had remarkable effects in alleviating hypoxia, promoting angiogenesis, lowering blood sugar, and eradicating pathogenic bacteria. This study not only provides an efficient multi-enzyme-like nanozyme solution for DFU treatment but also stands out as an all-in-one therapeutic option [84].

Hydrogels stand out as a promising platform for postoperative wound care; they keep the wound surface cool, reduce pain, control drug release, promote skin regeneration, and prevent infection. Surgical methods in cancer treatment are the primary choice for tumor removal, but completely clearing remaining cancer cells and ensuring wound healing presents significant challenges. Yin and colleagues have designed and developed a therapeutic hydrogel for such cases. The palladium nanoparticles (Pd NPs) used in this study are noted to exhibit effects such as photothermal therapy (PTT) and photodynamic therapy (PDT). PTT is a highly tumor-selective method with minimal risk to healthy tissues, while PDT eliminates cancer cells by generating reactive oxygen species (ROS) through light exposure. Yin's team developed a hydrogel made from natural polymers (agarose and chitosan) that provides both wound healing and photothermal and photodynamic therapies using Pd nanoparticles. This hydrogel, loaded with the chemotherapy drug doxorubicin (DOX), offers controlled drug release. In vitro studies with Pd/DOX@hydrogel observed that under 808 nm laser light, the temperature increased from 27 to 59 °C within 5 min, simultaneously producing ROS, which led to cancer cell death. The thermosensitive hydrogel structure softened under laser influence, enabling rapid drug release. The combination of photothermal therapy and laser-triggered chemotherapy effectively destroyed cancer cells while supporting wound healing. In vivo experiments on melanoma-model mice revealed that Pd/DOX@hydrogels had potential for postoperative tumor suppression, wound healing, and prevention of tumor recurrence. No signs of organ toxicity were observed in the treated mice, indicating that Pd nanoparticles and DOX remained localized at the target site with minimal systemic circulation. With its photothermal properties, pH-sensitive drug release, and natural components, Pd@hydrogel emerges as an innovative therapeutic tool with potential for clinical translation [85].

In recent years, the concept of integrating diagnostic sensors into wound dressings to support wound management has been developed. These "theranostic" platforms, which combine treatment and diagnosis, have the potential to not only enhance treatment efficacy but also provide a better understanding of the underlying mechanisms of chronic wounds, potentially reducing hospital stays [72]. The increased biomarker levels in chronic wounds indicate imbalances in the wound microenvironment and disruptions in the healing process. Therefore, accurately measuring biomarker levels provides valuable insights into the current condition of the wound, enabling personalized treatment approaches [86]. Hydrogel-based wound dressings not only support wound healing but also offer an effective platform for detecting and analyzing biomarker levels in the wound microenvironment. The formulation of hydrogels facilitates the absorption of wound exudate or the collection of materials released from the wound cavity via micro-sampling. This allows for the undisturbed collection of biomarkers and enables an accurate assessment of the wound's current status. The low viscosity of hydrogels at low temperatures ensures that exudate can be easily extracted without damaging the surrounding tissue during sampling. Additionally, hydrogels transform into a high-viscosity solid dressing at body temperature, providing advantages in both sample collection and wound closure processes. This dual functionality makes hydrogels an innovative tool in advanced wound care [72].

Matrix metalloproteinases (MMPs) play critical roles in wound healing, contributing to processes such as the removal of damaged tissue, angiogenesis, fibroblast and keratinocyte migration, epithelialization, wound matrix contraction, and tissue remodeling. However, increased MMP-9 concentration and elevated MMP-9/TIMP ratios in chronic wounds are significant indicators of imbalances in the wound microenvironment. In this context, Wang and colleagues have designed and developed a hydrogel that functions as a sensor for detecting MMP-9 and monitoring the wound healing process. This hydrogel is prepared using carboxymethyl chitosan (Fc-CMCS) modified with an MMP-9-specific Fc-Met-Gly peptide and dextran-aldehyde (DEX-ALD). The two components, combined via a Schiff base reaction, form stable chemical bonds, maintaining the structural integrity of the hydrogel. Methionine, cleaved by MMP-9, releases ferrosene, an electroactive component, while glycine enhances the flexibility of the peptide, facilitating the cleavage process. When the hydrogel comes into contact with an MMP-9-containing environment, the enzyme penetrates the hydrogel and breaks the short peptide bonds in the polymer backbone. The concentration of the released ferrosene segments is measured rapidly and accurately using differential pulse voltammetry (DPV), allowing for the determination of MMP-9 activity. In in vitro studies conducted on Raw264.7 cells, the hydrogel-based DPV method showed lower standard deviation and higher accuracy compared to ELISA. The hydrogel system developed in this study not only enables the precise detection of MMP-9 activity but also stands out as an innovative approach in biosensor technologies [86].

4 Advantages, Challenges and Limitations of Hydrogel Dressings

An ideal wound dressing material should maintain moisture in the wound area, facilitate gas exchange, protect the wound from infections, and provide physical support for wound healing [87, 88]. Additionally, it must be biocompatible and biodegradable [89]. Traditional dry dressings often fail to meet these requirements, whereas hydrogel dressings emerge as promising alternatives due to their remarkable properties, including mechanical characteristics compatible with biological tissues and exceptional water retention capacity, which maintains wound moisture and continuously absorbs exudate [90, 91]. Hydrogel dressings are fabricated using natural or synthetic polymers combined through various physical or chemical cross-linking techniques to provide diverse functionalities and properties [90].

Natural hydrogels address the limitations of conventional dressings by offering high water content, excellent moisture retention, superior biocompatibility, and a microstructure resembling the extracellular matrix (ECM) [92]. The transparency or translucency of natural hydrogels allows for easy monitoring of the wound healing process. Additionally, when combined with various drugs, they facilitate healing by enabling a gradual and sustained release at the wound site [93]. Compared to synthetic polymers, natural hydrogels offer significant opportunities for molecular construction and modification of three-dimensional cross-linked networks due to the abundance of active groups such as hydroxyl, carboxyl, and amino groups [94]. Despite their numerous advantages, natural hydrogels have certain limitations in clinical applications that require improvement. These include enhancing their mechanical properties to suit various wound types, improving their limited antibacterial capabilities to prevent infections, and increasing their biodegradability to eliminate the need for surgical removal after use [95].

Synthetic hydrogels have gained significant attention in biomedical applications due to their tunable mechanical properties and the ease of production [96]. These polymers exhibit high stability and inherent adaptability, enabling their use in various forms to support optimal tissue development. However, their limited biocompatibility remains a critical challenge [97]. To overcome this limitation, hybrid hydrogels have been designed by combining the tunable physical properties of synthetic materials with the superior biocompatibility and biodegradability of natural polymers [98–100]. For instance, Rasool and colleagues created hydrogels using a combination of chitosan, poly (N-vinyl-2-pyrrolidone), and polyacrylic acid (PAA). These hydrogels demonstrated thermal stability due to synthetic PAA, antibacterial activity against *E. coli* due to the inherent properties of chitosan, a natural polymer, and were biodegradable [101].

Natural and synthetic hydrogels with intrinsic antibacterial properties have demonstrated effectiveness in wound healing by exhibiting antibacterial activity against bacteria in direct contact. However, these hydrogels also have limitations, particularly their inability to target bacteria in the surrounding environment [95]. Furthermore, incorporating antibiotics, synthetic antibacterial agents, and antibacterial peptides into hydrogels has been shown to enhance their infection-resistant properties [102]. An additional strategy to combat bacterial infections involves integrating inorganic metal ions (such as Ag, Zn, Au, Cu, and Mg) and their metallic oxide nanoparticles, which are well-known for their antibacterial properties, into hydrogel wound dressings [103, 104]. Moreover, studies utilizing metal nanoparticles alone have highlighted the instability of these nanoparticles in aqueous solutions [105]. The incorporation of hydrogels addresses this issue by stabilizing the nanoparticles, leveraging the physiological and mechanical properties of the hydrogel matrix. This also allows for controlled release of antibacterial metal agents [45, 106, 107]. Silver, a widely recognized antibacterial material, is often the preferred choice due to its excellent bioactivity and biocompatibility [108, 109]. For instance, the antibacterial effect of cellulose hydrogels loaded with silver (Ag) nanoparticles has been demonstrated against both gram-positive and gram-negative bacteria [81]. In another study, Ag nanoparticles, embedded in biocompatible chitosan-polyvinyl alcohol hydrogels, exhibited significant antibacterial activity against *S. aureus* and *E. coli* without showing cytotoxic effects [110]. Additionally, the use of zinc oxide (ZnO) nanoparticles, another metallic oxide material, has been reported as safer and more cost-effective for wound healing, owing to its low cytotoxicity and superior antibacterial properties [111]. A nanohydrogel design consisting of methylcellulose (MC) and polyvinylpyrrolidone (PVP), encapsulating ZnO nanocomposites, has been shown to promote effective healing of burn wounds. This hydrophilic polymer design not only moisturizes the wound microenvironment but also accelerates healing by absorbing wound fluid, thanks to its high mechanical strength. Furthermore, the integration of nanoparticles with antibacterial properties significantly enhances wound healing efficiency by restricting bacterial growth [112].

The ability to produce nanoparticles (NPs) from various natural sources through green synthesis has significantly enhanced the potential for biomedical applications in the development of metallic nanocomposite hydrogels [113]. An exemplary study demonstrated that a wound dressing containing silver nanoparticles (LSH-AgNPs) incorporated into flaxseed hydrogels via green synthesis exhibited excellent antimicrobial properties. LSH-AgNP hydrogels were found to actively inhibit the growth of various bacteria and fungi, including *S. epidermidis*, *E. coli*, *S. aureus*, *B. subtilis*, and *A. niger*. Moreover, 100% wound closure was observed in rabbits by the 15th day after treatment with LSH-AgNPs [114]. NPs embedded in hydrogels can serve as effective antimicrobial agents by inhibiting the growth and formation of both bacterial and fungal species [115]. Although metal-based composite antibacterial hydrogels remain a major focus of research in biomaterials, further extensive investigation is required, particularly concerning the potential risks associated with the release of nanoparticles into the human circulatory system through wound sites [116].

In the context of cutaneous wound healing, antioxidant nanoparticle-loaded hydrogel systems have been widely used and have shown promising results [117]. Xu and colleagues introduced a multifunctional wound dressing called Alg-PDA-Se hydrogel, designed for treating infected wounds. This hydrogel demonstrated remarkable reactive oxygen species (ROS) scavenging ability by effectively alleviating oxidative stress, despite containing a low dose of selenium nanoparticles [118]. Additionally, it has been observed that highly charged nanoparticles can localize clotting factors and promote blood clotting. It has been proposed that nanoclays, which are self-assembling structures capable of dynamically breaking apart, could be utilized for minimally invasive therapies due to these properties [119]. By incorporating nanoclays into gelatin, a novel injectable bleeding dressing material with physiological stability has been developed, which solidifies rapidly in the wound area. This formulation significantly enhanced the hemostatic performance and clot strength of the hydrogel. In vitro studies showed a significant reduction in clotting time, with human blood-typically clotting in 5–6 min-demonstrating accelerated clotting when treated with this hydrogel [120].

Conventional hydrogel applications often lack the necessary functions for optimal wound healing and typically limit healing efficiency by passively contributing to the process, primarily through moisture retention in the wound area [88]. In contrast, mechanically active adhesive hydrogels and self-adapting smart hydrogels have been developed as advanced hydrogel dressing materials for wound healing applications. These innovative hydrogels actively support the wound healing process due to their unique functionalities. Mechanically active adhesive hydrogels were designed based on the contraction ability of embryonic wounds. These hydrogels can adhere to the wound and actively contract in response to external stimuli such as temperature, humidity, and changes in salinity [121]. Li et al. developed a mechanically active adhesive dressing (AAD) for skin wound healing, utilizing a double-network stiff hydrogel matrix of temperature-responsive PNIPAM and alginate, with alginate providing a tissue-adhering surface layer. When applied to an in vitro wound model, AAD was shown to contract the wound and reduce the wound area by 45% [122]. Similarly, in another study, a conductive and self-shrinking adhesive smart hydrogel was created through free radical polymerization of NIPAM using quaternized chitosan (QCS) and polydopamine-coated reductive graphene oxide (rGO-PDA). The synergy of the hydrogel's conductivity and self-shrinking ability significantly contributed to the skin wound healing process [123].

Self-adaptive hydrogels, on the other hand, can adjust to the local biological environment due to their dynamic network structure, allowing them to conform to wounds of various sizes, shapes, and thicknesses [124]. In an exemplary study, a self-adapting hydrogel was prepared using catechol-Fe^{3+} coordinated cross-linked poly(glycerol sebacate)-co-poly(ethylene glycol) (PEGSD) and quadruple hydrogen bond cross-linked ureido-pyrimidinone modified gelatin solutions. This self-healing hydrogel, designed with a dynamic cross-linking mechanism, exhibited excellent shape adaptation and self-healing capabilities. Furthermore, the hydrogel demonstrated antibacterial, antioxidant, and reversible tissue adhesive properties in response to near-infrared (NIR) irradiation or pH changes. In vivo skin trauma studies have shown that the hydrogel can regulate inflammation during the wound healing process, accelerate collagen deposition, promote granulation tissue formation and vascularization, and significantly support the healing of full-thickness skin defects [125]. Despite the various advantages of self-healing and mechanically active adhesive smart hydrogels, they also have limitations. Mechanically active hydrogels may not fully conform to the wound bed due to their stiffness, while self-adapting hydrogels may not remain in place for extended periods due to their high fluidity. These limitations could potentially be addressed by combining both hydrogel types in future studies [88]. By leveraging the synergy between the two types, the self-adapting hydrogel could initially be used to cover the wound bed, followed by the application of a mechanically active adhesive hydrogel patch to restrict the movement of the self-adapting hydrogel. This combination could enhance wound healing efficiency.

Another class of smart responsive hydrogels includes those designed with light-activated phototherapeutic agents, such as those used in photothermal therapy (PTT) and photodynamic therapy (PDT). Both PTT and PDT are characterized by high selectivity and broad-spectrum antibacterial activity, making them promising alternatives to antibiotics in wound care application [126, 127]. PTT has gained significant attention in antibacterial applications due to its remote control capability, excellent biocompatibility, deep tissue penetration, and non-invasive nature [128]. Antibacterial composite hydrogels with adhesive and self-healing properties have been designed for the treatment of infected wounds. The PTT method is effective in thermally inactivating bacteria, however, it carries the risk of damaging surrounding healthy tissues due to operating temperatures ranging from 55 to 60 °C. Additionally, the production of heat shock proteins in bacteria may confer resistance to PTT [95].

An alternative method, PDT, targets the absorption of a photosensitizer by bacteria, followed by stimulation of the photosensitizer with specific wavelengths of light. This activation produces reactive oxygen species (ROS), which damage the bacterial cell wall and internal structures, leading to bacterial death [129]. When encapsulated in hydrogel, the photosensitizer is slowly released into the infected area, enhancing therapeutic efficacy and minimizing potential damage to surrounding healthy tissue [130]. The incorporation of smart responsive hydrogels with phototherapeutic agents enables targeted delivery to wound sites and allows for controlled release of the agents. Moreover, since these agents can be activated by external stimuli as needed, this design offers the potential for single-application treatments without requiring frequent changes to the hydrogel dressing. However, optimizing the light parameters is essential to prevent damage to surrounding tissues, and potential side effects must be carefully considered. Additionally, the levels of reactive oxygen species (ROS) should be regulated during photodynamic therapy, as excessive ROS can negatively impact wound healing [131].

Since each stage of the wound healing process has its own distinct microenvironmental characteristics, it is essential to detect changes within the wound microenvironment and use dressings that can respond appropriately. To address this need, sensor-based smart hydrogels have been developed as an innovative method to monitor the overall progress of wound healing. These systems incorporate sensors that can provide critical information about wound status, including pH levels, humidity, temperature, oxygen levels, and inflammation intensity [132–135]. Such sensors enable the detection of wound parameters and allow for targeted interventions to enhance healing [136]. For example, hydrogel dressings with temperature-sensitive drug release systems, which are triggered by inflammation, have been developed for burn wounds [137]. It is important to ensure that the sensors used in these approaches are non-toxic, biocompatible, and sufficiently flexible to integrate with the hydrogel.

Researchers have developed nanocomposite hydrogel dressings by incorporating nanoparticles or nanorods into hydrogel networks, offering a novel approach that enhances not only the mechanical, self-healing, and antibacterial properties of these hydrogels, but also their ability to monitor and treat wounds [54, 138, 139]. For instance, a polymer-based wound dressing composed of polyacrylic acid (PAA)-grafted poly(N-isopropyl acrylamide) (PNIPAM), vinyl-based polyacrylamide (PAM), and silver nanowires (AgNWs) has been recently developed as a conductive, soft, temperature-responsive, antibacterial, and biocompatible hydrogel [140]. This hydrogel system was integrated with a Bluetooth module, enabling wireless transmission of temperature changes to a smart device. This innovative integration, along with the wireless transmission capabilities of the conductive hydrogel, allows for continuous and real-time monitoring of wound temperature, facilitating the early detection of potential infections. The development of these integrated hydrogel systems opens new possibilities for advancing wound care therapies (Table 1).

In summary, although these advanced hydrogel dressings show great potential in wound care, they also present limitations based on the features evaluated. Table 8.4.1

Table 1 Advantages and limitations of using advanced composite hydrogels as wound dressings

Features	Advantages	Limitations
Mechanical stability	Durability can be enhanced through the use of advanced synthetic polymers. Their high stability and internal adaptability contribute to the optimal development of tissues	However, materials that enhance stability may also reduce biodegradability and biocompatibility
Biocompatibility	Natural hydrogels, which can mimic the ECM structure, are compatible with human tissues	Long-term testing may be required to confirm their biocompatibility
Biodegradability	Biomaterials that naturally degrade in the wound area without causing environmental harm are used	The biocompatibility of the by-products released during biodegradation may raise concerns
Antibacterial properties	It prevents infections and supports wound healing by exhibiting antibacterial properties against both direct and surrounding bacteria	However, the dosing of antibacterial agents may lead to cytotoxicity or the development of bacterial resistance in the wound area
Wound moisturization	By providing a constantly moist environment, it prevents tissue dehydration and accelerates healing	Excess moisture may lead to softening of the surrounding skin and, in sensitive individuals, allergic reactions
Flexibility and compatibility	Mechanically adhesive and self-healing hydrogels conform easily to the wound without restricting skin movement, maintaining mobility	In cases of low flexibility, surface cracking may occur, or the hydrogel may fail to conform to the wound area due to high fluidity
Easy applicability	With lightweight and easily moldable structures, these hydrogels enable non-invasive or minimally invasive applications	Certain formulations may require gradual and precise application
Drug release	It supports the healing process by providing continuous and controlled drug release to the wound area	It is not suitable for all types of drugs; formulations need to be optimized based on the specific drug
Thermal tunability-detection	Sensor hydrogel systems that can solidify at the wound site in response to temperature sensitivity, detect wound temperature changes, and enable the development of appropriate treatments are utilized	In extreme hot or cold environments, the stability and detection sensitivity of hydrogels may be affected
Cost	Cost reduction can be achieved by advancing production methods with new technologies and materials	These hydrogels are more expensive compared to traditional wound dressings

summarizes the advantages and limitations of composite hydrogel dressing applications. Hydrogel dressings face challenges related to biocompatibility testing and clinical outcomes that must be addressed before they can be successfully translated into clinical practice [90]. Animal models used to evaluate these treatments cannot fully replicate the complexity and physiology of chronic wound healing in humans. Additionally, potential complications associated with hydrogel dressings must be carefully considered. These may include immune responses such as as softening of periarticular wound skin due to excessive moisture, as well as allergic reactions and autoimmune diseases in susceptible individuals [141]. Furthermore, the emphasis on developing accessible and cost-effective hydrogel dressings for use in healthcare settings with limited resources has been instrumental in their broader clinical adoption [95]. Making these treatments more affordable and accessible will help ensure their widespread use. Additionally, most studies to date have primarily focused on treating external skin wounds. Future research should place greater emphasis on developing advanced hydrogel dressing systems for internal wound healing. Addressing these multifaceted challenges presents promising opportunities for future research, with the potential to develop multifunctional hydrogel dressings that can simultaneously promote wound healing, provide infection control, and enable personalized drug delivery.

5 Conclusion

Ultimately, the evolution of hydrogel-based wound dressings represents a significant milestone in the support and enhancement of wound healing. These dressings, with their biocompatibility, moisture retention capabilities, and potential for customization, accelerate wound healing by reducing infection risks and improving pain management. Furthermore, innovative components integrated into hydrogel-based materials, such as growth factors, metallic nanoparticles, and sensors, offer multifunctional properties that enhance the efficiency of wound healing. Nanoparticles like gold nanoparticles provide significant benefits, such as neutralizing oxidative stress, alleviating inflammation, and promoting cell proliferation. Additionally, research has demonstrated the critical role of macrophage polarization in wound healing, particularly the transition of M1 (pro-inflammatory) macrophages to M2 (anti-inflammatory) macrophages, which plays a pivotal role in the healing process. Hydrogel-based dressings offer not only a promising solution for wound care but also versatile solutions that personalize treatment and optimize the healing environment [142]. However, challenges such as biocompatibility, clinical application, and accessibility persist. Nonetheless, significant strides are being made in overcoming these obstacles, and ongoing research is advancing the development of multifunctional and smart hydrogels, which hold great potential for providing more effective and personalized solutions for both external and internal wound healing. In conclusion, the future of hydrogel-based wound care is shaped by technological innovations and research, promising improved clinical outcomes and enhanced quality of life for patients.

References

1. Blackburn J, Kopecki Z, Ousey KJ (2024) Skin integrity, antimicrobial stewardship and infection control: a critical review of current best practice. Wound Pract Res 32. https://doi.org/10.33235/wpr.32.1.34-43
2. Farahani M, Shafiee A (2021) Wound healing: from passive to smart dressings. Adv Healthcare Mater 10:2100477
3. Daunton C (2012) A history of materials and practices for wound management. 20
4. Sussman G (2023) An update on wound management. Aust Prescr 46:29–35
5. Santos E, Cunha M (2022) Uma breve história da evolução do tratamento de feridas. Millenium-J Educ Technologies 9–11 Páginas
6. Lopes AI, Pintado MM, Tavaria FK (2024) Plant-based films and hydrogels for wound healing. Microorganisms 12:438
7. Bhattacharya S (2012) Wound healing through the ages. Indian J Plastic Surg 45:177–179
8. Dhivya S, Padma VV, Santhini E (2015) Wound dressings—a review. Biomedicine 5:22
9. Montaser AS, Rehan M, El-Senousy WM, Zaghloul S (2020) Designing strategy for coating cotton gauze fabrics and its application in wound healing. Carbohyd Polym 244:116479
10. Jones V, Grey JE, Harding KG (2006) Wound dressings. BMJ 332:777–780
11. Bryan J (2004) Moist wound healing: a concept that changed our practice. J Wound Care 13:227–228
12. Milestones in the History of Wound Healing (2005) Içinde Wound Healing and Ulcers of the Skin (ss. 19–29). Springer-Verlag, Berlin/Heidelberg
13. Butcher M, White R (2013) Reviewing the evidence for advanced dressings. Nurs Stand 27:51–62
14. Eaglstein WH (2001) Moist wound healing with occlusive dressings: a clinical focus. Dermatol Surg 27:175–182
15. Liang Z, Lai P, Zhang J, Lai Q, He L (2023) Impact of moist wound dressing on wound healing time: a meta-analysis. Int Wound J 20:4410–4421
16. Ousey K, Cutting KF, Rogers AA, Rippon MG (2016) The importance of hydration in wound healing: reinvigorating the clinical perspective. J Wound Care 25:122–130
17. Tang NFR, Heryanto H, Armynah B, Tahir D (2023) Bibliometric analysis of the use of calcium alginate for wound dressing applications: a review. Int J Biol Macromol 228:138–152
18. Queen D, Orsted H, Sanada H, Sussman G (2004) A dressing history. Int Wound J 1:59–77
19. Laurano R, Boffito M, Ciardelli G, Chiono V (2022) Wound dressing products: a translational investigation from the bench to the market. Eng Regenerat 3:182–200
20. Sood A, Granick MS, Tomaselli NL (2014) Wound dressings and comparative effectiveness data. Adv Wound Care 3:511–529
21. Fogh K, Nielsen J (2015) Clinical utility of foam dressings in wound management: a review. Chronic Wound Care Manag Res 31
22. Hargis A, Yaghi M, Bermudez NM, Gefen A (2024) Foam dressings for wound healing. Curr Dermatol Rep 13:28–35
23. Rybka M, Mazurek Ł, Konop M (2022) Beneficial effect of wound dressings containing silver and silver nanoparticles in wound healing—from experimental studies to clinical practice. Life 13:69
24. Arapoglou V, Katsenis K, Syrigos KN, Dimakakos EP, Zakopoulou N, Tsoutsos D, … Petersen B (2011) Analgesic efficacy of an ibuprofenreleasing foam dressing compared with local best practice for painful exuding wounds. J Wound Care 20:319–325

25. Ho TC, Chang CC, Chan HP, Chung TW, Shu CW, Chuang KP, … Tyan YC (2022a) Hydrogels: properties and applications in biomedicine. Molecules 27:2902
26. Rijns L, Baker MB, Dankers PYW (2024) Using chemistry to recreate the complexity of the extracellular matrix: guidelines for supramolecular hydrogel-cell interactions. J Am Chem Soc 146:17539–17558
27. Alshangiti DM, El-damhougy TK, Zaher A, Madani M, Mohamady Ghobashy M (2023) Revolutionizing biomedicine: advancements, applications, and prospects of nanocomposite macromolecular carbohydrate-based hydrogel biomaterials: a review. RSC Adv 13:35251–35291
28. Gyles DA, Castro LD, Silva JOC, Ribeiro-Costa RM (2017) A review of the designs and prominent biomedical advances of natural and synthetic hydrogel formulations. Eur Polymer J 88:373–392
29. Hahn S, Hennecke D (2023) What can we learn from biodegradation of natural polymers for regulation? Environ Sci Eur 35:50
30. Ho TT, Tran HA, Doan VK, Maitz J, Li Z, Wise SG, … Rnjak-Kovacina J (2024) Natural polymer-based materials for wound healing applications. Adv NanoBiomed Res 4:2300131
31. Vasile C, Pamfil D, Stoleru E, Baican M (2020) New developments in medical applications of hybrid hydrogels containing natural polymers. Molecules 25:1539
32. Afruzi FH, Abdouss M, Zare EN, Rezvani Ghomi E, Mahmoudi S, Neisiany RE (2025) Metal-organic framework-hydrogel composites as emerging platforms for enhanced wound healing applications: material design, therapeutic strategies, and future prospects. Coord Chem Rev 524:216330
33. Design Considerations, Formulation Approaches, and Strategic Advances of Hydrogel Dressings for Chronic Wound Management | ACS Omega (ty) Geliş tarihi 27 Kasım 2024, gönderen. https://pubs.acs.org/doi/https://doi.org/10.1021/acsomega.2c06806
34. Polysaccharide hydrogels for skin wound healing: Heliyon. (ty) Geliş tarihi 27 Kasım 2024, gönderen. https://www.cell.com/heliyon/fulltext/S2405-8440(24)11045-6?_returnURL=https%3A%2F%2Flinkinghub.elsevier.com%2Fretrieve%2Fpii%2FS2405844024110456%3Fshowall%3Dtrue
35. Yang Y, Du Y, Zhang J, Zhang H, Guo B (2022) Structural and Functional Design of Electrospun Nanofibers for Hemostasis and Wound Healing. Advanced Fiber Materials 4:1027–1057
36. Zheng Y, Wu J, Zhu Y, Wu C (2022) Inorganic-based biomaterials for rapid hemostasis and wound healing. Chem Sci 14:29–53
37. Park M, Shin HK, Kim B-S, Kim MJ, Kim I-S, Park B-Y, Kim H-Y (2015) Effect of discarded keratin-based biocomposite hydrogels on the wound healing process in vivo. Mater Sci Eng C 55:88–94
38. Dhandhi S, Yeshna V, Monika GB, Chauhan S, … Jhawat V (2024) The interplay of skin architecture and cellular dynamics in wound healing: Insights and innovations in care strategies. Tissue Cell 91:102578
39. Fernández-Guarino M, Hernández-Bule ML, Bacci S (2023) Cellular and molecular processes in wound healing. Biomedicines 11:2526
40. Sang F, Liu C, Yan J, Su J, Niu S, Wang S, … Dang Q (2024) Polysaccharide- and protein-based hydrogel dressings that enhance wound healing: a review. Int J Biologic Macromol 280:135482
41. Gao M, Guo H, Dong X, Wang Z, Yang Z, Shang Q, Wang Q (2024) Regulation of inflammation during wound healing: the function of mesenchymal stem cells and strategies for therapeutic enhancement. Front Pharmacol 15. https://doi.org/10.3389/fphar.2024.1345779
42. Verma P, Dixit J, Singh C, Singh AN, Singh A, Tiwari KN, … Mishra SK (2024) Preparation of hydrogel from the hydroalcoholic root extract of *Premna integrifolia* L. and its mediated green synthesis of silver nanoparticles for wound healing efficacy. Mater Today Commu 41:110228
43. Sharma A, Dheer D, Puri V, Alsayari A, Wahab S, Kesharwani P (2024) Insights of biopolymeric blended formulations for diabetic wound healing. Int J Pharm 656:124099

44. Wang Z, Qi F, Luo H, Xu G, Wang D (2022) Inflammatory microenvironment of skin wounds. Front Immunol 13. https://doi.org/10.3389/fimmu.2022.789274
45. Zhang W, Liu L, Cheng H, Zhu J, Li X, Ye S, Li X (2024) Hydrogel-based dressings designed to facilitate wound healing. Materials Adv 5:1364–1394
46. Gupta A, Kowalczuk M, Heaselgrave W, Britland ST, Martin C, Radecka I (2019) The production and application of hydrogels for wound management: a review. Eur Polymer J 111:134–151
47. El Ayadi A, Jay JW, Prasai A (2020) Current approaches targeting the wound healing phases to attenuate fibrosis and scarring. Int J Mol Sci 21:1105
48. Han A, Liu C, Wu Q, Gong Z, Liu M, Xu B, Su X (2024) External physical field-responsive nanocomposite hydrogels for wound healing applications. Adv Nanocomp. https://doi.org/10.1016/j.adna.2024.11.002
49. Qian W, Bai H, Yan Y, Bi Q (2024) Regulation of physicochemical properties of alginate-based hydrogels and preliminary applications in wound healing. Int J Biol Macromol 283:137617
50. Wang Y, Zhang N, Zhang J, Yao R, He J, Wu F (2025) Reinforced enzyme mineralized chitosan hydrogels with superior mechanical and osteogenic properties. Carbohyd Polym 349:123032
51. Shi C, Li S, Li C, Liu H, Wang Z, Li Y, … Han X (2024) Bovine serum albumin-based hydrogels: preparation, properties and biological applications. Chem Eng J 498:154651
52. El-Husseiny HM, Mady EA, Doghish AS, Zewail MB, Abdelfatah AM, Noshy M, … El-Dakroury WA (2024) Smart/stimuli-responsive chitosan/gelatin and other polymeric macromolecules natural hydrogels vs. synthetic hydrogels systems for brain tissue engineering: a state-of-the-art review. Int J Biologic Macromol 260:129323
53. Yadav P, Singh S, Jaiswal S, Kumar R (2024) Synthetic and natural polymer hydrogels: a review of 3d spheroids and drug delivery. Int J Biol Macromol 280:136126
54. Zhang YZ, El-Demellawi JK, Jiang Q, Ge G, Liang H, Lee K, … Alshareef HN (2020) MXene hydrogels: fundamentals and applications. Chem Soc Rev 49:7229–7251
55. Norahan MH, Pedroza-González SC, Sánchez-Salazar MG, Álvarez MM, Trujillo de Santiago G (2022) Structural and biological engineering of 3D hydrogels for wound healing. Bioactive Mater 24:197–235
56. Liu J, Jiang W, Xu Q, Zheng Y (2022) Progress in antibacterial hydrogel dressing. Gels 8:503
57. Las Heras K, Garcia-Orue I, Rancan F, Igartua M, Santos-Vizcaino E, Hernandez RM (2024) Modulating the immune system towards a functional chronic wound healing: a biomaterials and Nanomedicine perspective. Adv Drug Deliv Rev 210:115342
58. Gounden V, Singh M (2024) Hydrogels and wound healing: current and future prospects. Gels 10:43
59. Nuutila K, Eriksson E (2021) Moist wound healing with commonly available dressings. Adv Wound Care 10:685–698
60. Huang X, Hu B, Zhang X, Fan P, Chen Z, Wang S (2024) Recent advances in the application of clay-containing hydrogels for hemostasis and wound healing. Expert Opin Drug Deliv 21:457–477
61. Kharaziha M, Baidya A, Annabi N (2021) Rational design of immunomodulatory hydrogels for chronic wound healing. Adv Mater 33:2100176
62. Dam P, Celik M, Ustun M, Saha S, Saha C, Kacar EA, … Mandal AK (2023) Wound healing strategies based on nanoparticles incorporated in hydrogel wound patches. RSC Adv 13:21345–21364
63. Nunes D, Andrade S, Ramalho MJ, Loureiro JA, Pereira MC (2022) Polymeric nanoparticles-loaded hydrogels for biomedical applications: a systematic review on in vivo findings. Polymers 14:1010
64. Mirrahimi M, Beik J, Mirrahimi M, Alamzadeh Z, Teymouri S, Mahabadi VP, … Moustakis C (2020) Triple combination of heat, drug and radiation using alginate hydrogel co-loaded with gold nanoparticles and cisplatin for locally synergistic cancer therapy. Int J Biologic Macromol 158:617–626
65. Poustchi F, Amani H, Ahmadian Z, Niknezhad SV, Mehrabi S, Santos HA, Shahbazi M-A (2021) Combination therapy of killing diseases by injectable hydrogels: from concept to medical applications. Adv Healthcare Mater 10:2001571

66. Algi MP, Sarıgöl R (2024) Cross-linker engineered poly(hydroxyethyl methacrylate) hydrogel allows photodynamic and photothermal therapies and controlled drug release. Eur J Pharm Biopharm 202:114419
67. Gao S, He X, Liu H, Liu Y, Wang H, Zhou Z, … Xie J (ty) Multifunctional bioactive nanozyme systems for enhanced diabetic wound healing. https://doi.org/10.1002/adhm.202401580
68. Okawa M, Tanabe A, Ohta S, Nagatoishi S, Tsumoto K, Ito T (2022) Extracellular matrix-inspired hydrogel of hyaluronan and gelatin crosslinked via a Link module with a transglutaminase reactive sequence. Commun Mater 3:1–11
69. Alves PM, Fonseca DR, Bidarra SJ, Gomes A, Gomes P, Barrias CC, Martins MCL (2024) Norbornene-chitosan nanoparticles with and without a conjugated VEGF-peptide analog to promote vascularization. Mater Today Chemistr 36:101942
70. Qin P, Tang J, Sun D, Yang Y, Liu N, Li Y, … Yang X (2022) Zn2+ cross-linked alginate carrying hollow silica nanoparticles loaded with RL-QN15 peptides provides promising treatment for chronic skin wounds. ACS Appl Mater Interfaces 14:29491–29505
71. Shah S, Sasmal PK, Lee K-B (2014) Photo-triggerable hydrogel–nanoparticle hybrid scaffolds for remotely controlled drug delivery. J Mater Chemistr B 2:7685–7693
72. Bakshi S, Sahoo PK, Li K, Johnson S, Raxworthy MJ, Krauss TF (2023) Nanophotonic and hydrogel-based diagnostic system for the monitoring of chronic wounds. Biosens Bioelectron 242:115743
73. Wu Z, Wu W, Zhang C, Zhang W, Li Y, Ding T, … Huang F (2025) Enhanced diabetic foot ulcer treatment with a chitosan-based thermosensitive hydrogel loaded self-assembled multi-functional nanoparticles for antibacterial and angiogenic effects. Carbohydr Polym 347:122740
74. Li W, Su H, Ma Y, Ren H, Feng Z, Wang Y, … Zhu Z (2024) Multicargo-loaded inverse opal gelatin hydrogel microparticles for promoting bacteria-infected wound healing. Int J Biologic Macromol 260:129557
75. Kuan CH, Chang L, Ho CY, Tsai CH, Liu YC, Huang WY, … Wang TW (2025) Immunomodulatory hydrogel orchestrates pro-regenerative response of macrophages and angiogenesis for chronic wound healing. Biomaterials 314:122848
76. Wu J, Wu Y, Tang H, Li W, Zhao Z, Shi X, … Deng H (2024) Self-adapting biomass hydrogel embodied with miRNA immunoregulation and long-term bacterial eradiation for synergistic chronic wound therapy. ACS Nano (world). https://doi.org/10.1021/acsnano.4c02736
77. Li J, Zhao M, Liang J, Geng Z, Fan Y, Sun Y, Zhang X (2024) Hollow CuS immobilized in polysaccharide hydrogel promotes photothermally-driven vascularization. Chem Eng J 481:148287
78. Moorcroft SCT, Roach L, Jayne DG, Ong ZY, Evans SD (2020) Nanoparticle-loaded hydrogel for the light-activated release and photothermal enhancement of antimicrobial peptides. ACS Appl Mater Interfac (world). https://doi.org/10.1021/acsami.9b22587
79. Zhao X, Chang L, Hu Y, Xu S, Liang Z, Ren X, … Chen Z (2022) Preparation of photocatalytic and antibacterial MOF nanozyme used for infected diabetic wound healing. ACS Appl Mater Interfaces 14:18194–18208
80. Omidiyan M, Srinoi P, Tajalli P, Lee TR (2024) Review of light-activated antimicrobial nanoparticle-polymer composites for biomedical devices. ACS Appl Nano Mater (world). https://doi.org/10.1021/acsanm.3c05173
81. Gupta A, Briffa SM, Swingler S, Gibson H, Kannappan V, Adamus G, … Radecka I (2020) Synthesis of silver nanoparticles using curcumin-cyclodextrins loaded into bacterial cellulose-based hydrogels for wound dressing applications. Biomacromolecules 21:1802–1811
82. Fang Z, Chen P, Ji Q, Yan C, Gong A (2024) Stimuli-responsive hydrogel for disease therapy. Polym Bull 81:1981–2000
83. Algi MP, Sarıgöl R (2023) Cross-linker engineering of hydrogel enables photothermal therapy and controlled drug release. J Drug Deliv Sci Technol 89:104993
84. Shang L, Yu Y, Jiang Y, Liu X, Sui N, Yang D, Zhu Z (2023) Ultrasound-augmented multienzyme-like nanozyme hydrogel spray for promoting diabetic wound healing. ACS Nano (world). https://doi.org/10.1021/acsnano.3c04134

85. Yin X, Fan T, Zheng N, Yang J, Yan L, He S, … Hu J (2023) Palladium nanoparticle based smart hydrogels for NIR light-triggered photothermal/photodynamic therapy and drug release with wound healing capability. Nanoscale Adv 5:1729–1739
86. Wang J, Zhang H, Hu H, Hu S, Ma L (2024) An enzyme-responsive hydrogel of ferrocene-grafted carboxymethyl chitosan as a soft electrochemical sensor for MMP-9 detection. Int J Biol Macromol 268:131582
87. Koehler J, Brandl FP, Goepferich AM (2018) Hydrogel wound dressings for bioactive treatment of acute and chronic wounds. Eur Polymer J 100:1–11
88. Wang W, Ummartyotin S, Narain R (2023) Advances and challenges on hydrogels for wound dressing. Curr Opin Biomed Eng 26:100443
89. Zhang L, Liu M, Zhang Y, Pei R (2020) Recent progress of highly adhesive hydrogels as wound dressings. Biomacromol 21:3966–3983
90. Huang C, Dong L, Zhao B, Lu Y, Huang S, Yuan Z, … Qian W (2022) Anti-inflammatory hydrogel dressings and skin wound healing. Clinic Transl Med 12:e1094
91. Wang PH, Huang BS, Horng HC, Yeh CC, Chen YJ (2018) Wound healing. J Chinese Med Assoc 81. Geliş tarihi gönderen https://journals.lww.com/jcma/fulltext/2018/02000/wound_healing.3.aspx
92. Ho TC, Chang CC, Chan HP, Chung TW, Shu CW, Chuang KP, … Tyan YC (2022b) Hydrogels: properties and applications in biomedicine. Molecules 27. https://doi.org/10.3390/molecules27092902
93. Ahmad Raus R, Wan Nawawi WMF, Nasaruddin RR (2021) Alginate and alginate composites for biomedical applications. Asian J Pharm Sci 16:280–306
94. Varghese SA, Rangappa SM, Siengchin S, Parameswaranpillai J (2020) Chapter 2—natural polymers and the hydrogels prepared from them. In: İçinde YC (ed) Hydrogels based on natural polymers. Elsevier, ss 17–47
95. Liang X, Huang C, Liu H, Chen H, Shou J, Cheng H, Liu G (2024) Natural hydrogel dressings in wound care: design, advances, and perspectives. Chin Chem Lett 35:109442
96. Madduma-Bandarage USK, Madihally SV (2021) Synthetic hydrogels: synthesis, novel trends, and applications. J Appl Polym Sci 138:50376
97. Li Y, Zheng H, Liang Y, Xuan M, Liu G, Xie H (2022) Hyaluronic acid-methacrylic anhydride/polyhexamethylene biguanide hybrid hydrogel with antibacterial and proangiogenic functions for diabetic wound repair. Chin Chem Lett 33:5030–5034
98. Chen G, Yu Y, Wu X, Wang G, Ren J, Zhao Y (2018) Bioinspired multifunctional hybrid hydrogel promotes wound healing. Adv Func Mater 28:1801386
99. Masood N, Ahmed R, Tariq M, Ahmed Z, Masoud MS, Ali I, … Hasan A (2019) Silver nanoparticle impregnated chitosan-PEG hydrogel enhances wound healing in diabetes induced rabbits. Int J Pharmaceut 559:23–36
100. Zubik K, Singhsa P, Wang Y, Manuspiya H, Narain R (2017) Thermo-responsive poly(N-Isopropylacrylamide)-cellulose nanocrystals hybrid hydrogels for wound dressing. Polymers 9. https://doi.org/10.3390/polym9040119
101. Rasool A, Ata S, Islam A (2019) Stimuli responsive biopolymer (chitosan) based blend hydrogels for wound healing application. Carbohyd Polym 203:423–429
102. Kapanya A, Somsunan R, Molloy R, Jiranusornkul S, Leewattanapasuk W, Jongpaiboonkit L, Kong Y (2020) Synthesis of polymeric hydrogels incorporating chlorhexidine gluconate as antibacterial wound dressings. J Biomater Sci Polym Ed 31:895–909
103. Cao Z, Luo Y, Li Z, Tan L, Liu X, Li C, … Wu S (2021) Antibacterial hybrid hydrogels. Macromol Biosci 21:2000252
104. Li YP, Ben Fekih I, Chi Fru E, Moraleda-Munoz A, Li X, Rosen BP, … Rensing C (2021) Antimicrobial activity of metals and metalloids. Ann Rev
105. Teaima MH, Elasaly MK, Omar SA, El-Nabarawi MA, Shoueir KR (2022) Wound healing activities of polyurethane modified chitosan nanofibers loaded with different concentrations of linezolid in an experimental model of diabetes. J Drug Deliv Sci Technol 67:102982
106. Dzulkharnien NS, Rohani R (2022) A review on current designation of metallic nanocomposite hydrogel in biomedical applications. Nanomaterials 12. https://doi.org/10.3390/nano12101629

107. Khorasani MT, Joorabloo A, Adeli H, Mansoori-Moghadam Z, Moghaddam A (2019) Design and optimization of process parameters of polyvinyl (alcohol)/chitosan/nano zinc oxide hydrogels as wound healing materials. Carbohyd Polym 207:542–554
108. Hu T, Wu G-P, Bu H, Zhang H, Li W-X, Song K, Jiang G-B (2022) An injectable, adhesive, and self-healable composite hydrogel wound dressing with excellent antibacterial activity. Chem Eng J 450:138201
109. Schnaider L, Toprakcioglu Z, Ezra A, Liu X, Bychenko D, Levin A, ... Knowles TPJ (2020) Biocompatible hybrid organic/inorganic microhydrogels promote bacterial adherence and eradication in vitro and in vivo. Nano Lett 20:1590–1597
110. Nešović K, Janković A, Kojić V, Vukašinović-Sekulić M, Perić-Grujić A, Rhee KY, Mišković-Stanković V (2018) Silver/poly(vinyl alcohol)/chitosan/graphene hydrogels–synthesis, biological and physicochemical properties and silver release kinetics. Compos B Eng 154:175–185
111. Sharon Sofini PS, Biswas K, Mercy DJ, Girigoswami A, Girigoswami K (2024) Nanostructure-assisted wound dressing materials: a literature review. Macromol Res 32:1065–1087
112. Mondal MIH (2023) Development of methylcellulose/polyvinylpyrrolidone nanocomposites enriched with biogenic zinc oxide nanoparticles for wound healing applications. Biomed J Sci Techn Res 49:40624–40633
113. Aderibigbe BA (2022) Hybrid-based wound dressings: combination of synthetic and biopolymers. Polymers 14. https://doi.org/10.3390/polym14183806
114. Haseeb MT, Hussain MA, Abbas K, Youssif BG, Bashir S, Yuk SH, Bukhari SNA (2017) Linseed hydrogel-mediated green synthesis of silver nanoparticles for antimicrobial and wound-dressing applications. Int J Nanomed 2845–2855
115. Huang H, Zhang X, Dong Z, Zhao X, Guo B (2022) Nanocomposite conductive tough hydrogel based on metal coordination reinforced covalent Pluronic F-127 micelle network for human motion sensing. J Colloid Interface Sci 625:817–830
116. Zhang E, Zhao X, Hu J, Wang R, Fu S, Qin G (2021) Antibacterial metals and alloys for potential biomedical implants. Bioactive Mater 6:2569–2612
117. Wang M, Deng Z, Guo Y, Xu P (2023) Engineering functional natural polymer-based nanocomposite hydrogels for wound healing. Nanoscale Adv 5:27–45
118. Xu J, Younis MR, Zhang Z, Feng Y, Su L, Que Y, ... Yang C (2023) Mild heat-assisted polydopamine/alginate hydrogel containing low-dose nanoselenium for facilitating infected wound healing. ACS Appl Mater Interfaces 15:7841–7854
119. Gaharwar AK, Peppas NA, Khademhosseini A (2014) Nanocomposite hydrogels for biomedical applications. Biotechnol Bioeng 111:441–453
120. Gaharwar AK, Avery RK, Assmann A, Paul A, McKinley GH, Khademhosseini A, Olsen BD (2014) Shear-thinning nanocomposite hydrogels for the treatment of hemorrhage. ACS Nano 8:9833–9842
121. Dong R, Guo B (2021) Smart wound dressings for wound healing. Nano Today 41:101290
122. Blacklow SO, Li J, Freedman BR, Zeidi M, Chen C, Mooney DJ (t.y.) Bioinspired mechanically active adhesive dressings to accelerate wound closure. Sci Adv 5:eaaw3963
123. Li M, Liang Y, He J, Zhang H, Guo B (2020) Two-pronged strategy of biomechanically active and biochemically multifunctional hydrogel wound dressing to accelerate wound closure and wound healing. Chem Mater 32:9937–9953
124. Li Y, Wang X, Fu Y, Wei Y, Zhao L, Tao L (2018) Self-adapting hydrogel to improve the therapeutic effect in wound-healing. ACS Appl Mater Interfaces 10:26046–26055
125. Zhao X, Liang Y, Huang Y, He J, Han Y, Guo B (2020) Physical double-network hydrogel adhesives with rapid shape adaptability, fast self-healing, antioxidant and nir/ph stimulus-responsiveness for multidrug-resistant bacterial infection and removable wound dressing. Adv Func Mater 30:1910748
126. Cheng S, Pan M, Hu D, Han R, Li L, Bei Z, ... Qian Z (2023) Adhesive chitosan-based hydrogel assisted with photothermal antibacterial property to prompt mice infected skin wound healing. Chinese Chem Lett 34:108276

127. Wang Y, Zhang S, Wang J (2021) Photo-crosslinkable hydrogel and its biological applications. Chin Chem Lett 32:1603–1614
128. Yang K, Zhou X, Li Z, Wang Z, Luo Y, Deng L, He D (2022) Ultrastretchable, self-healable, and tissue-adhesive hydrogel dressings involving nanoscale tannic acid/ferric ion complexes for combating bacterial infection and promoting wound healing. ACS Appl Mater Interfaces 14:43010–43025
129. Piksa M, Lian C, Samuel IC, Pawlik KJ, Samuel ID, Matczyszyn K (2023) The role of the light source in antimicrobial photodynamic therapy. Chem Soc Rev 52:1697–1722
130. Ran P, Zheng H, Cao W, Jia X, Zhang G, Liu Y, Li X (2022) On-demand changeable theranostic hydrogels and visual imaging-guided antibacterial photodynamic therapy to promote wound healing. ACS Appl Mater Interfaces 14:49375–49388
131. Maleki A, He J, Bochani S, Nosrati V, Shahbazi M-A, Guo B (2021) Multifunctional photoactive hydrogels for wound healing acceleration. ACS Nano 15:18895–18930
132. Ahmed B, Reiche CF, Magda JJ, Solzbacher F, Körner J (2024) Smart hydrogel swelling state detection based on a power-transfer transduction principle. ACS Appl Polym Mater 6:5544–5554
133. Mishra SK, Zou B, Chiang KS (2017) Wide-range ph sensor based on a smart- hydrogel-coated long-period fiber grating. IEEE J Sel Top Quantum Electron 23:284–288
134. Wang K, Zhang J, Li H, Wu J, Wan Q, Chen T, … Luo Y (2024) Smart hydrogel sensors for health monitoring and early warning. Adv Sens Res 3:2400003
135. Zhao Y, Lei M, Liu S-X, Zhao Q (2018) Smart hydrogel-based optical fiber SPR sensor for pH measurements. Sens Actuat B Chem 261:226–232
136. Jin S, Newton MA, Cheng H, Zhang Q, Gao W, Zheng Y, … Zhu J (2023) Progress of hydrogel dressings with wound monitoring and treatment functions. Gels 9. https://doi.org/10.3390/gels9090694
137. Kim MH, Park H, Nam HC, Park SR, Jung J-Y, Park WH (2018) Injectable methylcellulose hydrogel containing silver oxide nanoparticles for burn wound healing. Carbohyd Polym 181:579–586
138. Choi S, Han SI, Kim D, Hyeon T, Kim D-H (2019) High-performance stretchable conductive nanocomposites: materials, processes, and device applications. Chem Soc Rev 48:1566–1595
139. Xu M-L, Guan L-Y, Li S-K, Chen L, Chen Z (2019) Stable gold graphitic nanocapsule doped hydrogels for efficient photothermal antibacterial applications. Chem Commun 55:5359–5362
140. Jiang J, Ding J, Wu X, Zeng M, Tian Y, Wu K, … Fan H (2023) Flexible and temperature-responsive hydrogel dressing for real-time and remote wound healing monitoring. J Mater Chemistr B 11:4934–4945
141. Fawzy A, Fortunata V (2023) Hydrogel dressings in wound management: advances, applications, and future directions. Int J Med Sci Clinic Res Stud 3:2674–2680
142. Altuntas S, Kaur Dhaliwal H, Radwan AE, Amiji M, Buyukserin F (2023) Local epidermal growth factor delivery using nanopillared chitosan–gelatin films for melanogenesis and wound healing. Biomater Sci 11:181–194

Hydrogels as Combinational Therapy with Antimicrobial Peptides for Wound Healing

Héllen Cristina Almeida de Castro Alves, Ralyson Xavier Oliveira da Silva, Thuanny Borba Rios, Kamila Botelho Sampaio de Oliveira, Mariana Rocha Maximiano, and Octávio Luiz Franco

Abstract Wound healing is a dynamic and intricate process involving a precisely coordinated sequence of cellular, molecular, and biochemical events that interact to facilitate tissue repair. A significant challenge in wound treatment is delayed closure. Therapeutic strategies to accelerate the healing process are critical for effective clinical management. This study provides an overview of wound healing, focusing on conventional treatments such as hydrogels enriched with antimicrobial peptides, resulting in a combinational therapy that has a high potential for wound healing. It further explores combination therapies, which add antimicrobial agents to drug delivery systems. Advanced hydrogel technologies are also examined, including responsive hydrogels, targeted therapeutic applications, and 3D-printed hydrogels. The study also addresses this field's challenges and future directions, detailing technical obstacles, regulatory considerations, and the approval processes involved with regulatory agencies.

Keywords Tissue engineering · Treatment wound · Chronic wound · Delivery system · NanoAMPs

1 Introduction

The skin can present multiple critical functions, including mechanical protection, prevention of water loss, temperature regulation, sensory perception, and immune modulation [84]. Additionally, the dermis contains a network of cutaneous blood

H. C. A. de Castro Alves · R. X. O. da Silva · T. B. Rios · K. B. S. de Oliveira ·
M. R. Maximiano · O. L. Franco (✉)
Centro de Análises Proteômicas e Bioquímicas, Pós-graduação em Ciências Genômicas e Biotecnologia, Universidade Católica de Brasília, Brasília, Brazil
e-mail: ocfranco@gmail.com

R. X. O. da Silva · K. B. S. de Oliveira · M. R. Maximiano · O. L. Franco
S-Inova Biotech, Pós-graduação em Biotecnologia, Universidade Católica Dom Bosco, Campo Grande, Mato Grosso do Sul, Brazil

A. K. Mandal et al. (eds.), *Functionalized Nanoparticles Hydrogels for Wound Healing*,
Smart Nanomaterials Technology, https://doi.org/10.1007/978-981-95-0720-7_10

vessels, which proliferate significantly during tissue repair, extending into the hypodermis, a subcutaneous layer rich in adipose tissue [65]. All living organisms have developed mechanisms for cellular tissue repair following injury or trauma [26]. Wounds are commonly sorted as acute or chronic, based on clinical presentation. Untreated wounds invariably result in the same outcome, including necrosis and cellular death of the epidermis, with the extent and severity, determined by the depth and scope of the injury, directly impacting the healing process, which can lead to a delay in wound closure [77].

Wound healing is a dynamic and complex process involving a precisely coordinated sequence of cellular, molecular, and biochemical events that interact to facilitate tissue reconstruction [24, 58, 81, 91, 92]. However, intrinsic factors such as diabetes and venous insufficiency can interfere with this process, and the results include an imbalance in the physiology of healing and the perpetuation of the inflammatory phase [73]. The evolution of tissue repair can thus be interrupted, resulting in a chronic wound [75]. Extrinsic factors can also contribute to a poor prognosis, such as infections by pathogens and the formation of biofilm in the wound bed [39]. Local infection can be shown clinically through purulent exudate, pain, opaque and brittle granulation tissue that is difficult to visualize, characteristic odor, and the presence of phlogistic signs in the perilesional area, characterizing inflammation due to an infectious process [50].

In recent years, a significant focus of wound care has been the development of dressings that maintain a moist environment to promote healing at the same time as treating the wound [34]. Traditional wound-healing biomaterials are temporary barriers, controlling bleeding and preventing infection. While various natural and synthetic biomaterials, such as proteins, gum arabic, and polyurethane, are designed for wound healing applications, many remain cost-inefficient and insufficiently suitable for widespread clinical use [92]. Several studies have shown that hydrogels can enhance cell properties and processes, such as bone formation, cell growth rate, bone formation, vascular anastomosis, and several others [29, 52]. Hydrogels constitute a class of water-swollen three-dimensional (3D) polymer networks featuring tunable physicochemical properties that are required to satisfy specific requirements under different conditions [36]. Hydrogels can be composed of a variety of polymers, which may be classified as natural, modified natural, or synthetic [64]. These polymers are selected based on their ability to form hydrogels with properties suited for diverse drug delivery, wound care, and tissue engineering applications [19, 88]. That is often implemented in medical applications because of their biocompatibility, biodegradability, and hydrophilicity [10, 40, 49].

However, hydrogels can present limitations, particularly when the composition is derived from natural materials [11]. This limitation stems from the inherent lack of control over the hydrogel's structure and functional properties, as it is devoid of adaptive self-regulation mechanisms capable of responding to abrupt variations in environmental conditions, including pH, temperature, and pathogen invasion [12]. Despite these limitations, hydrogels represent a promising candidate for addressing and preventing pathogenic factors, owing to their ability to function as physical barriers, sequester specific molecules, and deliver therapeutic agents [2, 28, 41].

They can also encapsulate cells, form scaffolds, and act as a drug carrier, which has made exploring and overcoming their limitations worthwhile [85].

In this context, antimicrobial peptides (AMPs) can be a promising combination with hydrogel polymers. AMPs were discovered in 1980 and can play a fundamental role in the innate immune response [22]. AMPs present up to 5 kDa, have hydrophilic and hydrophobic characteristics, and can be cationic or anionic, which allows them to have a better affinity and propensity to interact with the membranes of infectious agents [8] AMPs represent a promising class of therapeutic molecules that can be delivered using hydrogel systems [56]. Additionally, AMPs often disrupt microbial membranes, reducing the likelihood of resistance development [23]. This may result from alterations in the bacterial membrane surface charge, serving as an escape mechanism from the action of the peptide [47].

However, AMPs' clinical application is limited by their instability in physiological conditions, where proteolytic enzymes can rapidly degrade them, leading to different salt concentrations and pH values, interaction with other molecules, and rapid clearance [66]. Therefore, modification of the peptide sequence and the use of delivery systems extend the half-life of the peptide [46]. Hydrogels can offer a protective matrix that helps stabilize AMPs, ensuring their gradual and controlled release at the wound site [74].

Combining these elements in a single formulation can result in a hydrogel or a nanogel associated with antimicrobial peptides [20]. Nanogels can be considered a class of hydrogels, which combine the efficacy and stability of nanoparticles made of polymers with the ability to transport drugs that can be distributed over a long period at the insertion site [74]. This mechanism is a drug delivery system [20, 76]. This chapter explores the utilization of hydrogels in combination with antimicrobial peptides (AMPs), highlighting their potential to enhance wound healing alongside the technological advancements and challenges associated with their application and delivery.

2 Combinational Use of Antimicrobial Agents and Hydrogels for Accelerated Wound Healing

The incorporation of antimicrobial agents, including conventional antibiotics and AMPs, into hydrogels allows for the targeted treatment of infections directly at the wound site, which is especially beneficial compared to traditional systemic delivery methods [3]. For instance, antibiotics such as vancomycin, ciprofloxacin, and gentamicin have been successfully integrated into hydrogel matrices, providing a sustained release that ensures effective concentrations of these drugs at the site of infection [25, 27, 42, 44, 55, 72, 86]. By cross-linking chitosan with bifunctional PEG glyoxylic aldehyde, PEG-chitosan hydrogels were developed to allow the controlled release of ciprofloxacin to both prevent infections and hasten the healing process [69].

Hydrogels offered continuous ciprofloxacin release for up to 24 h and were effective against *Escherichia coli* while also demonstrating mechanical robustness and cytocompatibility with mammalian cells [69]. A hydrogel dressing with superior swelling ratios and thermally induced self-peeling properties was created using gelatin nanospheres loaded with vancomycin and poly ((isopropyl acrylamide)-co–N—(methylol acrylamide)). The hydrogel demonstrated the slowest drug release, long-term antibacterial characteristics, and balanced biocompatibility [72].

Local treatment of bacterial infection was achieved with a hydrogel wound dressing made of hyaluronic acid cross-linked with gentamicin via the 1-ethyl-3-(3-dimethyl aminopropyl) carbodiimide hydrochloride (EDC) associated with the N-hydroxysuccinimide (NHS) bioconjugation approach. An extended release of gentamicin with various adhesive and cohesive qualities lasting up to nine days required careful cross-linking density adjustment [80]. This localized delivery is valuable in treating infections caused by multidrug-resistant bacteria, such as methicillin-resistant *Staphylococcus aureus* (MRSA) [14, 70]. Moreover, the release kinetics of antibiotics from hydrogels can be precisely controlled through adjustments in polymer composition, cross-linking density, and responsiveness to environmental triggers like pH. This is crucial in infected wounds, where pH is often more acidic or alkaline, allowing for the release of antibiotics only when needed, thus minimizing the risk of systemic side effects and the development of resistance [6, 30].

LL-37, a multifunctional human cathelicidin that presents antimicrobial [15, 89], and immunomodulatory activity [48], was encapsulated in chitosan (CS) hydrogel to promote pressure ulcer healing. By decreasing TNF-α production from macrophages, the LL-37/CS hydrogel revealed immunomodulatory properties without causing any harm to fibroblasts. Additionally, using the same hydrogel in a mouse model of pressure ulcers demonstrated antibacterial activity against *S. aureus* and accelerated wound healing, indicating that LL-37/CS hydrogel may be a suitable biocompatible material for the treatment of pressure ulcers [57, 58].

Studies exploring peptide modifications have conjugated the modified peptide SAAP-148, a synthetic derivative of LL-37 [60], to the copolymer poly(HEAAm-co-BPAAm-co-PFPA-co-EH3A). This copolymer is composed of hydroxyethyl acrylamide (HEAAm) as the primary biocompatible component, BPAAm as a photocrosslinker that facilitates the aggregation of hydrogel networks upon exposure to ultraviolet (UV) light, EH3A as a balancing agent between hydrophilic and hydrophobic properties, and PFPA as a reactive ester monomer, which integrates post-polymerization modification (PPM). Following polymerization, these components enable the aggregation of the SAAP-148 peptide via PPM, enhancing its bactericidal and antibiofilm activities. Researchers have confirmed that this hydrogel exhibits low cytotoxicity when exposed to human fibroblasts while demonstrating significant antimicrobial effects against the tested strains of *P. aeruginosa*, *S. aureus*, *A. baumannii*, and *E. coli* [4].

Peptide DJK-5 presents antimicrobial and immunomodulatory activity [31, 57], and its association with a hyaluronic acid (HA)-based nanogel crosslinked with octenyl succinic anhydride (OSA-HA) was tested for its in vivo efficacy [45]. The peptide loaded in nanogel and administered subcutaneously demonstrated a fourfold reduction in cytotoxic effect, compared to the non-formulated peptide when tested against a murine abscess induced by *Pseudomonas aeruginosa* infection, and the antibacterial efficacy of DJK-5 was conserved [45].

Another analysis proves the effectiveness of combining peptides with nanogels, when peptide DPK-060 was encapsulated in a poly(ethyl acrylate-co-methacrylic acid) microgel, which protects the drug from infection-related proteases. It was discovered that microgels prevented peptides from degrading and minimized the hemolytic effects of DPK-060 while maintaining antibacterial activity against methicillin-resistant *P. aeruginosa, E. coli,* and *S. aureus* [62]. By protecting AMPs from premature degradation, this hydrogel ensures that the peptide remains active long enough to exert its antimicrobial action, thereby supporting faster and more effective wound healing [5, 43]. Hydrogels are versatile drug delivery systems (DDS) tailored for optimal therapeutic outcomes. The structure and composition of hydrogels can be modified to control the release of encapsulated drugs, whether through adjustments in cross-linking density or the creation of stimuli-responsive systems [5, 43]. For instance, pH-sensitive hydrogels can release antimicrobial agents in response to the altered pH levels characteristic of chronic wounds, where an alkaline environment is typical [60].

A promising new development is a small-molecule hydrogel formed from the caffeic acid (CA) associate tetrapeptide, as in the case of Nap-F3K (Nap-Phe-Phe-Phe-Lys) [38]. This strategy relies on the utilization of ultrashort peptides comprising between 3 and 7 amino acids that assemble the hydrogel nets through non-covalent interactions. These sequences are designed for intermolecular self-assembly, forming organized nanostructures that facilitate the creation of highly stable, chemically diverse, and cost-effective hydrogels [54]. In the case of Nap-F3K-CA, CA controls the ROS, essential in reducing inflammation through its anti-oxidant capacity, eradicating free radicals in the wound, and boosting antimicrobial activity in the case of burns [38].

This targeted release ensures that drugs are delivered precisely when and where needed, enhancing efficacy while minimizing unnecessary drug exposure. Furthermore, hydrogels composed of natural polymers like chitosan, alginate, and hyaluronic acid are particularly attractive for wound care due to their biocompatibility, biodegradability, and intrinsic biological activities [13]. Chitosan can interact with the negatively charged bacterial membranes with its positive charge, which characterizes its cationic portion, from this interaction, chitosan enters the organism's membrane and prevents RNA synthesis, which reinforces its antimicrobial activity [83]. When combined with other agents like antibiotics or AMPs, chitosan-based hydrogels can provide a synergistic effect, in which one component enhances the action of another, offering a more effective product that presents robust protection against infections while supporting tissue regeneration [51].

The design and application of hydrogels in combinational therapy for wound healing continue to be a vibrant area of research (Fig. 1). The development of hydrogels that respond to specific wound conditions, such as pH changes or enzymatic activity, offers new possibilities for creating more personalized and effective treatments [87, 95]. Hydrogels are a potential method for treating infections and fostering

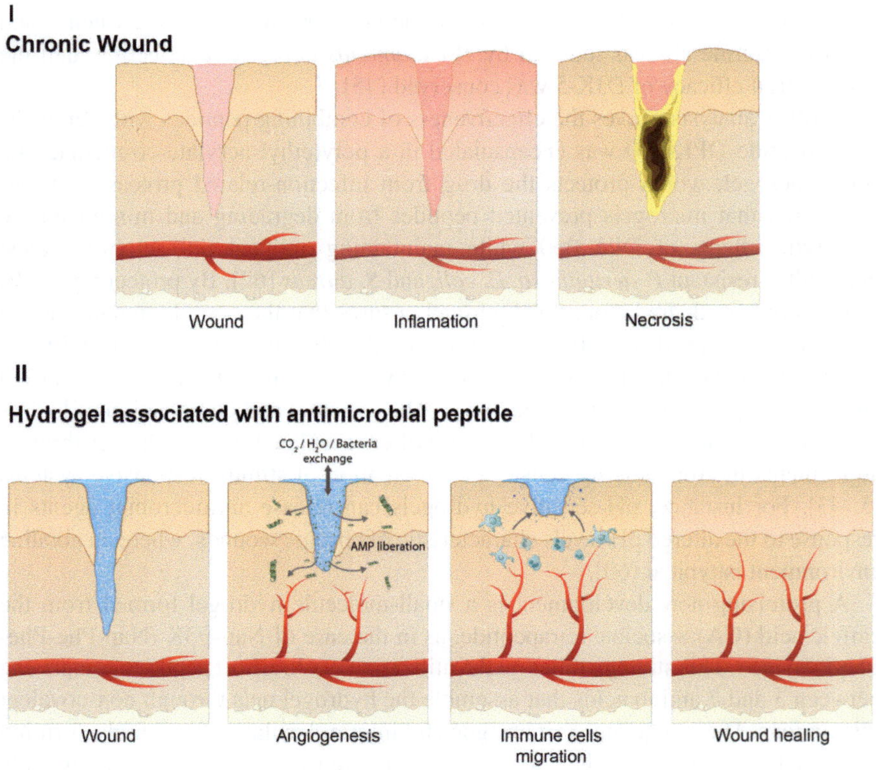

Fig. 1 **I** Wound-healing process the chronic wound. The delay scar occurs in the phase inflammatory with appearance the necrotic tissue. The NanoAMP is enhanced by the hydrogel associated with AMPs. **II** The NanoAMP enhances angioneses, allows immune system cells to arrive with the dressing to speed up the healing

tissue regeneration in complex wound settings, and they will continue to be at the forefront of creative therapeutic strategies as our understanding of the wound-healing process increases [95].

3 Advanced Hydrogel Technologies

Technological advances have increasingly facilitated the enhancement of nanostructure-based products. Notably, artificial intelligence (AI) strategies have driven significant progress in developing hydrogels for wound care, optimizing their properties and enabling the personalization of therapeutic solutions [78, 94]. Similarly, the employment of artificial intelligence for developing peptides has also been established, whereby predictions can be made by assessing the properties of antimicrobial peptides (AMPs), such as amino acid composition, sequence, secondary structure, and physicochemical attributes. These analyses extend to predicting the interactions of these structures concerning microbial control efficacy, as well as evaluating the potential cytotoxicity toward human cells [17, 71]. In this context, responsive hydrogels represent a key innovation, as they dynamically adapt to the conditions of the wound environment, such as detecting pH changes due to infectious processes and temperature, thereby facilitating more effective healing [59, 68].

Considering technological advancements and their potential as auxiliary tools in developing bioproducts, artificial intelligence is increasingly employed to establish patterns and enhance the understanding of molecular interactions within these polymerized networks [79]. This is exemplified by machine learning (ML), which facilitates the prediction of hydrogel behavior across various combinations in simulations featuring different microenvironments. This approach enables the analysis of substantial data sets more efficiently, optimizing both in vitro and in vivo experiments and consequently expediting the process of bringing the product on to the market [94]. The potential of these tools enables the development of innovative products, such as smart hydrogels for chronic wounds (Fig. 2).

Hydrogels' ability to adapt to each patient's characteristics further enhances wound-dressing personalization. Hydrogels produced by 3D printers are an example of this. They are molded according to the dimensions and depths of the wound, making them adaptable both in composition and in the arrangement of the dressing components. This confers an exceptional character, particularly concerning chronic

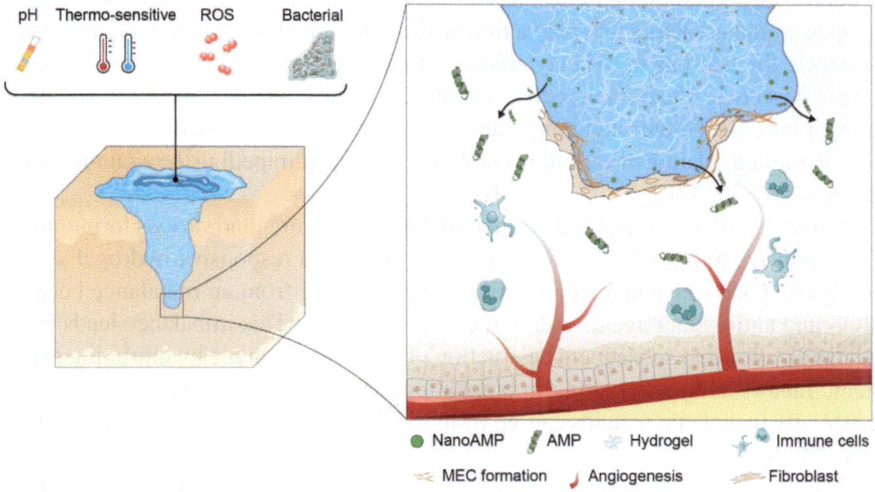

Fig. 2 Responsive hydrogel. Factorys how alterations pH, themperature, ROS and bacterial infection as possible stimuli to be detected by the hydrogel for the moving forward the process healing with the collagen production from fibroblast growth stimulation

wounds, as these devices can be tailored specifically to necrotic tissue, thereby preventing the stagnation of cellular proliferation [18, 33, 37, 67, 93].

Another advanced technology related to hydrogels is the development of innovative, responsive hydrogels; this term derives from the hydrogel's ability to adapt to its environment, including factors such as pH, temperature, presence of microorganisms, and glucose levels, making it possible to release medication progressively on demand from changes in pH or high amounts of ROS. The adoption of intelligent hydrogels makes it possible to demonstrate one or more of these indications of abnormalities in the healing process, resulting in a change in the physical and/or chemical conformation of these hydrogels as a result of the stimuli emitted by the wound, which allows the drug to be liberated [32]. Therefore, the nanoparticles can be released and affect the microbial load and effective healing. The result is a direct impact on reducing bacterial resistance, since high doses of medication are no longer needed in the wound [59, 68].

A vital characteristic of this type of hydrogel is its capacity for rearrangement based on specific needs, as well as its significant hydration potential, which subsequently facilitates the acceleration of wound closure [16]. By combining these characteristics with the use of antimicrobial peptides (AMPs), such as responsive and intelligent formulations, the prospect arises for a drug delivery system designed to protect peptides from potential degradation and reduced half-life while simultaneously promoting an environment conducive to fibroblast migration, collagen fiber deposition, and extracellular matrix (ECM) formation, thereby facilitating pathways for wound closure [18]

Responsive hydrogels exemplify a novel application designed to meet specific wound care requirements. Each wound encompasses a distinct microenvironment shaped by many factors that play a role in sustaining the inflammatory response. For instance, fluctuations in pH may arise due to infection or heightened blood glucose levels. Moreover, the immune cells' recruitment to the injury site frequently increases temperature. If the elevated temperature persists over an extended duration, it may lead to the degradation of various proteins, consequently impeding the wound-healing process [16, 33, 37].

Several studies have been performed on dressing features and various formulations of responsive hydrogels [67]. Among these, there is a responsive hydrogel specifically designed to counteract oxidative stress resulting from an imbalance between ROS and antioxidant agents within the wound [63, 93] This imbalance leads to the destabilization and subsequent destruction of the cell membrane by inhibiting angiogenic factors and exacerbating inflammation [7] caused a delay in wound closure, especially in diabetic wounds caused by hyperglycemia [9] and making the site more susceptible to infections [21].

Another factor that warrants exploration and that extends beyond the therapeutic aspect is the power to collect information for professionals to promote better care management [35, 90]. Researchers should investigate the potential of this type of intelligent hydrogel to detect significant alterations in the wound microenvironment, including excess free radicals, variations in temperature, infection indications, inflammatory process progression, and pH levels. Such information provides health-care professionals with a framework to optimize clinical management [37, 53]. One example is an intelligent hydrogel designed to detect alterations in pH and capable of transmitting this information to a smartphone. This allows for real-time assessment of the wound without the need for direct exposure, thereby reducing the risk of infection associated with excessive handling and environmental exposure [82].

Several examples illustrate that dressings possess high sensitivity for identifying changes within the wound microenvironment. This sensitivity through innovative responsive dressings is crucial for the early detection of injury and for addressing the underlying issues. Consequently, the permeability of these dressings to various factors and alterations within the wound enables the synthesis of increasingly specific and personalized products tailored to the pathophysiological conditions that give rise to particular manifestations. For instance, dressings that accommodate the stagnation of a chronic wound in the inflammatory phase or manage the acidity of an infected wound, while providing a controlled humidity environment conducive to cellular matrix deposition and subsequent tissue proliferation, exemplify the adaptability and efficacy of hydrogels.

4 Prospects and Concluding Remarks

Hydrogels combined with AMPs and antibiotics are extensively used as wound dressings, exerting antimicrobial, hemostatic, and proangiogenic activities, thus promoting wound healing. Despite considerable progress and promising results arising from hydrogels used in combination therapy for wound healing application, challenges related to clinical translation still need to be addressed. Issues related to stability, biodegradability, system administration route, and controlled release of incorporated antimicrobial factors are still barriers in vivo. As an example, the in vivo toxicity of hydrogel material is still a threat. Even though hydrogels are biocompatible and biodegradable, they depurate late, which leads to prolonged exposure in vivo and the ability to induce immune responses [1].

These challenges can be managed through accurate study, simulations, and design before developing the hydrogel system. The selection of polymers and crosslinking techniques based on the properties of components involved in the combination therapy are essential. Chemical properties such as hydrophilicity, geometrical properties such as shape and size, and the degree of cross-linking are critical for swelling capacity and biodegradability, influencing drug release and transport kinetics. Properties such as stability, solubility, dissolution, permeability, therapeutic activity, and clearance are impacted. Notably, the correct interaction between molecule and hydrogel, more precisely the loading of the molecule in the gel matrix, is of the utmost importance for controlled release over a sufficient time for the effectiveness of the combined system to be achieved [61]. Another essential factor to be carefully evaluated is the infection and the bacteria causing it. However, mimicking the complex wound environment in vivo is a considerable challenge. More significant efforts are needed from the research field so that hydrogel combination therapy can be widely applied clinically.

Overall, the prospects in this area are bright. Hydrogels' advantages, such as simple production, low cost, biodegradability, high compatibility, and their ability to protect sensitive drugs such as AMPs from severe environmental conditions, make them a promising platform for developing combination therapies for wound treatment. Therefore, advanced nanotechnology strategies and improvements in current technology should be applied to clinical translation.

Acknowledgements This work was supported by the Coordenação de Aperfeiçoamento de Pessoal de Nível Superior (CAPES).Conselho Nacional de Desenvolvimento Cienti´ fico e Tecnológico (CNPq). Fundação de Apoio à Pesquisa do Distrito Federal (FAPDF). Fundação de Apoio ao Desenvolvimento do Ensino,Ciência e Tecnologia do Estado de Mato Grosso do Sul (FUNDECT) and FINEP (Financiadora de Estudos e Projetos).

References

1. Ahmad N, Bukhari SNA, Hussain MA, Ejaz H, Munir MU, Amjad MW (2024) Nanoparticles incorporated hydrogels for delivery of antimicrobial agents: developments and trends. RSC Adv 14:13535–13564
2. Aliakbar Ahovan Z, Esmaeili Z, Eftekhari BS, Khosravimelal S, Alehosseini M, Orive G, Dolatshahi-Pirouz A, Pal Singh Chauhan N, Janmey PA, Hashemi A, Kundu SC, Gholipourmalekabadi M (2022) Antibacterial smart hydrogels: new hope for infectious wound management. Mater Today Bio 17:100499
3. Askari E, Seyfoori A, Amereh M, Gharaie SS, Ghazali HS, Ghazali ZS, Khunjush B, Akbari M (2020) Stimuli-responsive hydrogels for local post-surgical drug delivery. Gels 6:14
4. Atif M, Babuçcu G, Riool M, Zaat S, Jonas U (2024) Antimicrobial peptide SAAP-148-functionalized hydrogels from photocrosslinkable polymers with broad antibacterial activity. Macromol Rapid Commun
5. Attama AA, Nnamani PO, Onokala OB, Ugwu AA, Onugwu AL (2022) Nanogels as target drug delivery systems in cancer therapy: A review of the last decade. Front Pharmacol 13
6. Baek J, Song N, Yoo B, Lee D, Kim B-S (2024) Precisely programmable degradation and drug release profiles in triblock copolyether hydrogels with cleavable acetal pendants. J Am Chem Soc 146:13836–13845
7. Barbosa KBF, Costa NMB, Alfenas R de CG, De Paula SO, Minim VPR, Bressan J (2010) Estresse oxidativo: conceito, implicações e fatores modulatórios. Revista de Nutrição 23:629–643
8. Bechinger B, Gorr S-U (2017) Antimicrobial peptides: mechanisms of action and resistance. J Dent Res 96:254–260
9. Burgess JL, Wyant WA, Abdo Abujamra B, Kirsner RS, Jozic I (2021) Diabetic wound-healing science. Medicina (B Aires) 57:1072
10. Cao H, Duan L, Zhang Y, Cao J, Zhang K (2021) Current hydrogel advances in physicochemical and biological response-driven biomedical application diversity. Signal Transduct Target Ther 6
11. Catoira MC, Fusaro L, Di Francesco D, Ramella M, Boccafoschi F (2019) Overview of natural hydrogels for regenerative medicine applications. J Mater Sci Mater Med 30:115
12. Chaudhari A, Vig K, Baganizi D, Sahu R, Dixit S, Dennis V, Singh S, Pillai S (2016) Future prospects for scaffolding methods and biomaterials in skin tissue engineering: a review. Int J Mol Sci 17:1974

13. Chelu M, Calderon Moreno JM, Musuc AM, Popa M (2024) Natural regenerative hydrogels for wound healing. Gels 10:547
14. Chen B, Ponce Benavente L, Chittò M, Post V, Constant C, Zeiter S, Nylund P, D'Este M, González Moreno M, Trampuz A, Wagemans J, Lavigne R, Onsea J, Richards RG, Metsemakers W-J, Moriarty TF (2024) Combination of bacteriophages and vancomycin in a co-delivery hydrogel for localized treatment of fracture-related infections. NPJ Biofilms Microbiomes 10:77
15. Chen S-Y, Chang C-K, Lan C-Y (2023a) Antimicrobial peptide LL-37 disrupts plasma membrane and calcium homeostasis in *Candida albicans* via the Rim101 pathway. Microbiol Spectr 11
16. Chen Y, Wang X, Tao S, Wang Q, Ma P-Q, Li Z-B, Wu Y-L, Li D-W (2023) Research advances in smart responsive-hydrogel dressings with potential clinical diabetic wound healing properties. Mil Med Res 10:37
17. Chen Z, Wang R, Guo J, Wang X (2024) The role and future prospects of artificial intelligence algorithms in peptide drug development. Biomed Pharmacother 175:116709
18. Copling A, Akantibila M, Kumaresan R, Fleischer G, Cortes D, Tripathi RS, Carabetta VJ, Vega SL (2023) Recent advances in antimicrobial peptide hydrogels. Int J Mol Sci 24:7563
19. de Moura MR, Rubia AF, Muniz EC (2008) Hidrogéis Semi-IPN Baseados em Rede de Alginato-Ca2+ com PNIPAAm Entrelaçado: Propriedades Hidrofílicas, Morfológicas e Mecânicas. Polímeros: Ciência e Tecnologia 18:132–137
20. de Oliveira KBS, Leite ML, Melo NTM, Lima LF, Barbosa TCQ, Carmo NL, Melo DAB, Paes HC, Franco OL (2024) Antimicrobial peptide delivery systems as promising tools against resistant bacterial infections. Antibiotics 13:1042
21. Dong Y, Wang Z (2023) ROS-scavenging materials for skin wound healing: advancements and applications. Front Bioeng Biotechnol 11
22. Duarte-Mata DI, Salinas-Carmona MC (2023) Antimicrobial peptides' immune modulation role in intracellular bacterial infection. Front Immunol 14
23. Dwivedi M, Parmar MD, Mukherjee D, Yadava A, Yadav H, Saini NP (2024) Biochemistry, mechanistic intricacies, and therapeutic potential of antimicrobial peptides: an alternative to traditional antibiotics. Curr Med Chem 31:6110–6139
24. Ferreira ACGR, Mohamed YI, da Silva ALFS, de Souza CRS, Ferreira AC (2024) Cicatrização cutânea: uma revisão da literatura. J Arch Health 5:e2004
25. Florczyk A, Krajcer A, Wójcik K, Lewandowska-Łańcucka J (2024) Innovative vancomycin-loaded hydrogel-based systems—New opportunities for the antibiotic therapy. Int J Nanomedicine 19:3991–4005
26. Freedman BR, Hwang C, Talbot S, Hibler B, Matoori S, Mooney DJ (2023) Breakthrough treatments for accelerated wound healing. Sci Adv 9
27. Gao Y, Zhan X, Huo S, Fu L, Tang Z, Qi K, Lv C, Liu C, Zhu Y, Ding S, Lv Y (2022) A gentamicin-thioctic acid multifunctional hydrogel for accelerating infected wound healing. J Mater Chem B 10:2171–2182
28. Ghaly T, Rabadi MM, Weber M, Rabadi SM, Bank M, Grom JM, Fallon JT, Goligorsky MS, Ratliff BB (2011) Hydrogel-embedded endothelial progenitor cells evade LPS and mitigate endotoxemia. Am J Physiol-Renal Physiol 301:F802–F812
29. Gibbs DMR, Black CRM, Dawson JI, Oreffo ROC (2016) A review of hydrogel use in fracture healing and bone regeneration. J Tissue Eng Regen Med 10:187–198
30. Haidari H, Kopecki Z, Sutton AT, Garg S, Cowin AJ, Vasilev K (2021) PH-Responsive "smart" hydrogel for controlled delivery of silver nanoparticles to infected wounds. Antibiotics 10:49
31. Hancock REW, Alford MA, Haney EF (2021) Antibiofilm activity of host defence peptides: complexity provides opportunities. Nat Rev Microbiol 19:786–797
32. Hao F, Wang L, Chen B, Qiu L, Nie J, Ma G (2021) Bifunctional smart hydrogel dressing with strain sensitivity and NIR-responsive performance. ACS Appl Mater Interfaces 13:46938–46950
33. Hao M, Wang D, Duan M, Kan S, Li S, Wu H, Xiang J, Liu W (2023) Functional drug-delivery hydrogels for oral and maxillofacial wound healing. Front Bioeng Biotechnol 11

34. Harding KG, Morris HL, Patel GK (2002) Clinical review science, medicine, and the future Healing chronic wounds. 324
35. Haririan Y, Asefnejad A (2024) Biopolymer hydrogels and synergistic blends for tailored wound healing. Int J Biol Macromol 279:135519
36. Ho T-C, Chang C-C, Chan H-P, Chung T-W, Shu C-W, Chuang K-P, Duh T-H, Yang M-H, Tyan Y-C (2022) Hydrogels: properties and applications in biomedicine. Molecules 27:2902
37. Jin S, Newton MAA, Cheng H, Zhang Q, Gao W, Zheng Y, Lu Z, Dai Z, Zhu J (2023) Progress of hydrogel dressings with wound monitoring and treatment functions. Gels 9:694
38. Ji Q, Chen K, Yi H, He B, Jiang T (2024) A paintable small-molecule hydrogel with antimicrobial and ROS scavenging activities for burn wound healing. Gels 10:621
39. Kalan LR, Brennan MB (2019) The role of the microbiome in nonhealing diabetic wounds. Ann N Y Acad Sci 1435:79–92
40. Kass LE, Nguyen J (2022) Nanocarrier-hydrogel composite delivery systems for precision drug release. Wiley Interdiscip Rev Nanomed Nanobiotechnol 14
41. Kaur H, Gogoi B, Sharma I, Das DK, Azad MA, Das PD, Pramanik A (2024) Hydrogels as a potential biomaterial for multimodal therapeutic applications. Mol Pharm 21:4827–4848
42. Khalil IA, Saleh B, Ibrahim DM, Jumelle C, Yung A, Dana R, Annabi N (2020) Ciprofloxacin-loaded bioadhesive hydrogels for ocular applications. Biomater Sci 8:5196–5209
43. Khan F, Mohd A, Haseen M, Kamal S, Khan MS, Shahid S, Nami SAA (2022) Synthesis, classification and properties of hydrogels: their applications in drug delivery and agriculture. J Mater Chem B 10:170–203
44. Khumalo SM, Bakare BF, Rathilal S (2024) Single and multicomponent adsorption of amoxicillin, ciprofloxacin, and sulfamethoxazole on chitosan-carbon nanotubes hydrogel beads from aqueous solutions: Kinetics, isotherms, and thermodynamic parameters. J Hazardous Mater Adv 13:100404
45. Kłodzińska SN, Pletzer D, Rahanjam N, Rades T, Hancock REW, Nielsen HM (2019) Hyaluronic acid-based nanogels improve in vivo compatibility of the anti-biofilm peptide DJK-5. Nanomedicine 20:102022
46. Kumar P, Kizhakkedathu J, Straus S (2018) Antimicrobial peptides: diversity, mechanism of action and strategies to improve the activity and biocompatibility in vivo. Biomolecules 8:4
47. Lee T-H, Hofferek V, Separovic F, Reid GE, Aguilar M-I (2019) The role of bacterial lipid diversity and membrane properties in modulating antimicrobial peptide activity and drug resistance. Curr Opin Chem Biol 52:85–92
48. Leite ML, Duque HM, Rodrigues GR, da Cunha NB, Franco OL (2023) The LL-37 domain: a clue to cathelicidin immunomodulatory response? Peptides (NY) 165:171011
49. Li J, Mooney DJ (2016) Designing hydrogels for controlled drug delivery. Nat Rev Mater 1
50. Li S, Renick P, Senkowsky J, Nair A, Tang L (2021) Diagnostics for Wound Infections. Adv Wound Care (New Rochelle) 10:317–327
51. Liu H, Wang C, Li C, Qin Y, Wang Z, Yang F, Li Z, Wang J (2018) A functional chitosan-based hydrogel as a wound dressing and drug delivery system in the treatment of wound healing. RSC Adv 8:7533–7549
52. Li W, Hu J, Chen C, Li X, Zhang H, Xin Y, Tian Q, Wang S (2023) Emerging advances in hydrogel-based therapeutic strategies for tissue regeneration. Regen Ther 24:459–471
53. Li X, Xue X, Xie P (2024) Smart dressings and their applications in chronic wound management. Cell Biochem Biophys 82:1965–1977
54. Loo Y, Wong Y-C, Cai EZ, Ang C-H, Raju A, Lakshmanan A, Koh AG, Zhou HJ, Lim T-C, Moochhala SM, Hauser CAE (2014) Ultrashort peptide nanofibrous hydrogels for the acceleration of healing of burn wounds. Biomaterials 35:4805–4814
55. Mahmoudian M, Ganji F (2017) Vancomycin-loaded HPMC microparticles embedded within injectable thermosensitive chitosan hydrogels. Prog Biomater 6:49–56
56. Malik E, Dennison S, Harris F, Phoenix D (2016) PH dependent antimicrobial peptides and proteins, their mechanisms of action and potential as therapeutic agents. Pharmaceuticals 9:67
57. Mann K, Aveyard J, Dallos Ortega M, Chen T, Koduri MP, Fothergill JL, Schache AG, Curran JM, Poole RJ, D'Sa RA (2025) Gelatin emulsion gels loaded with host defence peptides for the treatment of antibiotic-resistant infections. Biomater Adv 166:214071

58. de Mendonça RJ, Coutinho-Netto J (2009) Aspectos celulares da cicatrização. An Bras Dermatol 84:257–262
59. Merino S, Martín C, Kostarelos K, Prato M, Vázquez E (2015) Nanocomposite hydrogels: 3D polymer-nanoparticle synergies for on-demand drug delivery. ACS Nano 9:4686–4697
60. Miranda-Calderon L, Yus C, Landa G, Mendoza G, Arruebo M, Irusta S (2022) Pharmacokinetic control on the release of antimicrobial drugs from pH-responsive electrospun wound dressings. Int J Pharm 624:122003
61. Nassar N, Kasapis S (2023) Fundamental advances in hydrogels for the development of the next generation of smart delivery systems as biopharmaceuticals. Int J Pharm 633:122634
62. Nordström R, Nyström L, Andrén OCJ, Malkoch M, Umerska A, Davoudi M, Schmidtchen A, Malmsten M (2018) Membrane interactions of microgels as carriers of antimicrobial peptides. J Colloid Interface Sci 513:141–150
63. Olufunmilayo EO, Gerke-Duncan MB, Holsinger RMD (2023) Oxidative stress and antioxidants in neurodegenerative disorders. Antioxidants 12:517
64. Patel DK, Jung E, Priya S, Won S-Y, Han SS (2024) Recent advances in biopolymer-based hydrogels and their potential biomedical applications. Carbohydr Polym 323:121408
65. Peña OA, Martin P (2024) Cellular and molecular mechanisms of skin wound healing. Nat Rev Mol Cell Biol 25:599–616
66. Selvaraj SP, Chen J-Y (2023) Conjugation of antimicrobial peptides to enhance therapeutic efficacy. Eur J Med Chem 259:115680
67. Serpico L, Dello Iacono S, Cammarano A, De Stefano L (2023) Recent Advances in stimuli-responsive hydrogel-based wound dressing. Gels 9:451
68. Shao Z, Yin T, Jiang J, He Y, Xiang T, Zhou S (2023) Wound microenvironment self-adaptive hydrogel with efficient angiogenesis for promoting diabetic wound healing. Bioact Mater 20:561–573
69. Sharma PK, Halder M, Srivastava U, Singh Y (2019) Antibacterial PEG-chitosan hydrogels for controlled antibiotic/protein delivery. ACS Appl Bio Mater 2:5313–5322
70. Sun M, Zhu C, Long J, Lu C, Pan X, Wu C (2020) PLGA microsphere-based composite hydrogel for dual delivery of ciprofloxacin and ginsenoside Rh2 to treat *Staphylococcus aureus* -induced skin infections. Drug Deliv 27:632–641
71. Szymczak P, Szczurek E (2023) Artificial intelligence-driven antimicrobial peptide discovery. Curr Opin Struct Biol 83:102733
72. Tang Y, Shu X, He G, Zhang Y, Zhao Y, Yuan H, Yu J, Guo J, Chen Q (2024) Vancomycin-loaded hydrogels with thermal-responsive, self-peeling, and sustainable antibacterial properties for wound dressing. J Mater Chem B 12:752–761
73. Tapiwa Chamanga E (2018) Clinical management of non-healing wounds. Nursing Standard 32:48–63
74. Thapa RK, Winther-Larsen HC, Ovchinnikov K, Carlsen H, Diep DB, Tønnesen HH (2021) Hybrid hydrogels for bacteriocin delivery to infected wounds. Eur J Pharm Sci 166:105990
75. Uberoi A, McCready-Vangi A, Grice EA (2024) The wound microbiota: microbial mechanisms of impaired wound healing and infection. Nat Rev Microbiol 22:507–521
76. Wang C, Hong T, Cui P, Wang J, Xia J (2021) Antimicrobial peptides towards clinical application: delivery and formulation. Adv Drug Deliv Rev 175:113818
77. Wang P-H, Huang B-S, Horng H-C, Yeh C-C, Chen Y-J (2018) Wound healing. J Chin Med Assoc 81:94–101
78. Wang Y, Wallmersperger T, Ehrenhofer A (2023a) Application of back propagation neural networks and random forest algorithms in material research of hydrogels. PAMM 23
79. Wang Y, Wallmersperger T, Ehrenhofer A (2023b) Application of back propagation neural networks and random forest algorithms in material research of hydrogels. PAMM 23
80. Watson AL, Eckhart KE, Wolf ME, Sydlik SA (2022) Hyaluronic acid-based antibacterial hydrogels for use as wound dressings. ACS Appl Bio Mater 5:5608–5616
81. Wilkinson HN, Hardman MJ (2020) Wound healing: cellular mechanisms and pathological outcomes. Open Biol 10

82. Xie X, Lei Y, Li Y, Zhang M, Sun J, Zhu M-Q, Wang J (2023) Dual-crosslinked bioadhesive hydrogel as NIR/pH stimulus-responsiveness platform for effectively accelerating wound healing. J Colloid Interface Sci 637:20–32
83. Yan D, Li Y, Liu Y, Li N, Zhang X, Yan C (2021) Antimicrobial properties of chitosan and chitosan derivatives in the treatment of enteric infections. Molecules 26:7136
84. Yanez DA, Lacher RK, Vidyarthi A, Colegio OR (2017) The role of macrophages in skin homeostasis. Pflugers Arch 469:455–463
85. Yang J, Chen Z, Pan D, Li H, Shen J (2020) Umbilical cord-derived mesenchymal stem cell-derived exosomes combined pluronic F127 hydrogel promote chronic diabetic wound healing and complete skin regeneration. Int J Nanomedicine 15:5911–5926
86. Yan T, Kong S, Ouyang Q, Li C, Hou T, Chen Y, Li S (2020) Chitosan-gentamicin conjugate hydrogel promoting skin scald repair. Mar Drugs 18:233
87. Yao H, Wu M, Lin L, Wu Z, Bae M, Park S, Wang S, Zhang W, Gao J, Wang D, Piao Y (2022) Design strategies for adhesive hydrogels with natural antibacterial agents as wound dressings: status and trends. Mater Today Bio 16:100429
88. Yuan N, Shao K, Huang S, Chen C (2023) Chitosan, alginate, hyaluronic acid and other novel multifunctional hydrogel dressings for wound healing: a review. Int J Biol Macromol 240:124321
89. Zarei-Mehrvarz E, Fahimirad S, Ghaznavi-rad E, Abbasian SS, Abtahi H (2023) The LL-37 antimicrobial peptide as a treatment for systematic infection of *Acinetobacter baumannii* in a mouse model. Protein Pept Lett 30:44–53
90. Zhang M, Zhao X (2020) Alginate hydrogel dressings for advanced wound management. Int J Biol Macromol 162:1414–1428
91. Zhang X, Kang X, Ji L, Bai J, Liu W, Wang Z (2018) Stimulation of wound healing using bioinspired hydrogels with basic fibroblast growth factor (bFGF). Int J Nanomedicine 13:3897–3906
92. Zhang X, Kang X, Jin L, Bai J, Liu W, Wang Z (2018) Stimulation of wound healing using bioinspired hydrogels with basic fibroblast growth factor (bFGF). Int J Nanomed 13:3897–3906
93. Zhao H, Huang J, Li Y, Lv X, Zhou H, Wang H, Xu Y, Wang C, Wang J, Liu Z (2020) ROS-scavenging hydrogel to promote healing of bacteria infected diabetic wounds. Biomaterials 258:120286
94. Zhao Y, Mulder RJ, Houshyar S, Le TC (2023) A review on the application of molecular descriptors and machine learning in polymer design. Polym Chem 14:3325–3346
95. Zheng Y, Pan C, Xu P, Liu K (2024) Hydrogel-mediated extracellular vesicles for enhanced wound healing: the latest progress, and their prospects for 3D bioprinting. J Nanobiotechnol 22:57

Hydrogels Against Wound Causing Opportunistic Pathogens

Snigdha Majumder and Sankar N. Sinha

Abstract In spite of modern developments in the area of health care and biomedical field, different infectious wound causing by pathogens became a major public health problems which directly affect our socio-economic status. Wounds are sometimes fatal because of multidrug-resistant bacterial infections. One productive and effective alternate way free from the resistant mechanisms is immediately desired. Hydrogels are soft three-dimensional insoluble crosslinking polymer structure that can absorb a lot of water and provide a favourable environment for cell growth. It also provide necessary nutrition and oxygen supply to the cell and fight against different wound causing pathogens. This studies focuses on advances of using modern hydrogels against different skin wound infection causing bacteria. In that regard, a precise introduction on skin wound healing and smart approaches of hydrogels are presented in their allied sections. These review studies can open the door to produce a new biomaterials against wound causing bacteria with therapeutic uses.

Graphical Abstract

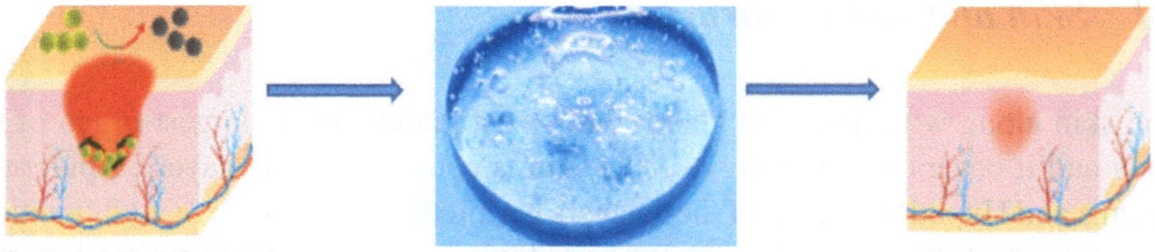

Keywords Hydrogels · Wound healing · Opportunistic pathogens · Antimicrobial hydrogels · Infectious wound management · Nano-hydrogels · Controlled drug delivery

S. Majumder · S. N. Sinha (✉)
Department of Botany, University of Kalyani, Kalyani, West Bengal, India
e-mail: sinhasn62@yahoo.co.in

A. K. Mandal et al. (eds.), *Functionalized Nanoparticles Hydrogels for Wound Healing*, Smart Nanomaterials Technology, https://doi.org/10.1007/978-981-95-0720-7_11

1 Introduction

Wound infection became a major public health problems which directly affect our socio-economic status due to the extensive production of antibiotic-resistant bacteria. When the reliability and function of the human skin are interrupted, wounds occur [33]. Wounds can be classified into two major groups: Acute wounds including surgical wounds, burns, minor cuts and Traumatic wounds including diabetes, pressure ulcers and gunshot wound etc. [67]. According to WHO, about 265,000 deaths are noticed annually just due to burn wounds and around 300,000–500,000 surgical wound infections are reported annually [49, 53]. Moreover, overuse of antibiotics and modern developments in the area of health care and biomedical field leads to increased multidrug resistance [30, 55]. In this situation, Wound management is become a crucial global concern that needs effective treatments. Over the past several decades, a great effort has been given to develop a strategy with antibacterial properties that deals with wound healing process [65]. Infection at the wound site due to bacteria like *Staphylococcus aureus, Staphylococcus epidermidis* are very common. So, using broad spectrum antimicrobial agents are safe and work effectively against these bacterial infections. Recently hydrogels have become more attractive and promote wound healing by maintaining moisture, preventing desiccation and supporting cell growth [81]. Hydrogels also have intrinsic antibacterial properties.

In light of this, the present review focused on characteristics of hydrogels and their advantages in wound healing applications with antimicrobial activity against wound causing opportunistic pathogens.

1.1 Skin and Wound Structure

The skin along with hairs, oil glands and sweat glands, is the largest organ of the body and it covers the entire body. The skin is consists of three layers: Epidermis, Dermis and Hypodermis. All of these layers permit the skin to achieve vital functions including (1) Protection: It protects the body from external factors including bacteria, chemicals, temperature and UV radiations. It also protects from different injury; (2) Temperature regulation: In cold the skin regulates temperature by constricting blood vessels in the dermis; (3) Sensation: The skin allows the body to feel different sensations such as warmth, cold, itching and pain; (4) Vitamin D production: It also produces vitamin D; (5) Hormone Production: Skin also produces many important body hormones; and (6) Wound healing: After skin injuries, the blood supply to the skin increases to help Wound heal and protect from infections [26, 66].

A break or opening resulting from an injury or surgical incision on the skin defines a wound, which can be prolonged to subcutaneous tissues and can damage nerves [5, 39, 91]. Based on the pathogenesis and consequences, the skin wounds are classified into acute and chronic wounds [87]. Acute wounds go through a chains of molecular events and lead to recovering of structural integrity of the skin. On the other hand,

Chronic wounds omit to resolve and go through by pathogenic processes, such as constant inflammation, infections and necrosis. We can consider four overlapping phases during wound healing such as coagulation, inflammation, fibroplasia, and remodelling [22, 44, 80]. In case of acute wound, skin recovers within an 8- to 12-week period and in chronic wounds such as diabetes and pressure ulcers, the skin treatments goes for longer than 12 weeks [5, 68].

1.2 Wound Healing Process

During an injury, the first step hemostasis is happened. During the hemostasis, the primary phase is an outpouring of blood and lymphatic fluid. Both the coagulation (extrinsic and intrinsic) pathways are triggered and perform a crucial role for stopping bleeding. Then the arterial vasoconstriction in the injured lymphatic endothelium and platelet clumping take place. The release of ADP begans the process of thrombosis. This vasoconstriction minimize the blood loss, followed by vasodilation to allow increased blood flow facilitating the transport of immune cells for healing [92].

Just after hemostasis, the inflammatory phase begins with the release of chemicals injured cells at the wound site. Then the white cells and thrombocytes releases mediators and cytokines to speed up the process. During this phase, platelet aggregation and infiltration of leucocytes are occurred at the wound site. The inflammatory neutrophil cells able to phagocytosis of cellular waste and bacteria and maintain the sterility of the wound [59] (Fig. 1).

The proliferation process is always going on in the background. It is the phase in which new tissue, known as granulation tissue, fills in the wound. Proliferation events involves several stages: Stabilization-proteoglycans form the wound core and stabilize the wound site; Reepithelialization-is the movement of cells from periphery and adjacent ends of the wound; Neovascularization-occurs through both angiogenesis (new blood vessels originates from existing one) and vasculogenesis (new blood vessels originates from endothelial progenitor cells). Then the wound starts to mature and begins to contract [92].

The remodeling phase, also called maturation phase begins around week 3 and can last for a year. At this stage the newly formed tissue is refined, strengthened and reorganized initially by replacing excess collagen fibres with type I collagen which lids to formation of scar tissue and restoration of tissue functions. The final resulting scar tissue will never have 100% of the original functional capability as the original tissue of the wound and only have about 80% of the workable strength [7].

1.3 Microbiology of Wounds

A wound becomes infected or contaminated with bacteria that inhibit wound healing and also can influence the host's immune system. Chronic wounds are a suitable

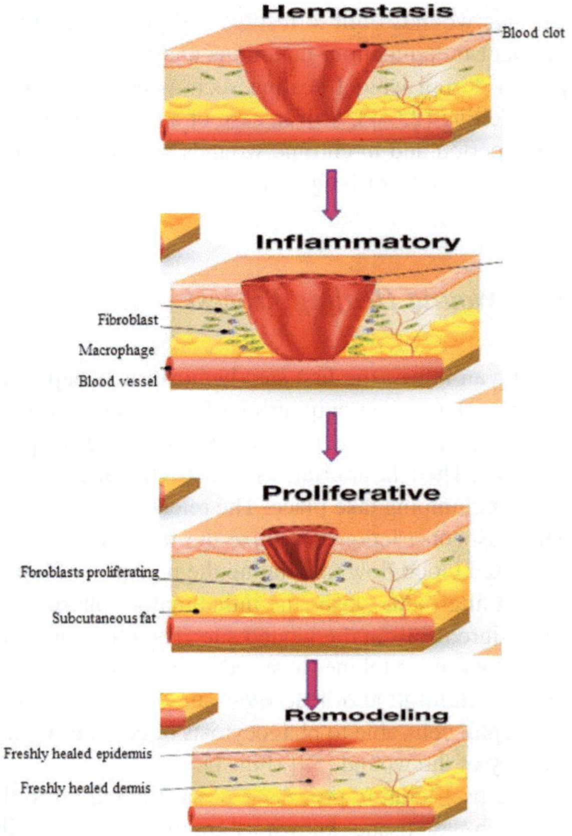

Fig. 1 Healing phases of a wound [51]

place to inhabit some pathogenic bacteria, including *Staphylococcus aureus*, *Pseudomonas aeruginosa*, *Peptoniphilus*, *Enterobacter*, *Stenotrophomonas*, *Escherichia coli, Acinetobacter baumannii*, *Finegoldia*, and *Serratia* [69]. They produce biofilms that lead to a delay in wound healing. Acute wounds become chronic due to the resistance of bacteria to antibiotics because of this mature biofilms [25]. Conversely, the diversity of microorganisms in any wound site also influenced by Wound type, site, depth, the level of tissue damage and the antibacterial efficacy of the host defense mechanism.

Wound contaminants are classified into three groups based on sources: (a) the environment-the exogenous microorganisms found in the air, (b) the surrounding skin-concerning the normal skin microflora including skin diphtheroids, *Staphylococcus epidermidis*, and micrococci, propionibacteria, and (iii) Internal sources including mucous membranes [20]. The normal body microfloras of the gut, oral cavity, and the vagina are aslo provide the bacteria colonize wounds. A list of microorganisms are presented in Table 1 [9] with references.

Due to the overuse of antibiotics, a variety of multi-drug resistant (MDR) bacteria have appeared in last few years [15]. This MDR bacteria are difficult to treat and can

Table 1 List of wound microorganisms

Wound study description	Principle isolates	References
Microbiology of 20 diabetic ulcers (scrapings of ulcer bed and wound edge)	*Bacteroides*spp., *Peptostreptococcus* spp., *Proteus* spp.	Louie et al. [48]
392 specimens taken from burn sites in 180 children	*S. epidermidis, S. aureus,* alpha-hemolytic streptococci, *Propionibacterium acnes, Peptostreptococcus* spp., *Bacteroides* spp.	Brook and Randolph [14]
49 specimens from 25 pressure sores	*Bacteroides* spp., coliforms, *P. aeruginosa*	Sapico et al. [76]
Aspirates from human and animal bite wounds in 39 children	*S. aureus, Peptostreptococcus* spp., *Bacteroides* spp.	Brook [10]
Analysis of 584 wounds from many sites	*Bacteroides* spp., *Peptostreptococcus* spp., *S. aureus, Clostridium* spp., *Fusobacterium* spp.	Brook and Frazier [12]
Analysis of decubitus ulcers in 58 children	*S. aureus, Peptostreptococcus* spp., *B. fragilis* group, *P. aeruginosa*	Brook [11]
Analysis of leg ulcers without clinical signs of infection in 58 patients	*S. aureus, Enterococcus faecalis, Enterobacter cloacae, Peptostreptococcus magnus*	Hansson et al. [31]
43 swab specimens from chronic leg ulcers in 41 patients	*S. aureus, Peptostreptococcus* spp., *B. fragilis* group	Brook and Frazier [13]
Swab specimens from 30 noninfected leg ulcers without clinical signs of infection	*S. aureus*, coliforms, coagulase-negative staphylococci, fecal streptococci, *Peptostreptococcus* spp.	Bowler and Davis [8]

cause serious infection. Therefore, it become a great concern to design novel tactics and technologies for preventing antibiotic resistance and decreasing mortality.

1.4 Hydrogels

Wound healing is generally along with bacterial infection. The overuse of artificial antibiotics results in multidrug resistance, causing a threat to social health. In this situation, Hydrogel-based dressings of wound focused on reducing the bacterial infections have developed as an active wound treatment. Therefore, several approaches and strategies have been established to make the hydrogels very much potent with extensive antibacterial spectrum [103]. Hydrogel is a net of water-loving polymers that can absorb and hold lot of water. Hydrogel may be reversible hydrogel i.e.

Fig. 2 Classification of hydrogels [42]

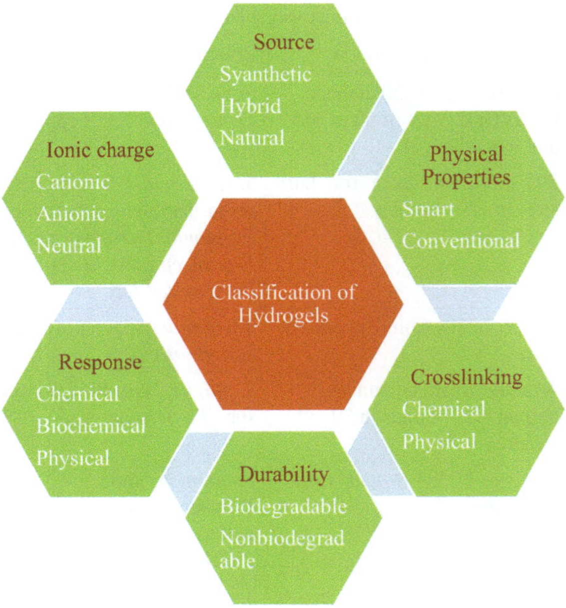

cross-linked by physical entanglement or permanent hydrogel that are cross-linked through covalent bonds. It is a gel like semi-solid material containing of large quantities of liquid and small quantities of solids. Hydrogels are three-di-mensional (3D) networks consisting hydrophilic functional groups such as amino (NH_2), carboxyl (–COOH), amide, and hydroxyl group (−OH). Hydrogels shows response under the presence of exterior stimuli such as temperature, magnetic field, pH, light, ionic strength and therefore hydrogel called as a 'smart material' [71]. Based on several aspects, hydrogels can be classified into different groups which is stated briefly in Fig. 2.

1.5 Antibacterial Hydrogel

Hydrogels have noteworthy advantages in antibacterial resources because of their effective loading, release of drugs and several antibacterial agents. Thus hydrogel can reduce the toxic effects of several infectious bacteria [35, 52]. Consequently, the hydrogels have fascinated growing attention as substitute biomaterials to treat several bacterial infections. The hydrogel-based antibacterial therapy concise into following groups: (1) Antibiotic loaded hydrogels, (2) Antibiotic-free loaded hydrogels, (3) The stimuli-responsive smart antibacterial hydrogels, and (4) Light-mediated antibacterial hydrogels [83].

2 Role of Hydrogels in Combating Opportunistic Pathogens

2.1 Antimicrobial Properties

Intrinsic Antimicrobial Polymers Hydrogels can be synthesized from different polymers of natural, semisynthetic, and synthetic origin. Some hydrogels are synthesized from natural polymers origin with inherent antimicrobial properties. Among them, Carrageenan, chitosan, and gels from plant extracts are become very popular [90].

Incorporation of Antimicrobial Agents Incorporated antimicrobial-agent hydrogels are those in which antimicrobial substances are inserted to control microbial infection [1]. On the basis of incorporation hydrogels can grouped into following classes: antibiotics incorporated hydrogels, metal nanoparticle incorporated hydrogels, antimicrobial peptides (AMPs) incorporated hydrogels, and biological (plant or organism) extracts incorporated hydrogels (Fig. 3).

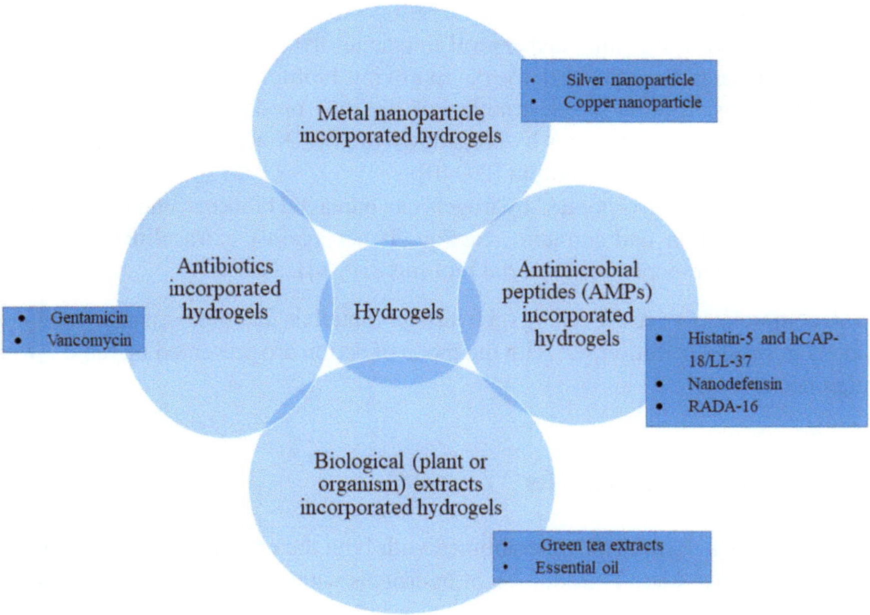

Fig. 3 Classification of hydrogels on the basis of incorporation

2.2 Pathogen-Specific Targeting

Pathogen-specific targeting using hydrogels relating to the design of hydrogels that can specifically bind to and bring therapeutic agents directly to definite pathogens, such as *Pseudomonas aeruginosa* and *Staphylococcus aureus*, at the place of wound infection [23, 95]. These bacteria form biofilms that play a role as a physical and biochemical blockade, keeping protected them from several antimicrobial agents and the human's immune system. Smart hydrogels can be planed to target bacterial biofilms that protect these bacteria [16]. Diverse strategies are obtainable to interrupt the biofilm such as:

2.2.1 Enzymatic Destruction of Biofilms

Biofilms are highly resistant to drug due to their shielding extracellular polymeric substance (EPS) medium [93, 94]. Enzymatic disruption of biofilms through hydrogel-based methods is an inventive strategy that combines the **targeted act of enzymes** with the **adaptableness of hydrogels** to destroyed the EPS and increase antimicrobial activity [85].

a. **Dispersin B**: Integrating dispersin B into smart hydrogels empower the process for precise and restricted delivery, approving biofilm interruption lacking any side effects. It aims the destruction of biofilms produced by *Staphylococcus aureus* and other pathogens by breakdown of poly-N-acetylglucosamine, a vital component of bacterial biofilm [29, 40].
b. **DNase**: DNase-incorporated hydrogels can damaged biofilms, enhancing antibiotic penetration and immune reaction. It act against extracellular DNA, an important component of bacterial biofilms [19, 84].

A comprehensive list focusing on hydrogel materials and their specific pathogen targets in wound infection site with the mode of action are presented in Table 2 with references.

2.2.2 Combination Therapies

In this approach, hydrogels are combined with both the enzymes and antibiotics or antimicrobial peptides. The addition of hydrogels with **enzymes** and **antibiotics** or **antimicrobial peptides (AMPs)** signifies an influential strategy to tackle bacterial biofilms [17]. These combine therapies destroys bacteria enhancing their efficacy. It provide broad-spectrum antimicrobial activity. **Enzymes** degrade the extracellular polymeric substance, distracting the biofilm's structure and exposing bacteria to antimicrobials and **Antibiotics/AMPs** kill bacteria within the bacterial film [3]. A smart hydrogel incorporated with both DNase and gentamicin disrupts the biofilms and shows pathogen-killing action.

Table 2 List of used hydrogel materials and their specific pathogen targets

Hydrogel material	Pathogen target in wound site	Mode of action	Application	Examples	References
Chitosan	*E. coli*, *S. aureus*	Disruption of microbial membranes, pH sensitivity	Wound healing, drug delivery	Chitosan-based hydrogels with silver ions	Aziz et al. [2], Goy et al. [27]
Alginate	*Pseudomonas aeruginosa*, *Candida*	Enzyme-responsive degradation, drug release	Antibacterial coatings, burn wounds	Alginate-antibiotic hydrogel films	Morais et al. [56], Osmokrovic et al. [63], Raj et al. [70]
Polyethylene glycol (PEG)	SARS-CoV-2, *S. aureus*	Antibody or ligand functionalization for targeting	Viral inhibition, injectable hydrogels	PEG-based nanogel for virus binding	Yang et al. [102] Mallakpour et al. [50]
Hyaluronic acid (HA)	*Helicobacter pylori*, *E. coli*	Mucosal adhesion, enzyme-responsive degradation	Gastrointestinal infections	HA hydrogels for mucosal application	Snetkov et al. [79], Wekwejt et al. [96]
PHEMA (poly hydroxyethyl methacrylate)	*S. aureus*, *Candida*	Sustained drug release	Antifungal implants	PHEMA-based hydrogels for slow drug delivery	Laverty et al. [45]
Polyvinyl alcohol (PVA)	*Salmonella*, *E. coli*	pH-triggered antimicrobial release	Food packaging, biosensors	PVA hydrogel coatings	Boggione et al. [6], Wikanta et al. [97]
Gelatin	*Staphylococcus epidermidis*, MRSA	Enzyme degradation, antibiotic conjugation	Wound dressings, tissue scaffolds	Gelatin hydrogels with antibiotics	Ren et al. [73], Hasan et al. [32], Bhattacharjee et al. [4]
Carboxymethylcellulose (CMC)	*Candida albicans*, *S. aureus*	Controlled release of antifungal agents	Oral infections, wound healing	CMC hydrogels loaded with antifungal drugs	Turky et al. [88], Pagano et al. [64]

(continued)

Table 2 (continued)

Hydrogel material	Pathogen target in wound site	Mode of action	Application	Examples	References
Collagen	*Pseudomonas aeruginosa*, MRSA	Antimicrobial peptide delivery	Skin grafts, wound healing	Collagen-based hydrogel dressings	Jarman et al. [38]
Silk fibroin	*Klebsiella pneumoniae*, *E. coli*	Biofilm inhibition and antibacterial delivery	Wound healing, tissue repair	Silk-based nanocomposite hydrogels	Kaptan et al. [41], Khosravimelal et al. [43]

2.2.3 pH-Sensitive Hydrogels

Many times biofilms produce acidic microenvironments. PH-sensitive smart hydrogels convey significant variations in their physical or chemical properties in response to alterations in the pH. PH-sensitive smart hydrogels can release enzymes or medicines specifically in these situations, confirming targeted activity. It has risen as a encouraging tactic for aiming and distracting biofilms because of their ability to respond to the variation in pH, which often differs in biofilm-associated wound infections [18, 24].

2.3 Controlled Drug Delivery

Hydrogels have become a keystone in progressive wound care due to their capability to deliver **controlled release of antimicrobial agents due to because** these polymeric systems can be chemically and physically adjusted endowing new and remarkable properties [58, 77, 89]. This method confirms sustained therapeutic action at the wound infection site, reduces systemic toxicity, and improves the wound-healing process. Hydrogels can encapsulate several antimicrobials, such as **antibiotics, antimicrobial peptides (AMPs), metallic nanoparticles** (e.g., silver or zinc oxide), or **natural extracts** (e.g., honey or essential oil) [101]. Encapsulation shields the active agents from degradation and keep it stable. It provide a **diffusion-controlled discharge** mechanism, in which the antimicrobials are slowly released into the wound site over time. It also can be designed to release antimicrobial agents in response to precise stimuli, such as **pH, temperature,** or **enzyme activity** [104]. Sustained release of drugs through hydrogel-based controlled delivery reduces the frequency of dressing changes, improving patient relaxation and compliance. It also restricts drug action to the wound and reduces systemic side effects, such as nephrotoxicity or hepatotoxicity (Wang et al. 2023). Hydrogel-based controlled drug delivery represent an exciting frontline in wound care system, offering an effective and confined solution to combat infections even though supportive to the natural healing process.

2.4 Moisture Retaining and Absorption of Exudate

Hydrogels are extensively used in wound infection treatment by reason of their capability to preserve a **humid curing atmosphere** simultaneously absorbing additional exudate [46]. This bifunctionality is essential for wound treatment, preventing the infection risk affected by pathogens. A humid environment make easier the keratinocyte migration and fibroblast proliferation, i.e. essential for re-epithelialization and cell rejuvenation [61, 74, 82]. Moisture retaining capability of hydrogels prevents the wound from drying out, reducing scab formation. On the other hand, it provide a blockade against exterior contaminants while maintaining optimum moisture levels that diminishes bacterial growth. By tricking and isolating wound exudates, hydrogels decrease the risk of bacterial propagation and biofilm construction. Innovative hydrogels regulate their properties (e.g., swelling capacity) based on wound circumstances, ensuring optimum hydration and exudate controlling [28, 62]. For diabetic ulcers, venous leg ulcers, and pressure sores, it aid to manage high levels of exudate although maintaining hydration.

3 Stimuli-Responsive Hydrogels

Stimuli-responsive hydrogels are also called as **smart hydrogels.** These are polymer-based tools that can devotedly respond to several stimuli and able to modify their structure, properties, or functionality [54]. These are smart tools designed to antimicrobials delivery in a specific way when activated through infection's conditions, such as variations in pH, heat, or enzyme function [36]. These hydrogels improve therapeutic efficacy, decrease systemic toxicity, and reduce the development of microbes [98]. Infection sites often show acidic microenvironments (pH ~ 5–6) because of improved metabolic activity and inflammatory processes [18]. Similarly, temperature-responsive hydrogels swell or shrink, discharging encapsulated antimicrobials when the temperature increases [37]. Likewise hydrolysis of Enzyme-responsive hydrogels can deliver antimicrobials directly at the infection site.

4 Dual-Functional Hydrogels

Dualfunctional hydrogels describe an advanced technique in biotherapy, including antimicrobial activity with corresponding properties for example anti-inflammatory and tissue repairing abilities. This synergy increases healing outcomes of infected wounds or chronic injuries. It afford antimicrobials together with anti-inflammatory drugs or growth stimulating factors. It release antibiotics or AMPs upon exposure to infection-specific triggers like acidic pH or bacterial enzymes and Nanoparticles interrupt bacterial membranes. These hydrogels combines antimicrobial, anti-inflammatory, and reformative activity and reduces the necessity of regular dressing changes and facilitate painless wound healing [78, 99, 100].

5 Nano-hydrogels

The integration of nanotechnology or nanoparticle with hydrogels has an abundant effect on wound care by improving their antimicrobial activity, drug release ability. Several nanoparticles (silver or gold) embedded in hydrogels can combat against wound infection causing pathogens by preventing biofilm development, and thus recover wounds in challenging environments [47, 57]. Nanoparticles provided broad-spectrum antimicrobial activity and shows antimicrobial efficacy against bacteria, fungi, and viruses. NPs within hydrogels penetrate biofilms, breaking them apart and exposing bacteria to antimicrobials. Metallic nanoparticles (e.g., silver, zinc) kill pathogens by disrupting cell membranes and generating ROS [60]. Carbon-based and silica nanoparticles penetrate biofilms, disrupting their protective matrix and killing embedded bacteria. Nanoparticles can modify the inflammatory activity, promoting a stable immune response beneficial to healing process [75, 86].

6 Drawbacks and Considerations

6.1 Antimicrobial Resistance Development

Though antibiotics loaded hydrogels offer noteworthy rewards in wound treatment, **misuse** of antibiotics result in **antimicrobial resistance (AMR)**. This creates a foremost health concerns as resistant bacteria become more challenging.

6.2 Cytotoxicity of Hydrogel

However several antimicrobial agents (nanoparticles, antibiotics, or antimicrobial peptides) enrich the antimicrobial activity in wound infection treatment, over dosage of these agents can also present a threat to **human cells**. These poisonous effects are a leading attention in the design of hydrogel formulations. To diminish this, careful optimization of doses, surface adaptations, and use of decomposable, non-toxic nanoparticles are vital to treat wounds.

6.3 Expensive

Raw materials (such as **eco-friendly polymers, nanoparticles**, and **antimicrobial agents)** necessary for advanced hydrogels making are expensive, which significantly increase the final price of the smart hydrogel.

7 Emerging Pathogen-Focused Applications

Hydrogels get significant attention in the field of medicine, particularly for their unique capability to **treat infections** caused by pathogens. As a result of the growing prevalence of **evolving pathogens,** hydrogels are being planned with **innovative properties** to combat specific microbial threats [34]. These advanced hydrogels offer **customizable antimicrobial activities** and **targeted delivery** to combat emerging pathogens effectively while decreasing the side effects. Broad-spectrum antimicrobial hydrogels have emerged as a promising solution to combat these **multi-pathogen infections** effectively, offering a **localized** and **sustained release** of antimicrobial agents to the wound site. Having the ability to effectively kill a broad range of pathogens, **Silver nanoparticle-incorporated hydrogels** are used in wound treatment. Similarly, **copper nanoparticles** have showed the antimicrobial activity against **drug-resistant bacteria** and fungi [72]. Several broad-spectrum antimicrobial hydrogels can combat against multi-pathogen wound infections [21].

The complication of chronic wounds, mainly **caused by several types of microorganisms**, requires a **designed way** for wound treatment. It enriches the **efficacy** and **safety** of wound repair. As technology develops and becomes available, customized hydrogel formation could revolutionize the wound management.

8 Conclusion

Smart antibiotic loaded Hydrogels with their unique properties, are an effective resourceful solution for reducing wound infections and stimulating healing process. Hydrogels uphold a moist environment which is necessary for quick tissue repair and reducing scarring. It incorporated with effective antimicrobial agents (nanoparticles, chemical or natural antibiotics) providing antimicrobial activity against bacterial infections, including drug-resistant strains. Moreover, their exclusive ability to carry these agents in a targeted area in a controlled way confirms sustained protection and decreases the risk of side effects. Biocompatible and adaptable hydrogel also give a protective barrier over the wound, inhibiting external contamination although it minimizes pain during wound dressing on the skin offering a auspicious approach to modern wound treatment.

References

1. Ahmad N, Bukhari NA, Hussain MA, Ejaz H, Munir MU, Amjad MW (2024) Nanoparticles incorporated hydrogels for delivery of antimicrobial agents: developments and trends. RSC Adv 14(19):13535–13564

2. Aziz MA, Cabral JD, Brooks HJ, Moratti SC, Hanton LR (2012) Antimicrobial properties of a chitosan dextran-based hydrogel for surgical use. Antimicrob Agents Chemother 56(1):280–287
3. Barman S, Kurnaz LB, Leighton R, Hossain MW, Decho AW, Tang C (2024) Intrinsic antimicrobial resistance: molecular biomaterials to combat microbial biofilms and bacterial persisters. Biomaterials 122690
4. Bhattacharjee B, Ghosh S, Mukherjee R, Haldar J (2020) Quaternary lipophilic chitosan and gelatin cross-linked antibacterial hydrogel effectively kills multidrug-resistant bacteria with minimal toxicity toward mammalian cells. Biomacromol 22(2):557–571
5. Boateng JS, Matthews KH, Stevens HN, Eccleston GM (2008) Wound healing dressings and drug delivery systems: a review. J Pharm Sci 97(8):2892–2923
6. Boggione MG, Santos JB, de Souza SM, Mendonça CS (2021) Preparation of polyvinyl alcohol hydrogel containing bacteriophage and its evaluation for potential use in the healing of skin wounds. J Drug Delivery Sci Technol 63:102484
7. Bowden LG, Byrne HM, Maini PK, Moulton DE (2016) A morphoelastic model for dermal wound closure. Biomech Model Mechanobiol 15:663–681
8. Bowler PG, Davies BJ (1999) The microbiology of infected and noninfected leg ulcers.
9. Bowler PG, Duerden BI, Armstrong DG (2001) Wound microbiology and associated approaches to wound management. Clin Microbiol Rev 14(2):244–269
10. Brook I (1987) Microbiology of human and animal bite wounds in children. Pediatr Infect Dis J 6(1):29–32
11. Brook I (1991) Microbiological studies of decubitus ulcers in children. J Clin Pediatr Surg 26(2):207–209
12. Brook I, Frazier EH (1990) Aerobic and anaerobic bacteriology of wounds and cutaneous abscesses. Arch Surg 125(11):1445–1451
13. Brook I, Frazier EH (1998) Aerobic and anaerobic microbiology of chronic venous ulcers. Int J Dermatol 37(6):426–428
14. Brook I, Randolph JG (1981) Aerobic and anaerobic bacterial flora of burns in children. J Trauma Acute Care Surg 21(4):313–318
15. Carpa R, Remizovschi A, Culda CA, Butiuc-Keul AL (2022) Inherent and composite hydrogels as promising materials to limit antimicrobial resistance. Gels 8(2):70
16. Chen Z, Wang Z, Ren J, Qu X (2018) Enzyme mimicry for combating bacteria and biofilms. Acc Chem Res 51(3):789–799
17. Copling A, Akantibila M, Kumaresan R, Fleischer G, Cortes D, Tripathi RS, Vega SL (2023) Recent advances in antimicrobial peptide hydrogels. Int J Mol Sci 24(8):7563
18. Das IJ, Bal T (2024) pH factors in chronic wound and pH-responsive polysaccharide-based hydrogel dressings. Int J Mol Sci 135118
19. Deng W, Lei Y, Tang X, Li D, Liang J, Luo J, Chen Z (2022) DNase inhibits early biofilm formation in *Pseudomonas aeruginosa*-or *Staphylococcus aureus*-induced empyema models. Front Cell Infect Microbiol 12:917038
20. Duerden BI (1994) Virulence factors in anaerobes. Clin Infect Dis 18(Supplement_4):S253–S259
21. Dundas AA, Mikulskis P, Hook A, Dundas A, Irvine D, Sanni O, Winkler DA (2018) Prediction of broad-spectrum pathogen attachment to coating materials for biomedical devices. ACS Appl Mater Interfaces 10(1)
22. Ellis S, Lin EJ, Tartar D (2018) Immunology of wound healing. Curr Dermatol Rep 7:350–358
23. El-Naggar MY, Gohar YM, Sorour MA, Waheeb MG (2016) Hydrogel dressing with a nano-formula against methicillin-resistant *Staphylococcus aureus* and *Pseudomonas aeruginosa* diabetic foot bacteria. J Microbiol Biotechnol 26(2):408–420
24. Fan D, Xie R, Liu X, Li H, Luo Z, Li Y, Zeng W (2024) A peptide-based pH-sensitive antibacterial hydrogel for healing drug-resistant biofilm-infected diabetic wounds. J Mater Chem B 12(22):5525–5534
25. Fazli M, Bjarnsholt T, Kirketerp-Møller K, Jørgensen B, Andersen AS, Krogfelt KA, Tolker-Nielsen T (2009) Nonrandom distribution of *Pseudomonas aeruginosa* and *Staphylococcus aureus* in chronic wounds. J Clin Microbiol 47(12):4084–4089

26. Gilaberte Y, Prieto-Torres L, Pastushenko I, Juarranz Á (2016) Anatomy and function of the skin. In: Nanoscience in dermatology. Academic Press, pp 1–14
27. Goy RC, Morais ST, Assis OB (2016) Evaluation of the antimicrobial activity of chitosan and its quaternized derivative on *E. coli* and *S. aureus* growth. Rev Bras Farmacogn 26(1):122–127
28. Gupta A, Kowalczuk M, Heaselgrave W, Britland ST, Martin C, Radecka I (2019) The production and application of hydrogels for wound management: A review. Eur Polym J 111:134–151
29. Hagen CR, Singh A, Weese JS, Marshall Q, Zur Linden A, Gibson TW (2020) In vitro elution of amikacin and Dispersin B from a polymer hydrogel. Vet Surg 49(5):1035–1042
30. Hall CW, Mah TF (2017) Molecular mechanisms of biofilm-based antibiotic resistance and tolerance in pathogenic bacteria. FEMS Microbiol Rev 41(3):276–301
31. Hansson C, Hoborn J, Möller A, Swanbeck G (1995) The microbial flora in venous leg ulcers without clinical signs of infection. Repeated culture using a validated standardised microbiological technique. Acta Derm-Venereol 75(1):24–30
32. Hasan N, Cao J, Lee J, Kim H, Yoo JW (2021) Development of clindamycin-loaded alginate/pectin/hyaluronic acid composite hydrogel film for the treatment of MRSA-infected wounds. J Pharm Invest 51(5):597–610
33. Hernández-Rangel A, Martin-Martinez ES (2021) Collagen based electrospun materials for skin wounds treatment. J Biomed Mater Res, Part A 109(9):1751–1764
34. Hu B, Owh C, Chee PL, Leow WR, Liu X, Wu YL, Chen X (2018) Supramolecular hydrogels for antimicrobial therapy. Chem Soc Rev 47(18):6917–6929
35. Hu J, Quan Y, Lai Y, Zheng Z, Hu Z, Wang X, Cheng Y (2017) A smart aminoglycoside hydrogel with tunable gel degradation, on-demand drug release, and high antibacterial activity. J Control Rel 247:145–152
36. Hu Y, Ding M, Lv X, Jiang J, Zhang J, Yang D (2024) Stimuli-responsive NO delivery platforms for bacterial infection treatment. Adv Healthc Mater 2402240
37. Huang H, Qi X, Chen Y, Wu Z (2019) Thermo-sensitive hydrogels for delivering biotherapeutic molecules: a review. Saudi Pharm J 27(7):990–999
38. Jarman E, Burgess J, Sharma A, Hayashigatani K, Singh A, Fox P (2024) Human-derived collagen hydrogel as an antibiotic vehicle for topical treatment of bacterial biofilms. PLoS ONE 19(5):e0303039
39. Jiang T, Li Q, Qiu J, Chen J, Du S, Xu X, Chen T (2022) Nanobiotechnology: applications in chronic wound healing. Int J Nanomed 17:3125
40. Kaplan JÁ (2010) Biofilm dispersal: mechanisms, clinical implications, and potential therapeutic uses. J Dent Res 89(3):205–218
41. Kaptan Y, Karal-Yilmaz O, Izbudak B, Giray B, Yilmaz B, Bal-Ozturk A (2023) Preparation of tetracycline hydrochloride loaded chitosan/silk fibroin/ZnO antibacterial biocomposite hydrogel sponges for wound healing application. J Polym Res 30(2):49
42. Kasai RD, Radhika D, Archana S, Shanavaz H, Koutavarapu R, Lee DY, Shim J (2023) A review on hydrogels classification and recent developments in biomedical applications. Int J Polym Mater Polym Biomater 72(13):1059–1069
43. Khosravimelal S, Chizari M, Farhadihosseinabadi B, Moosazadeh MM, Gholipourmalekabadi M (2021) Fabrication and characterization of an antibacterial chitosan/silk fibroin electrospun nanofiber loaded with a cationic peptide for wound-dressing application. J Mater Sci Mater Med 32:1–11
44. Larouche J, Sheoran S, Maruyama K, Martino MM (2018) Immune regulation of skin wound healing: mechanisms and novel therapeutic targets. Adv Wound Care 7(7):209–231
45. Laverty G, Gorman SP, Gilmore BF (2012) Antimicrobial peptide incorporated poly (2-hydroxyethyl methacrylate) hydrogels for the prevention of Staphylococcus epidermidis-associated biomaterial infections. J Biomed Mater Res Part A 100(7):1803–1814
46. Li J, Yu F, Chen G, Liu J, Li XL, Cheng B, Pan JF (2020) Moist-retaining, self-recoverable, bioadhesive, and transparent in situ forming hydrogels to accelerate wound healing. ACS Appl Mater Interfaces 12(2):2023–2038

47. Liu Y, Li F, Guo Z, Xiao Y, Zhang Y, Sun X, Wang J (2020) Silver nanoparticle-embedded hydrogel as a photothermal platform for combating bacterial infections. Chem Eng J 382:122990
48. Louie TJ, Bartlett JG, Tally FP, Gorbach SL (1976) Aerobic and anaerobic bacteria in diabetic foot ulcers. Ann Intern Med 85(4):461–463
49. Mai B, Jia M, Liu S, Sheng Z, Li M, Gao Y, Wang P (2020) Smart hydrogel-based DVDMS/bFGF nanohybrids for antibacterial phototherapy with multiple damaging sites and accelerated wound healing. ACS Appl Mater Interfaces 12(9):10156–10169
50. Mallakpour S, Azadi E, Hussain CM (2021) Recent breakthroughs of antibacterial and antiviral protective polymeric materials during COVID-19 pandemic and after pandemic: coating, packaging, and textile applications. Curr Opin Colloid Interface Sci 55:101480
51. Maynard J, Collings C (2015) How wounds heal: the 4 main phases of wound healing. Ostomy and Wound Care Shield HealthCare
52. Mensah A, Rodgers AM, Larrañeta E, McMullan L, Tambuwala M, Callan JF, Courtenay AJ (2023) Treatment of periodontal infections, the possible role of hydrogels as antibiotic drug-delivery systems. Antibiotics 12(6):1073
53. Mingmalairak C (2011) Antimicrobial sutures: new strategy in surgical site infections. Sci Against Microbial Pathogens: Commun Curr Res Technol Adv: Formatex Res Center 13(23):313–323
54. Mohamed MA, Fallahi A, El-Sokkary AM, Salehi S, Akl MA, Jafari A, Cheng C (2019) Stimuli-responsive hydrogels for manipulation of cell microenvironment: From chemistry to biofabrication technology. Prog Polym Sci 98:101147
55. Moosazadeh MM, Eftekhary M, Erfanimanesh S, Hashemi A, Fallah Omrani V, Farhadi-hosseinabadi B, Gholipourmalekabadi M (2018) Comparison of the antibacterial effects of a short cationic peptide and 1% silver bioactive glass against extensively drug-resistant bacteria, *Pseudomonas aeruginosa* and *Acinetobacter baumannii*, isolated from burn patients. Amino acids 50:1617-1628.
56. Morais DS, Rodrigues MA, Lopes MA, Coelho MJ, Maurício AC, Gomes R, Botelho CM (2013) Biological evaluation of alginate-based hydrogels, with antimicrobial features by Ce(III) incorporation, as vehicles for a bone substitute. J Mater Sci Mater Med 24:2145–2155
57. Moreno Ruiz YP, de Almeida Campos LA, Alves Agreles MA, Galembeck A, Macário Ferro Cavalcanti I (2023). Advanced hydrogels combined with silver and gold nanoparticles against antimicrobial resistance. Antibiotics 12(1):104
58. Nazlı AB, Açıkel YS (2019) Loading of cancer drug resveratrol to pH-Sensitive, smart, alginate-chitosan hydrogels and investigation of controlled release kinetics. J Drug Delivery Sci Technol 53:101199
59. Ninan N, Thomas S, Grohens Y (2015) Wound healing in urology. Adv Drug Delivery Rev 82:93–105
60. Nisar P, Ali N, Rahman L, Ali M, Shinwari ZK (2019) Antimicrobial activities of biologically synthesized metal nanoparticles: an insight into the mechanism of action. JBIC J Biol Inorg Chem 24:929–941
61. Olteanu G, Neacşu SM, Joiţa FA, Musuc AM, Lupu EC, Ioniţă-Mîndrican CB, Mititelu M (2024) Advancements in regenerative hydrogels in skin wound treatment: a comprehensive review. Int J Mol Sci 25(7):3849
62. Op't Veld RC, Walboomers XF, Jansen JA, Wagener FA (2020) Design considerations for hydrogel wound dressings: strategic and molecular advances. Tissue Eng Part B 26(3):230–248
63. Osmokrovic A, Jancic I, Vunduk J, Petrovic P, Milenkovic M, Obradovic B (2018) Achieving high antimicrobial activity: composite alginate hydrogel beads releasing activated charcoal with an immobilized active agent. Carbohydr Polym 196:279–288
64. Pagano C, Calarco P, Di Michele A, Ceccarini MR, Beccari T, Primavilla S, Perioli L (2021) Development of sodium carboxymethyl cellulose based polymeric microparticles for in situ hydrogel wound dressing formation. Int J Pharm 602:120606

65. Pan F, Giovannini G, Zhang S, Altenried S, Zuber F, Chen Q, Ren Q (2022) PH-responsive silica nanoparticles for the treatment of skin wound infections. Acta Biomater 145:172–184
66. Pereira RF, Barrias CC, Granja PL, Bartolo PJ (2013) Advanced biofabrication strategies for skin regeneration and repair. Nanomedicine 8(4):603–621
67. Phaechamud T, Yodkhum K, Charoenteeraboon J, Tabata Y (2015) Chitosan–aluminum monostearate composite sponge dressing containing asiaticoside for wound healing and angiogenesis promotion in chronic wound. Mater Sci Eng, C 50:210–225
68. Pyun DG, Choi HJ, Yoon HS, Thambi T, Lee DS (2015) Polyurethane foam containing rhEGF as a dressing material for healing diabetic wounds: synthesis, characterization, in vitro and in vivo studies. Colloids Surf, B 135:699–706
69. Rahim K, Saleha S, Zhu X, Huo L, Basit A, Franco OL (2017) Bacterial contribution in chronicity of wounds. Biomech Model Mechanobiol 73:710–721
70. Raj V, Raorane CJ, Shastri D, Kim SC, Lee S (2024) Engineering a self-healing grafted chitosan–sodium alginate based hydrogel with potential keratinocyte cell migration property and inhibitory effect against fluconazole resistance Candida albicans biofilm. Int J Biol Macromol 261:129774
71. Ranjit E, Hamlet S, George R, Sharma A, Love RM (2022) Biofunctional approaches of wool-based keratin for tissue engineering. J Sci Adv Mater Devices 7(1):100398
72. Rasool U, Sah SK, Hemalatha S (2019) Effect of biosynthesized copper nanoparticles (CUNPS) on the growth and biofilm formation of fluconazole-resistant Candida albicans. J Microbiol Biotechnol Food Sci 9(1)
73. Ren L, He G, Zhou Y, Dai J, Miao W, Ouyang C, Chen G (2022) A hydrogel based on nanocellulose/polydopamine/gelatin used for the treatment of MRSA infected wounds with broad-spectrum antibacterial and antioxidant properties and tissue suitability. Biomater Sci 10(12):3174–3187
74. Ribeiro M, Simões M, Vitorino C, Mascarenhas-Melo F (2024) Hydrogels in cutaneous wound healing: insights into characterization, properties, formulation therapeutic potential. Gels 10(3):188
75. Salvo J, Sandoval C (2022) Role of copper nanoparticles in wound healing for chronic wounds: literature review. Int J Burns Trauma 10:tkab047
76. Sapico FL, Ginunas VJ, Thornhill-Joynes M, Canawati HN, Capen DA, Klein NE, Montgomerie JZ (1986) Quantitative microbiology of pressure sores in different stages of healing. Diagn Microbiol Infect Dis 5(1):31–38
77. Shafiee S, Ahangar HA, Saffar A (2019) Taguchi method optimization for synthesis of Fe_3O_4@ chitosan/Tragacanth gum nanocomposite as a drug delivery system. Carbohydr Polym 222:114982
78. Shi X, Wu Y, Tang L, Yin Z, Shi J, Wu X, Xu Y (2024) Dual-functional composite hydrogel platform: A "kill two birds with one stone" strategy for anti-infection and osseointegration in the treatment of infectious bone defects. Chem Eng J 155337
79. Snetkov P, Rogacheva E, Kremleva A, Morozkina S, Uspenskaya M, Kraeva L (2022) In-vitro antibacterial activity of curcumin-loaded nanofibers based on hyaluronic acid against multidrug-resistant ESKAPE pathogens. Pharmaceutics 14(6):1186
80. Sullivan JV, Myers S (2022) Skin structure and function, wound healing and scarring. In: Plastic surgery-principles and practice. Elsevier, pp 1–14
81. Surjushe A, Vasani R, Saple D (2008) Aloe vera: a short review. Ind J Dermatol 53(4):163–166
82. Tan ST, Dosan R (2019) Lessons from epithelialization: the reason behind moist wound environment. Open Dermatol J 13(1)
83. Tang Y, Xu H, Wang X, Dong S, Guo L, Zhang S, Xu C (2023) Advances in preparation and application of antibacterial hydrogels. J Nanobiotechnol 21(1):300
84. Tetz GV, Artemenko NK, Tetz VV (2009) Effect of DNase and antibiotics on biofilm characteristics. Antimicrob Agents Chemother 53(3):1204–1209
85. Thorn CR, Howell PL, Wozniak DJ, Prestidge CA, Thomas N (2021) Enhancing the therapeutic use of biofilm-dispersing enzymes with smart drug delivery systems. Adv Drug Delivery Rev 179:113916

86. Tian J, Wong KK, Ho CM, Lok CN, Yu WY, Che CM, Tam PK (2007) Topical delivery of silver nanoparticles promotes wound healing. ChemMedChem: Chem Enabling Drug Discovery 2(1):129–136
87. Tottoli EM, Dorati R, Genta I, Chiesa E, Pisani S, Conti B (2020) Skin wound healing process and new emerging technologies for skin wound care and regeneration. Pharmaceutics 12(8):735
88. Turky G, Moussa MA, Hasanin M, El-Sayed NS, Kamel S (2021) Carboxymethyl cellulose-based hydrogel: dielectric study, antimicrobial activity and biocompatibility. Arabian J Sci Eng 46:17–30
89. Ullah K, Sohail M, Buabeid MA, Murtaza G, Ullah A, Rashid H, Khan SA (2019) Pectin-based (LA-co-MAA) semi-IPNS as a potential biomaterial for colonic delivery of oxaliplatin. Int J Pharm 569:118557
90. Vanić Ž, Škalko-Basnet N (2020) Hydrogels as intrinsic antimicrobials. In: Hydrogels based on natural polymers. Elsevier, pp 309–328
91. Velnar T, Bailey T, Smrkolj V (2009) The wound healing process: an overview of the cellular and molecular mechanisms. J Int Med Res 37(5):1528–1542
92. Wallace HA, Basehore BM, Zito PM (2017) Wound healing phases
93. Wang K, Zhu K, Zhu Z, Shao F, Qian R, Wang C, Zhao J (2023) Triptolide with hepatotoxicity and nephrotoxicity used in local delivery treatment of myocardial infarction by thermosensitive hydrogel. J Nanobiotechnol 21(1):227
94. Wang X, Liu M, Yu C, Li J, Zhou X (2023) Biofilm formation: mechanistic insights and therapeutic targets. Mol Biomed 4(1):49
95. Wang Y, Guan A, Isayeva I, Vorvolakos K, Das S, Li Z, Phillips KS (2016) Interactions of Staphylococcus aureus with ultrasoft hydrogel biomaterials. Biomaterials 95:74–85
96. Wekwejt M, Małek M, Ronowska A, Michno A, Pałubicka A, Zasada L, Kaczmarek-Szczepańska B (2024) Hyaluronic acid/tannic acid films for wound healing application. Int J Biol Macromol 254:128101
97. Wikanta T, Erizal M, Tjahyono M, Sugiyono M (2012) Synthesis of polyvinyl alcohol-chitosan hydrogel and study of its swelling and antibacterial properties. Squalen Bull Marine Fisher Postharvest Biotechnol 7(1):1–10
98. Xin H, Naficy S (2022) Drug delivery based on stimuli-responsive injectable hydrogels for breast cancer therapy: a review. Gels 8(1):45
99. Xu J, Chang L, Xiong Y, Peng Q (2024) Chitosan-based hydrogels as antibacterial/antioxidant/anti-inflammation multifunctional dressings for chronic wound healing. Adv Healthcare Mater 13(30):2401490
100. Xu Z, Liang B, Tian J, Wu J (2021) Anti-inflammation biomaterial platforms for chronic wound healing. Biomater Sci 9(12):4388–4409
101. Yang K, Han Q, Chen B, Zheng Y, Zhang K, Li Q, Wang J (2018) Antimicrobial hydrogels: promising materials for medical application. Int J Nanomed pp. 2217–2263.
102. Yang T, Li D, Yan Y, Ettoumi FE, Wu RA, Luo Z, Lin X (2023) Ultrafast and absolute quantification of SARS-CoV-2 on food using hydrogel RT-LAMP without pre-lysis. J Hazard Mater 442:130050
103. Zhong Y, Xiao H, Seidi F, Jin Y (2020) Natural polymer-based antimicrobial hydrogels without synthetic antibiotics as wound dressings. Biomacromol 21(8):2983–3006
104. Zhou Q, Si Z, Wang K, Li K, Hong W, Zhang Y, Li P (2022) Enzyme-triggered smart antimicrobial drug release systems against bacterial infections. J Control Rel 352:507–526

Future Scopes of Hydrogels-Based Wound Healing

Mediha Gurel

Abstract Wound healing is a complex and dynamic biological process characterized by a series of highly coordinated events aimed at repairing skin integrity following injury. Despite recent advancements in wound management technologies, traditional methods continue to face significant challenges, including subjective assessments of healing, the diverse etiologies of wounds, and limitations in existing technologies. While significant strides have been made in understanding and managing complexities of wound healing, challenges persist in achieving effective, personalized care. The future of wound healing presents a paradigm shift driven by innovative advancements in material science, biotechnology, and regenerative medicine. One of the prominent areas of development includes smart dressings equipped with biosensors, which allow real-time monitoring of wound conditions. In the realm of biotechnology, the integration of growth factors and stems cell therapies provides promising alternative for enhancing wound healing. Ultimately, the intersection of these multidisciplinary innovations, from the use of smart bandages incorporating biosensor technology to advanced synthetic materials, are revolutionizing how wounds are monitored and treated. The incorporation of nanotechnology has markedly advanced the fields of cell proliferation and tissue regeneration, while controlled drug delivery systems are paving the way for personalized, on-demand therapies. These advancements not only promise to accelerate healing but also improve patient comfort, reduce healthcare costs, and address the underlying biological impairments in chronic wounds. As wound care continues to evolve, the wound healing field that increasingly necessitates the integration of personalized medicine and smart technologies to improve patient outcomes and elevate standards of care.

Keywords Wound healing · Skin integrity · Injury repair · Biological process · Healing challenges

M. Gurel (✉)
Biotechnology Research and Application Center, Cukurova University, Adana, Turkey
e-mail: medihagurell@gmail.com

Electronic and Automation Department, Bitlis Eren University, Bitlis, Turkey

A. K. Mandal et al. (eds.), *Functionalized Nanoparticles Hydrogels for Wound Healing*, Smart Nanomaterials Technology, https://doi.org/10.1007/978-981-95-0720-7_12

1 Introduction

Wounds, defined as injuries to the epithelial lining of the skin resulting from physical damage can have profound implications not only on physical health but also on mental well-being. Wound healing is a multifaceted and complex biological process that involves a series of dynamic coordinated cellular events aimed at restoring tissue integrity following injury. This process can be divided into distinct but overlapping phases: hemostasis, inflammation, proliferation, and remodeling. Each phase is characterized by specific cellular activities, signaling pathways, and extracellular matrix (ECM) interactions for restoring the integrity and functionality of the damages tissue. The interdependence of these phases means that any disturbance or imbalance at any stage can significantly hinder the overall healing process and potentially lead to failures in healing and chronic wounds.

In recent years, the field of wound healing has witnessed remarkable advancements. The integration of technology, evidence-based practices, and standardized assessment tools has improved the accuracy and consistency of wound evaluations that further facilitates decision-making in wound dressing selection, thereby promoting adherence to best practices [4, 42, 90]. Despite these recent progresses, the field of wound healing has consistently posed challenges for the medical community which can be categorized into several domains including, subjectivity in wound assessment, complexity of wound types, and technological limitations. Wound assessment has relied heavily on the clinician's subjective interpretation of wound characteristics and variability in assessment criteria can lead to inconsistent treatment approaches and patient outcomes [6]. While wounds can be classified into various categories, each type presents unique assessment and healing challenges, particularly chronic wounds, which often involve underlying comorbidities such as diabetes and vascular disease [97].

The future of wound healing is poised for transformative advancements, driven by innovations in materials science, biotechnology, regenerative medicine, and advanced therapies. As the field evolves, the integration of novel technologies and methodologies will significantly enhance the speed and efficacy of wound healing while simultaneously addressing complications and improving the overall quality of healing. This synthesis of emerging trends and technologies is crucial for understanding the trajectory of wound care in the coming decades.

Wound dressings are integral to the healing process, serving multiple functions that not only also protect the wound but facilitate the regeneration of tissue. One of the primary roles of wound dressings is to act as a barrier against external contaminants, thereby preventing infection. Studies have shown that dressings can significantly reduce the risk of wound contamination by providing a protective layer that inhibits the entry of bacteria and other pathogens [86, 92]. In addition to protection, wound dressings play a critical role in creating an optimal environment for healing with contributing the moist wound healing theory that posits to be maintaining a proper environment for tissue regeneration and repair [9, 78]. Research indicates that modern dressings, such as hydrocolloids and hydrogels, can maintain moisture levels that are conducive to healing, thereby reducing necrosis and promoting faster recovery [78, 110]. Hydrocellular foam dressings have been shown to decrease inflammation and enhance granulation tissue formation, which is vital for the healing process [110]. This moist environment not only aids in the healing of the wound but also minimizes pain and discomfort associated with dressing changes, as moist dressings are less likely to adhere to the wound bed compared to traditional dry dressings [76, 112].

The most promising developments in wound healing are encouraged using advanced technologies for dressings that incorporate with sensors which can provide real-time data on wound moisture levels, temperature, and degree of infection, allowing for timely dressing changes and minimizing the risk of infection and complications associated with prolonged exposure to exudates. Milne et al. demonstrated that dressings designed to effectively manage exuding wounds can lead to better treatment efficiencies, as the healing environment remains undisturbed for longer periods [67]. This approach not only optimizes the healing process but also enhances patient comfort and reduces healthcare costs associated with frequent dressing changes. The development of synthetic materials that mimic the extracellular matrix (ECM) has shown great promise in enhancing wound healing. Abicht et al. evaluated a synthetic hybrid-scale fiber matrix designed to facilitate cellular infiltration and neovascularization, leading to improved healing rates in diabetic foot ulcers [1]. Their study reported that 75% of wounds achieved complete healing within 12 weeks, underscoring the potential of engineered materials to revolutionize wound care. Such advancements not only provide a scaffold for tissue regeneration but also address the underlying biological processes that contribute to impaired healing. The exploration of biological therapies, particularly stem cell applications, represents another frontier in wound healing. Li et al. demonstrated that mesenchymal stem cells (MSCs) modified with angiopoietin-1 gene significantly promote wound healing by enhancing angiogenesis and tissue regeneration [54]. This approach addresses chronic wounds that are notoriously difficult to heal, such as diabetic ulcers and radiation-induced injuries. The ability to harness the regenerative potential of stem cells offers a powerful strategy for overcoming the limitations of traditional wound care methods. Moreover, the integration of nanotechnology into wound care is paving the way for innovative therapeutic solutions. Kiiskinen et al. explored the use of wood-derived nanofibrillar cellulose as a wound dressing material that supports the growth of human MSCs without the need for cell adhesion coatings [44]. This biocompatible material not only enhances cell viability but also promotes effective healing by providing a conducive environment for cell proliferation and tissue repair. The application of nanotechnology in wound dressings exemplifies how interdisciplinary approaches can lead to significant advancements in patient care. The role of growth factors in wound healing is also gaining attention, with research indicating that localized delivery of these factors can enhance healing rates. Mostafalu et al. discussed the potential of textile dressings for controlled drug delivery, allowing for the on-demand release of anti-inflammatory agents and growth factors [72]. This targeted approach minimizes systemic side effects while maximizing therapeutic efficacy, representing a significant shift in how wound care is approached. As the field continues to evolve, the importance of personalized medicine in wound healing cannot be overstated.

Tailoring therapies to individual patient needs, particularly in the context of chronic wounds, is essential for improving outcomes. The work of Liu and Lu underscores the importance of re-evaluating therapeutic strategies for diabetic patients, integrating advancements from biotechnology and molecular biology that contribute to enhanced personalized approaches to wound care [60]. This focus on individualized care aligns with broader trends in healthcare, where patient-centered approaches are increasingly prioritized.

As we look to the future, the convergence of various disciplines—ranging from materials science to biotechnology—will continue to shape the landscape of wound healing. Several breakthrough treatments are entering the market, including advanced wound dressings and targeted therapies that address the underlying causes of impaired healing [25]. These innovations are set to redefine the standards of care and improve patient outcomes significantly (Fig. 1).

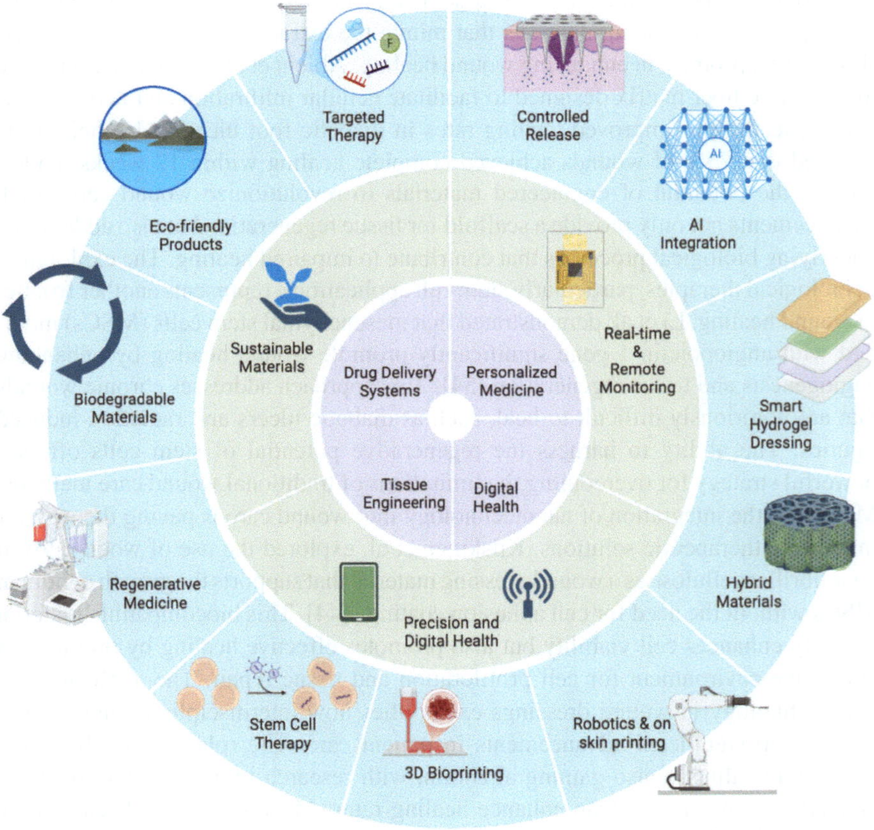

Fig. 1 Overall schematic of the future scopes of hydrogel dressings for wound management (Figure was created with Biorender)

2 Advanced Drug Delivery Systems of Hydrogel-Based Wound Healing for Future Applications

Advanced drug delivery systems (ADDS) leverage innovative materials and technologies to ensure localized, sustained, and controlled release of therapeutic agents, thereby addressing the multifaceted challenges associated with wound management. In addressing the challenges of wound healing, it needs to consider the dynamic nature of the wound environment and future ADDS must be designed to adapt to changes in the wound's biochemical and physical conditions. The importance of these systems was emphasized as they can deliver active factors in appropriate dosages to the healing tissue in a controlled manner, which helps enhancing therapeutic outcomes [88].

Precision Delivery and Targeted Therapy for Personalized Medicine Treatments

The integration of ADDS with personalized medicine represents a significant shift in healthcare, moving from a traditional "one-size-fits-all" model to a more nuanced, patient-centered approach. This transformation is driven by the need to tailor therapeutic interventions to individual patient profiles. ADDS facilitate this personalization by enabling precise control over the timing, location, and dosage of drug administration, therefore enhancing therapeutic efficacy, and minimizing adverse effects based on individual patient characteristics, including genetic makeup, disease state, and environmental factors. This approach is particularly relevant in the context of chronic wounds, where traditional treatment modalities often fall short due to the complex interplay of factors that impede healing [94]. As research progresses, it is expected that new materials and technologies will emerge, further enhancing the capabilities of wound care products. The development of injectable self-healing hydrogels that can release bioactive factors on demand represents a promising avenue for future research [89]. Such innovations will not only improve the efficiency of drug delivery but also address the challenges associated with chronic wounds, ultimately leading to better patient outcomes.

ADDS represent an important leap in the field of pharmacology, enabling targeted drug delivery to specific tissues, cells, or molecules. This capability is particularly crucial in the context of personalized medicine for wound healing technologies, which utilizes genetic, proteomic, or metabolomic data to tailor treatments to individual patients and the use of biocompatible materials that can be engineered to release drugs in a controlled manner. Studies demonstrated a flexible transient bioelectronic system capable of multifunctional active-controlled drug delivery, which can be tailored to promote endothelial repair while inhibiting smooth muscle proliferation, a common issue in vascular repair processes [61, 65]. This dual-action capability is essential in preventing complications and highlighting the importance of targeted drug delivery systems in managing complex medical conditions. Hydrogel microneedles (HMNs) is the technology to be considered as future of wound healing techniques that can be combined with smart devices to achieve precise control over drug release rates, thereby addressing the issue of sudden drug release that can hinder healing processes [38]. The ability to regulate drug delivery based on the specific requirements of the wound represents a significant advancement in wound care technology [38]. Furthermore, hydrogels can be engineered to respond to physiological stimuli, enhancing their functionality as drug delivery systems [89].

The integration of electronic devices into drug delivery systems has also opened new avenues for selective drug administration and the development of programmable systems that allow for the controlled release of multiple drugs simultaneously which has significantly improved the precision and effectiveness of drug administration in wound healing scenarios for individuals. One of the most promising advancements in this field is the development of battery-free and wireless smart wound dressings. These innovative dressings not only facilitate on-demand drug delivery but also integrate monitoring and diagnostic capabilities, creating a closed-loop medical system. Xu et al. highlight that such systems can utilize specific sensor parameters to automate drug delivery, enhancing the treatment of wound infections and promoting healing processes through multimodal electrically controlled therapies [109]. This new integrated systems with personalized medicine technologies allows for precise real-time monitoring of wound conditions, enabling timely interventions that can significantly improve patient outcomes.

By employing nanoparticle-mediated drug delivery, researchers have been trying to be able to overcome some of the limitations of traditional drug delivery methods. The use of nanomedicine offers a promising approach to optimize drug absorption in specific areas of the body while minimizing exposure to healthy tissues [69]. This technology allows for a more precise targeting mechanism, ensuring that therapeutic agents are delivered directly to the site of injury or infection, which is particularly beneficial in wound healing applications. The development of solid lipid nanoparticles (SLNs) has emerged as a promising approach for combinational drug therapy in cancer and wound healing contexts. Lala et al. note that SLNs can enhance drug bioavailability while reducing systemic toxicity, which is particularly important in managing wounds that may be complicated by underlying conditions such as cancer [48]. The ability of these functional nanoparticles to encapsulate multiple therapeutic agents allows for synergistic effects, potentially improving healing outcomes in complex wound scenarios. The use of polymer nanoparticles and other nanocarriers has also been extensively studied for their ability to enhance drug delivery precision. The advantages of using aptamer-modified polymer nanoparticles, which can selectively target specific tissues with high performance, thereby improving the therapeutic index of the delivered drugs [71]. This targeted approach is particularly beneficial in wound healing, where the goal is to concentrate therapeutic agents at the site of injury while minimizing exposure to surrounding healthy tissues. Moreover, the exploration of peptide-drug conjugates has opened new pathways for targeted drug delivery systems. Zhu et al. discuss how these conjugates can enhance the accumulation of drugs in tumor tissues, which is relevant for wound healing in cancer patients [116]. By leveraging the unique properties of peptides, researchers can design drug delivery systems that are not only effective but also tailored to the specific needs of patients with complex wound healing challenges.

The role of biomaterials in drug delivery systems cannot be overstated. Barrs et al. discuss the importance of using biocompatible materials that can facilitate the controlled release of therapeutic agents while also supporting tissue regeneration [5]. The dynamic interactions between cells and the matrix provided by these biomaterials can significantly influence the healing process. By adjusting the physical properties of the materials used, researchers can create environments that mimic natural tissue characteristics, thereby enhancing the efficacy of drug delivery systems in wound healing applications. In addition to traditional drug delivery methods, the emergence of stimuli-responsive systems has opened new avenues for precision medicine. These systems can release drugs in response to specific environmental triggers, such as pH changes or temperature variations. Feng et al. illustrate how such platforms can be utilized for postoperative treatments, allowing for controlled drug release that aligns with the body's healing processes [23]. This sustained release is essential for managing chronic wounds, where prolonged drug action is often required to combat infections and promote healing. The ability to tailor the release profiles of these systems enhances their effectiveness in delivering therapeutic agents precisely and locally when and where they are needed [100]. This adaptability not only improves therapeutic outcomes but also minimizes the risk of systemic side effects, making these systems particularly suitable for sensitive applications. Yang et al. present a novel MXene-based hydrogel system that exhibits intelligent drug delivery capabilities, allowing for targeted and controlled release of therapeutic agents [111]. This system is designed to respond to environmental cues, thereby enhancing the healing process while minimizing the risk of adverse effects associated with systemic drug administration.

Advanced imaging techniques of drug delivery systems have also been considered to improve the precision of drug targeting with emphasizing the importance of understanding the biodistribution of nanoparticle formulations to improve their therapeutic efficacy [12]. By employing multimodal imaging approaches, researchers can track the accumulation of drugs at the target site, ensuring that the delivery system is functioning as intended, and optimize treatment regimens and ensuring that patients receive the most effective therapies for their wounds.

3 Incorporating of Biologically Active Molecules

The incorporation of biologically active molecules into these hydrogels enhances their therapeutic potential, promoting faster and more efficient healing through various mechanisms, including anti-inflammatory effects, antimicrobial activity, tissue regeneration, and angiogenesis. Hydrogels can mimic the extracellular matrix (ECM), facilitating cell–hydrogel interactions that are vital for tissue repair [58]. This ECM-like structure not only provides mechanical support but also enhances the proliferation and migration of cells involved in wound healing, as incorporating bioactive molecules into these gels can significantly enhance their healing capabilities collaborating with the cells [53, 51, 99].

Growth factors in ECM stimulates proliferation, differentiation, and tissue repair which are mainly Epidermal Growth Factor (EGF), Vascular Endothelial Growth Factor (VEGF), and Platelet-Derived Growth Factor (PDGF) and particularly significant for the stages of wound healing. EGF is known to accelerate skin cell proliferation and migration, thereby enhancing wound closure [14, 33]. VEGF is essential for angiogenesis, facilitating the formation of new blood vessels in the wound area, which is critical for supplying nutrients and oxygen necessary for healing [33, 34]. PDGF promotes cell migration and collagen production, both of which are vital for effective wound closure and tissue regeneration [45, 117].

Biodegradable hydrogels have been developed that can release growth factors over an extended period, thereby increasing their therapeutic efficacy [26, 36]. This sustained release is particularly important as it allows for prolonged exposure of cells to growth factors, which can lead to improved outcomes in wound healing [33, 41]. Additionally, the physical properties of hydrogels, such as their porosity and water content, can be tailored to optimize the release rates of the encapsulated growth factors [39]. Studies have demonstrated that hydrogels functionalized with EGF can improve the proliferation of epidermal cells, thereby promoting faster healing [14, 37] Similarly, VEGF-loaded hydrogels have been shown to effectively stimulate angiogenesis, which is crucial for the healing of larger wounds [34, 41]. Furthermore, PDGF has been incorporated into keratin hydrogels, which not only supports the sustained release of growth factors but also enhances the overall healing efficacy [68, 117].

The future of wound healing is increasingly intertwined with the development of advanced biomaterials, particularly hydrogels that incorporate antimicrobial agents. These innovations aim to address the significant challenge of wound infections, which can severely impede the healing process. Among these agents, silver nanoparticles (AgNPs) and antibiotics are prominent due to their efficacy in combating microbial infections and promoting tissue regeneration. They have garnered attention for their broad-spectrum antimicrobial properties. Studies have demonstrated that AgNPs can be effectively integrated into hydrogels to enhance their wound healing capabilities. Liu et al. reported that a hydrogel composed of silver nanoparticles, aloe vera, and silk fibroin significantly accelerated wound closure in rat models, achieving a healing rate of 97.02% by promoting wound contraction rather than re-epithelialization [62]. Similarly, Chinnasamy et al. highlighted the effectiveness of AgNPs synthesized from *Azadirachta indica* in stimulating wound healing processes, including remodeling and re-epithelialization, thereby validating their use in wound dressings [13]. Furthermore, Papaioannou et al. emphasized that AgNPs are widely utilized in various biomedical products to resist infections and expedite healing, showcasing their versatility and importance in wound care [82]. In addition to nanoparticles, the incorporation of traditional antibiotics into hydrogels presents another effective strategy for infection control. The development of antimicrobial collagen hydrogels that release antibiotics like levofloxacin has shown promise in inhibiting biofilm formation, a common barrier to effective in vitro wound healing [108]. This is particularly crucial as biofilms can significantly delay recovery by protecting bacteria from conventional treatments [80]. Moreover, studies have indicated that the use of antibiotic-loaded hydrogels can lead to improved healing outcomes compared to standard treatments, as seen in the clinical trial conducted by Pathi, which demonstrated enhanced wound healing with silver nanoparticle formulations compared to traditional antibiotics [83].

The design of hydrogels also plays a critical role in their effectiveness. The ability of hydrogels to swell and maintain moisture is vital for creating an optimal healing environment. Research by Chuah et al. showed that hybrid hydrogels with enhanced swelling capabilities can significantly improve the biological properties necessary for effective wound healing [15]. Additionally, the incorporation of natural polymers, such as chitosan and collagen, into hydrogel formulations has been shown to enhance antimicrobial properties while promoting tissue regeneration [21, 22]. These natural polymers not only provide structural support but also possess inherent antimicrobial properties that can further aid in infection control.

4 Smart and Responsive Hydrogels Integration with Digital Health

Smart and responsive hydrogels, especially when integrated with biosensors, offer an innovative approach to enhancing wound care in biomedical applications, by providing **real-time monitoring** of key physiological markers. These hydrogels can be integrated with digital health technology to monitor critical parameters such as **pH, oxygen levels, and specific biomarkers**, providing valuable feedback for wound management. The ability of hydrogels to respond to environmental stimuli is a fundamental characteristic that enhances their functionality in biomedical applications. For instance, **pH-**responsive hydrogels can undergo significant volume changes in response to variations in pH levels, which can be crucial for monitoring wound healing environments where pH can indicate infection or healing status [32, 107]. The swelling behavior of these hydrogels can be finely tuned to achieve rapid equilibrium, allowing for timely monitoring of wound conditions [107]. Moreover, the integration of conductive materials within hydrogels has been shown to enhance their responsiveness, making them suitable for real-time biosensing applications [10, 18].

In addition to pH, other physiological markers such as **oxygen** levels can also be monitored using smart hydrogels. Hydrogels that incorporate oxygen-sensitive dyes can provide visual feedback on the oxygenation status of a wound, which is critical for assessing healing progress [10, 32]. The incorporation of biosensors into hydrogels allows for the development of multifunctional platforms that can simultaneously monitor multiple parameters, thereby providing a comprehensive overview of the wound healing [32]. Certain hydrogels can be engineered to release therapeutic agents in response to elevated levels of inflammatory cytokines, thereby allowing for targeted treatment of infections [95]. The integration of nanosensors within these hydrogels enhances their functionality, also enabling continuous monitoring of oxidative stress and other biochemical markers [87]. This capability is crucial, as it allows healthcare providers to detect infections early and adjust treatment protocols accordingly. Recent advancements in hydrogel technology have led to the development of systems that can not only sense but also respond to the wound environment dynamically. Hydrogels that incorporate lipase-responsive components can identify bacterial infections by detecting specific lipid profiles associated with pathogens [93]. Moreover, the use of hybrid hydrogels that combine natural and synthetic polymers is gaining traction. These materials leverage the beneficial properties of both types of polymers, enhancing biocompatibility and functionality [102]. The incorporation of proteins and polysaccharides can improve the hydrogels' ability to mimic the extracellular matrix, thereby facilitating better cell adhesion and proliferation [8]. This biomimetic approach not only aids in wound healing but also allows for the incorporation of sensors that can detect specific biomarkers, such as elevated levels of inflammatory cytokines, providing real-time feedback on the wound's status [73].

The mechanical properties of **smart hydrogels**, such as their flexibility and biocompatibility, make them particularly suitable for applications in tissue engineering and wound care. These properties enable the hydrogels to conform to the dynamic environments such as wounds that occur on the movable parts of the body, ensuring effective monitoring and support for healing tissues [105, 115]. A critical aspect of wound dressings for movable parts, particularly in areas prone to flexion and extension, must include flexibility and strength to adapt to the irregular contours of wounds. Silk fibroin/ polycaprolactone nanofibrous membranes have been shown to provide the necessary mechanical support while maintaining flexibility, which is crucial for areas with high mobility [49]. Additionally, the incorporation of materials such as hydrocolloids has demonstrated improved tensile strength and elongation at break, outperforming traditional dressings in terms of skin adhesion and exudate management [40]. Additionally, the integration of flexible electronics into these dressings would revolutionize their functionality, where traditional dressings may fail to provide adequate monitoring due to their static nature.

The development of hydrogels that can respond to multiple stimuli—such as temperature, light, and mechanical stress—further enhances their utility in real-time monitoring applications [103]. The integration of biosensors into smart and responsive hydrogels represents a significant advancement in the field of wound care. These hydrogels not only provide real-time monitoring of critical healing markers but also possess the mechanical and chemical properties necessary for effective application in biomedical settings. The ongoing research and development in this area promise to enhance the capabilities of hydrogels, making them indispensable tools in modern medicine.

The future of wound healing is poised for significant advancements through the integration of smart hydrogel systems and artificial intelligence (AI). These technologies promise to enhance remote monitoring of wounds and optimize treatment strategies tailored to individual patient needs. Smart hydrogels represent a transformative approach in wound care, particularly through their capability to enable remote monitoring. These hydrogels can be designed to transmit real-time healing data to healthcare providers, enabling timely interventions. For instance, hydrogels that maintain a moist environment at the wound site are crucial for promoting tissue regeneration and preventing infections, which are vital for effective healing [11]. The incorporation of sensors within these hydrogels can allow for the continuous assessment of wound conditions, such as pH levels and temperature, which are indicators of healing progress [114]. This capability not only enhances patient care but also empowers patients by involving them in their own healing process, as they can receive feedback and guidance remotely [4]. Moreover, the application of AI in conjunction with these smart hydrogels can further refine the monitoring process. AI algorithms can analyze data collected from the hydrogels to predict healing outcomes and identify potential complications early [55, 52]. This predictive capability is essential for chronic wounds, which often require prolonged management and can lead to significant healthcare costs if not monitored effectively [27, 77].

The role of AI in the design of hydrogels is equally promising. AI-driven methodologies can optimize hydrogel formulations tailored to specific wound types or patient demographics. By leveraging machine learning techniques, researchers can analyze vast datasets to identify the most effective combinations of materials and properties for different wound healing scenarios [75]. AI can assist in developing hydrogels that not only provide healing environment but also release therapeutic agents in a controlled manner, enhancing the healing process. Additionally, AI can facilitate the classification of wounds based on their characteristics, such as the presence of infection or the type of tissue involved, which is critical for determining the appropriate treatment strategy [35, 63]. This capability can significantly reduce the burden on healthcare professionals by streamlining the assessment process and allowing for more personalized care plans [63] (Fig. 2).

5 Tissue Engineering and Hybrid Materials

The wound healing is increasingly being shaped by the integration of advanced technologies, particularly through the use of bioengineered and hybrid materials that leverage 3D bioprinting and sophisticated biomaterial composites. These innovations are enhancing the efficacy of wound healing processes and addressing the complexities associated with various wound types, including chronic and diabetic wounds.

3D bioprinting technology has emerged as a transformative approach in the field of wound healing and allows for the precise placement of cells and biomaterials, creating scaffolds that closely mimic the natural extracellular matrix (ECM). Studies have demonstrated that 3D bioprinted patches can significantly enhance wound healing dynamics in situ by promoting angiogenesis, reducing inflammation, and minimizing fibrosis with using robotic printers [50, 106]. The integration of robotics and on-skin printing technologies for wound healing highlights the potential of in situ bioprinting systems that can replicate the layered structure of skin with delivering autologous skin cells directly to wound sites. Precise in situ delivery of autologous dermal fibroblasts and epidermal keratinocytes in a hydrogel carrier with

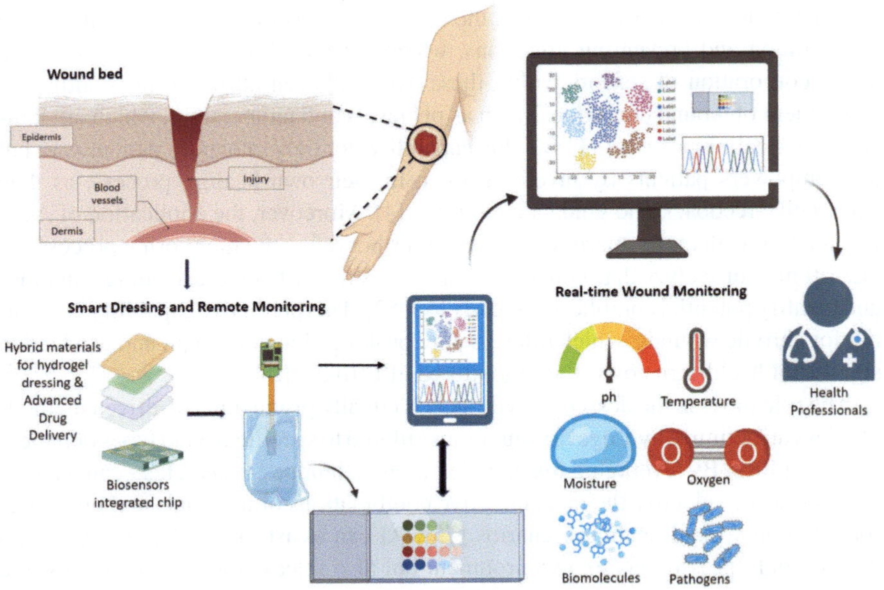

Fig. 2 Smart and responsive wound dressing with integrated components of digital health for real-time microenvironment monitoring (Figure was created with Biorender)

using custom skin bioprinter resulted in rapid wound closure and reduced contraction in extensive excisional wounds [3]. The ability to customize these dressings with bioactive molecules further enhances their effectiveness, as they can be designed to release growth factors that stimulate tissue regeneration [79].

The incorporation of various biomaterials into these 3D-printed constructs is crucial for optimizing their properties. Materials such as hydrogels, cellulose, and alginate have been shown to possess excellent biocompatibility and moisture-retaining capabilities, which are essential for maintaining a conducive healing environment [96]. The use of hybrid materials that combine different biomaterials can also lead to improved mechanical properties and bioactivity, thereby enhancing the overall performance of the wound dressings [30]. The development of polysaccharide-peptide cryogels has been highlighted as a promising strategy for managing infected wounds while promoting hemostasis [31]. These innovations, including the combinations of 3D bioprinting and the development of synthetic hybrid-scale fiber matrices, are revolutionizing the wound management, especially for those challenging-to-heal.

Hybrid materials, such as collagen and silk fibroin composites, have shown significant promise in enhancing wound healing. These materials create a moisture-rich microenvironment that is crucial for effective healing. Research indicates that maintaining moisture at the wound site accelerates angiogenesis and collagen deposition, which are vital for tissue repair [64]. The use of synthetic hybrid-scale fiber matrices has also been demonstrated to significantly improve healing outcomes in chronic wounds, including diabetic foot ulcers. A study found that 75% of patients achieved complete healing within 12 weeks when treated with these advanced materials, highlighting their effectiveness in managing complex wound scenarios [1]. Moreover, the application of negative pressure wound therapy (NPWT) has been a game-changer in chronic wound management. NPWT promotes healing by removing exudate and infectious materials while enhancing blood flow and granulation tissue formation [28]. This therapy, combined with hybrid materials, can lead to synergistic effects that further enhance healing rates and reduce complications associated with chronic wounds.

The incorporation of nanotechnology into wound care is another frontier that holds great potential. Nanomaterials, such as silver nanoparticles and MXene nanosheets, have been shown to possess antimicrobial properties and can be integrated into hydrogels to create multifunctional dressings that not only promote healing but also prevent infection [43, 56]. Furthermore, the development of injectable hydrogels that are adhesive and hemostatic represents a significant advancement in wound care. These hydrogels can bond to damaged tissues, providing a protective barrier while facilitating the healing process [57]. Their ability to deliver therapeutic agents in a controlled manner enhances their utility in treating various wound types, including those associated with diabetes and other chronic conditions [70].

6 Integration with Regenerative Medicine

The integration of regenerative medicine into wound healing represents a transformative advancement in medical science, leveraging innovative technologies and biological insights to enhance healing processes. Regenerative medicine encompasses a variety of strategies, including the use of stem cells, biomaterials, and growth factors, all aimed at promoting tissue repair and regeneration.

One of the most promising approaches in regenerative medicine is the application of mesenchymal stem cells (MSCs) and their derivatives. Research has shown that conditioned media derived from MSCs can significantly enhance wound healing by promoting cellular migration and proliferation, as evidenced by in vitro studies utilizing scratch assays and transepithelial electrical resistance (TEER) analysis [2]. MSCs have been demonstrated to facilitate cutaneous wound healing through mechanisms such as collagen synthesis and angiogenesis, which are critical for tissue repair [113]. This aligns with findings that highlight the role of MSCs in modulating inflammation and promoting tissue regeneration, particularly in chronic wound scenarios [19]. The development of advanced biomaterials combined with stem cell application has emerged as a pivotal aspect of regenerative medicine in wound healing. While these hydrogels can be engineered to release bioactive factors, they possess unique properties that allow for controlled drug delivery and enhanced healing environments with enhanced cell differentiation capabilities [89]. Additionally, the versatility of hydrogels enables their use as scaffolds for tissue engineering, providing structural support while facilitating cellular interactions necessary for regeneration [104]. The exploration of natural polymers, such as collagen and keratin, has gained traction in regenerative medicine and these biomaterials not only provide structural integrity but also possess bioactive properties that can enhance cellular responses during the healing process [47]. Keratin, in particular, has been highlighted for its role in promoting keratinocyte differentiation, epithelial cell migration and proliferation, which are essential for effective wound closure [20, 46].

7 Sustainability and Biodegradability

The principles of sustainability and biodegradability, as researchers strive to develop advanced wound dressings that not only promote healing but also minimize environmental impact. Biodegradable wound dressings are designed to break down naturally in the body, which can enhance the healing process by reducing the need for frequent dressing changes and associated patient discomfort. Biodegradable materials such as poloxamer and mastic gum have been shown to facilitate drug release and improve healing outcomes, thereby addressing both clinical and patient-centered needs [7]. The introduction of fully biodegradable dressings has been reported to promote wound healing and mitigate environmental pollution, aligning with sustainable development goals [55, 52]. Polysaccharide-based hydrogels represent a significant advancement in the field of wound dressings due to their inherent properties such as high-water retention, biocompatibility, and tunable biodegradability. These materials can be tailored to meet the specific requirements of various wound types, making them versatile options for both acute and chronic wounds [16]. Additionally, the incorporation of bioactive compounds into these hydrogels can enhance their therapeutic efficacy, as evidenced by studies demonstrating improved healing rates when herbal extracts are included [53, 51]. The challenge remains, however, to ensure that these materials maintain consistent mechanical properties and prevent microbial contamination while optimizing their biodegradability [85]. Recent innovations in wound dressing technologies, such as sprayable biodegradable dressings with antimicrobial properties, have shown promise in enhancing healing outcomes for larger wounds [17]. The support of biodegradable scaffolds, like those made from silk fibroin, has been highlighted for their ability to enhance tissue regeneration while being biocompatible and environmentally friendly [66]. These materials not only facilitate healing but also contribute to a more sustainable approach to wound management by reducing waste associated with traditional dressings.

In addition to the materials used, the design of wound dressings is critical for their effectiveness. For instance, the development of Janus sponges with biodegradable amphiphilic polyurethanes (PUs) that can absorb exudates while maintaining a moist environment has been shown to improve healing outcomes without causing additional trauma to the wound site [98]. The integration of eco-friendly bio-nanocomposites into these sustainable wound dressings has also been explored, providing enhanced properties such as controlled drug release and improved antibacterial activity, which are essential for preventing infections and promoting faster healing [74]. The future of wound healing is increasingly leaning towards the use of eco-friendly materials which hydrogels derived from natural sources, have shown promising results in enhancing wound healing processes while being sustainable.

One other significant area of research in this field focuses on self-healing hydrogels. These materials are designed to mimic the natural healing properties of skin, providing a support environment that is conducive to cellular growth and tissue regeneration. Chen et al. demonstrated that injectable self-healing hydrogels can possess antibacterial and angiogenic properties, crucial for diabetic skin wound repair, thereby addressing both infection control and promoting blood vessel formation, which is vital for effective healing [59]. Similarly, Tu et al. highlighted the multifunctional properties of hydrogels, which can absorb exudates and maintain moisture while delivering therapeutic agents to the wound site, thus facilitating rapid healing [101]. The incorporation of biocompatible materials such as alginate and hyaluronic acid has been shown to enhance the healing process. Shafei et al. reported that exosome-loaded alginate hydrogels significantly promoted tissue regeneration in full-thickness skin wounds, indicating the potential of these materials to accelerate reepithelialization and improve healing outcomes [91]. Hadisi et al. further supported this by showing that hyaluronic acid-based dressings could effectively reduce wound debris and accelerate healing, underscoring the importance of biocompatibility in wound care [29].

The use of natural polymers and herbal extracts in wound dressings is another promising avenue. Recent studies have explored the incorporation of plant-derived compounds, such as quercetin, which has been shown to enhance the wound healing process through its biological activities [84]. Additionally, the development of electrospun nanofibers from polysaccharides and herbal constituents has been noted for their ability to provide a supportive matrix for cell migration and proliferation, which is essential for effective wound healing [81]. These materials not only promote healing but also reduce the risk of infection due to their inherent antimicrobial properties. The integration of biodegradable and sustainable materials into wound dressings is gaining traction with the use of plant-based microfibers loaded with bioactive compounds, such as vitamin C, has been proposed as an eco-friendly alternative that supports skin healing while being environmentally responsible [24]. The potential of these materials to degrade without releasing toxic residues aligns with the growing emphasis on sustainability in medical applications.

8 Conclusion

The future of hydrogel-based wound healing is promising and driven by advancements in biotechnology, materials science, tissue engineering and digital health. As the understanding of the complex biological processes involved in wound healing deepens, hydrogel formulations are anticipated to evolve significantly and interdisciplinary research incorporating more advanced features gain more prominence containing enhanced bioactivity, controlled drug release, bio-inspired structures that better mimic the natural healing environment, and real-time monitoring that also supports personalized treatments.

The integration and different combinations of innovative materials, nanotechnology, and smart monitoring systems has the potential to revolutionize the management of wounds, particularly in challenging cases particularly for complex and chronic wounds. As research continues to advance in this field, the development of multifunctional drug delivery systems that can address the multifaceted nature of wound healing will be crucial in improving patient outcomes and reducing the burden of wound-related complications.

Future research in the field of hydrogels is poised to focus on the optimization of appropriate treatments for wound healing applications integrating cutting-edge technologies mentioned in this chapter such as stem cell therapies, robotics and on-skin-printing, and precision medicine to accelerate tissue regeneration considering the complex wounds of individuals. Collaborative efforts across interdisciplinary fields, that improve adaptability of applications in clinical settings where varying wound conditions are encountered, will be key to realizing the full potential of hydrogels in wound healing, making them an indispensable tool in modern medical care.

References

1. Abicht B, Deitrick G, MacEwan M, Jeng L (2021) Evaluation of wound healing of diabetic foot ulcers in a prospective clinical trial using a synthetic hybrid-scale fiber matrix. Foot Ankle Surgery: Tech Rep Cases 2:100135. https://doi.org/10.1016/j.fastrc.2021.100135
2. Akkoç T, Aksu MB (2023) Investigation of the efficacy of mesenchymal stem cell-derived conditioned medium on wound healing through scratch test and Teer analysis. 27(5):2135–2144. https://doi.org/10.29228/jrp.492
3. Albanna M, Binder KW, Murphy SV, Kim J, Qasem SA, Zhao W, Tan J, El-Amin IB, Dice DD, Marco J, Green J, Xu T, Skardal A, Holmes JH, Jackson JD, Atala A, Yoo JJ (2019) In situ bioprinting of autologous skin cells accelerates wound healing of extensive excisional full-thickness wounds. Sci Rep 9(1):1856. https://doi.org/10.1038/s41598-018-38366-w
4. Barakat-Johnson M, Jones A, Burger M, Leong T, Frotjold A, Randall S, Kim B, Fethney J, Coyer F (2022) Reshaping wound care: evaluation of an artificial intelligence app to improve wound assessment and management amid the COVID-19 pandemic. Int Wound J 19(6):1561–1577. https://doi.org/10.1111/iwj.13755. Epub 2022 Feb 25. PMID: 35212459; PMCID: PMC9111327
5. Barrs RW, Jia J, Silver SE, Yost M, Mei Y (2020) Biomaterials for bioprinting microvasculature. Chem Rev 120(19):10887–10949. https://doi.org/10.1021/acs.chemrev.0c00027
6. Bhutani M, Gupta D, Shikha M, Gupta D, Settia M (2024) AI-driven wound assessment: leveraging wearable sensors for clinical innovation, pp 45–70. https://doi.org/10.58532/nbennuraich3
7. Buyana B, Aderibigbe B, Sinha Ray S, Ndinteh D, Thierry F (2019) Development, characterization, and in vitro evaluation of water soluble poloxamer/pluronic-mastic gum-gum acacia-based wound dressing. J Appl Polym Sci 137:48728. https://doi.org/10.1002/app.48728
8. Chafran L, Carfagno A, Altalhi A, Bishop B (2022) Green hydrogel synthesis: emphasis on proteomics and polymer particle-protein interaction. Polymers 14:4755. https://doi.org/10.3390/polym14214755

9. Chen Y, Fan X, Lu J, Liu X, Chen J, Chen Y (2022) Mussel-inspired adhesive, antibacterial and stretchable composite hydrogel for wound dressing. Macromol Biosci 23:e2200370. https://doi.org/10.1002/mabi.202200370
10. Chen J, Peng Q, Peng X, Han L, Wang X, Wang J, Zeng H (2020) Recent advances in mechano-responsive hydrogels for biomedical applications. ACS Appl Polym Mater XXXX. https://doi.org/10.1021/acsapm.0c00019
11. Chen T, Chen Y, Rehman HU, Chen Z, Yang Z, Wang M, Li H, Liu H (2018) Ultratough, self-healing, and tissue-adhesive hydrogel for wound dressing. ACS Appl Mater Interfaces 10(39):33523–33531. https://doi.org/10.1021/acsami.8b10064. Epub 2018 Sep 24. PMID: 30204399
12. Cheng S-H, Groseclose M, Mininger C, Bergstrom M, Zhang L, Lenhard S, Skedzielewski T, Comroe D, Hong H, Cui H, Hoover J, Rittenhouse S, Castellino S, Jucker B, Alsaid H (2022) A multimodal imaging approach to assess the quantitative biodistribution of a liposomal nanoformulation with the antibiotic GSK2485680 (Lipo680). https://doi.org/10.21203/rs.3.rs-1479300/v1
13. Chinnasamy G, Chandrasekharan S, Koh TW, Bhatnagar S (2021) Synthesis, characterization, antibacterial and wound healing efficacy of silver nanoparticles from *Azadirachta indica*. Front Microbiol 12:611560. https://doi.org/10.3389/fmicb.2021.611560. PMID: 33679635; PMCID: PMC7932996
14. Choi JS, Yoo HS (2010) Pluronic/chitosan hydrogels containing epidermal growth factor with wound-adhesive and photo-crosslinkable properties. J Biomed Mater Res Part A 95A(2):564–573. https://doi.org/10.1002/jbm.a.32848
15. Chuah C, Wang J, Tavakoli J, Tang Y (2018) Novel bacterial cellulose-poly (acrylic acid) hybrid hydrogels with controllable antimicrobial ability as dressings for chronic wounds. Polymers 10:1323. https://doi.org/10.3390/polym10121323
16. Cui R, Zhang L, Ou R, Xu Y, Xu L, Zhan X-Y, Li D (2022) Polysaccharide-based hydrogels for wound dressing: design considerations and clinical applications. Front Bioeng Biotechnol 10:845735. https://doi.org/10.3389/fbioe.2022.845735
17. Daristotle JL, Lau LW, Erdi M, Hunter J, Djoum A Jr, Srinivasan P, Wu X, Basu M, Ayyub OB, Sandler AD, Kofinas P (2020) Sprayable and biodegradable, intrinsically adhesive wound dressing with antimicrobial properties. Bioeng Transl Med 5(1):e10149. https://doi.org/10.1002/btm2.10149
18. Deng Z, Hu T, Lei Q, He J, Ma PX, Guo B (2019) Stimuli-responsive conductive nanocomposite hydrogels with high stretchability, self-healing, adhesiveness, and 3D printability for human motion sensing. ACS Appl Mater Interfaces 11(7):6796–6808. https://doi.org/10.1021/acsami.8b20178. Epub 2019 Feb 7. PMID: 30673228
19. Domingues R, Rautenfeld M, Kavalco C, Caliari C, Dellagiustina C, Fonseca L, Costa F, Reis A, Santos G, Azzini G, Faria A, Santos N, Pires L, Huber S, Mahmood A, Dallo I, Everts P, Lana J (2024) The role of orthobiologics in chronic wound healing. Int Wound J 21:e14854. https://doi.org/10.1111/iwj.14854
20. Ehrlich F, Fischer H, Langbein L, Praetzel-Wunder S, Ebner B, Figlak K, Weissenbacher A, Sipos W, Tschachler E, Eckhart L (2018) Differential evolution of the epidermal keratin cytoskeleton in terrestrial and aquatic mammals. Mol Biol Evol 36:328–340
21. Farasati Far B, Naimi-Jamal MR, Jahanbakhshi M, Hadizadeh A, Dehghan S, Hadizadeh S (2024) Enhanced antibacterial activity of porous chitosan-based hydrogels crosslinked with gelatin and metal ions. Sci Rep 14(1):7505. https://doi.org/10.1038/s41598-024-58174-9
22. Farasati Far B, Naimi-Jamal MR, Jahanbakhshi M, Rostamani H, Karimi M, Keihankhadiv S (2023) Synthesis and characterization of chitosan/collagen/polycaprolactone hydrogel films with enhanced biocompatibility and hydrophilicity for artificial tendon applications. Int J Biol Macromol 253:127448. https://doi.org/10.1016/j.ijbiomac.2023.127448
23. Feng C, Ouyang J, Tang Z, Kong N, Liu Y, Fu L, Ji X, Xie T, Farokhzad OC, Tao W (2020) Germanene-based theranostic materials for surgical adjuvant treatment: inhibiting tumor recurrence and wound infection. Matter 3(1):127–144. https://doi.org/10.1016/j.matt.2020.04.022

24. Fiorentini F, Suarato G, Summa M, Miele D, Sandri G, Bertorelli R, Athanassiou A (2023) Plant-based, hydrogel-like microfibers as an antioxidant platform for skin burn healing. ACS Appl Bio Mater 6(8):3103–3116. https://doi.org/10.1021/acsabm.3c00214. Epub 2023 Jul 26. PMID: 37493659; PMCID: PMC10445266
25. Freedman B, Hwang C, Talbot S, Hibler B, Matoori S, Mooney D (2023) Breakthrough treatments for accelerated wound healing. Sci Adv 9:eade7007. https://doi.org/10.1126/sciadv.ade7007
26. Fu, A., Gwon, K., Kim, M., Tae, G., & Kornfield, J. (2015). Visible-Light-Initiated Thiol−Acrylate Photopolymerization of Heparin-Based Hydrogels. Biomacromolecules, 16.
27. Gilmore J, McClendon J, Madisetti M, Kelechi T (2022) Predictive chronic wound monitoring protocol for healing assessment via intelligent biosensing technology and machine learning. Biomed J Sci Tech Res 45(5). https://doi.org/10.26717/BJSTR.2022.46.007429 PDF
28. Gopal SV, Solomon A (2019) An inexpensive method of negative pressure wound therapy for extremities. Int Wound J 16(1):131–133. https://doi.org/10.1111/iwj.13002
29. Hadisi Z, Farokhi M, Bakhsheshi-Rad HR, Jahanshahi M, Hasanpour S, Pagan E, Dolatshahi-Pirouz A, Zhang YS, Kundu SC, Akbari M (2020) Hyaluronic acid (HA)-based silk fibroin/zinc oxide core-shell electrospun dressing for burn wound management. Macromol Biosci 20(4):1900328. https://doi.org/10.1002/mabi.201900328
30. Han W, Zhou B, Yang K, Xiong X, Luan S, Wang Y, Xu Z, Lei P, Zhengshan L, Gao J, Zhan Y, Chen G, Liang L, Wang R, Li S, Xu H (2020) Biofilm-inspired adhesive and antibacterial hydrogel with tough tissue integration performance for sealing hemostasis and wound healing. Bioactive Mater 5:768–778. https://doi.org/10.1016/j.bioactmat.2020.05.008
31. Hou S, Liu Y, Feng F, Zhou J, Feng X, Fan Y (2020) Polysaccharide-peptide cryogels for multidrug-resistant-bacteria infected wound healing and hemostasis. Adv Healthc Mater 9(3):1901041. https://doi.org/10.1002/adhm.201901041
32. Hu C, Yang L, Wang Y (2022) Recent advances in smart-responsive hydrogels for tissue repairing. MedComm Biomater Appl 1(2):e23. https://doi.org/10.1002/mba2.23
33. Hu L, Zhou J, He Z, Zhang L, Du F, Nie M, Zhou Y, Hao H, Zhang L, Yu S, Zhang J, Chen Y (2020) In situ-formed fibrin hydrogel scaffold loaded with human umbilical cord mesenchymal stem cells promotes skin wound healing. https://doi.org/10.21203/rs.3.rs-136523/v1
34. Huang L-C, Wang H-C, Chen L-H, Ho C-Y, Hsieh P-H, Huang J, Wu H-C, Wang T-W (2019) Bioinspired self-assembling peptide hydrogel with proteoglycan-assisted growth factor delivery for therapeutic angiogenesis. Theranostics 9:7072–7087. https://doi.org/10.7150/thno.35803
35. Hüsers J, Hafer G, Heggemann J, Wiemeyer S, Przysucha M, Dissemond J, Moelleken M, Erfurt-Berge C, Hübner U (2022) Automatic classification of diabetic foot ulcer images – a transfer-learning approach to detect wound maceration. In: Studies in health technology and informatics, vol 289. https://doi.org/10.3233/SHTI210919
36. Jeon O, Sok MC, Solorio L, Krebs M, Alsberg E (2011) Affinity-based growth factor delivery using biodegradable, photocrosslinked heparin-alginate hydrogels. J Control Rel: Off J Control Rel Soc 154:258–266. https://doi.org/10.1016/j.jconrel.2011.06.027
37. Jeong S, Kim B, Park M, Ban E, Lee SH, Kim A (2020) Improved diabetic wound healing by EGF encapsulation in gelatin-alginate coacervates. Pharmaceutics 12(4):334. https://doi.org/10.3390/pharmaceutics12040334
38. Ji M, Zhan F, Qiu X, Liu H, Liu X, Bu P, Zhou B, Serda M, Feng Q (2024) Research progress of hydrogel microneedles in wound management. ACS Biomater Sci Eng 10(8):4771–4790. https://doi.org/10.1021/acsbiomaterials.4c00972. Epub 2024 Jul 10. PMID: 38982708; PMCID: PMC11322915
39. Jiang F, Chi Z, Ding Y, Quan M, Tian Y, Shi J, Song F, Liu C (2021) Wound dressing hydrogel of enteromorpha prolifera polysaccharide-polyacrylamide composite: a facile transformation of marine blooming into biomedical material. ACS Appl Mater Interfaces 13(12):14530–14542. https://doi.org/10.1021/acsami.0c21543. Epub 2021 Mar 17. PMID: 33729756

40. Jin SG, Yousaf AM, Jang SW, Son M-W, Kim KS, Kim D-W, Li DX, Kim JO, Yong CS, Choi H-G (2015) In vivo wound-healing effects of novel benzalkonium chloride-loaded hydrocolloid wound dressing. Drug Dev Res 76(3):157–165. https://doi.org/10.1002/ddr.21253
41. Johnson, Wang N (2012) Controlled delivery of heparin-binding EGF-like growth factor yields fast and comprehensive wound healing. journal of controlled release . Off J Control Rel Soc 166. https://doi.org/10.1016/j.jconrel.2012.11.004
42. Jordan S, Mcswiggan J, Parker J, Halas G, Friesen M (2018) An mHealth app for decision-making support in wound dressing selection (WounDS): protocol for a user-centered feasibility study. JMIR Res Protocols 7:e108. https://doi.org/10.2196/resprot.9116
43. Khalid A, Bai D, Abraham AN, Jadhav A, Linklater D, Matusica A, Nguyen D, Murdoch BJ, Zakhartchouk N, Dekiwadia C, Reineck P, Simpson D, Vidanapathirana AK, Houshyar S, Bursill CA, Ivanova EP, Gibson BC (2020) Electrospun nanodiamond-silk fibroin membranes: a multifunctional platform for biosensing and wound-healing applications. ACS Appl Mater Interfaces 12(43):48408–48419. https://doi.org/10.1021/acsami.0c15612. Epub 2020 Oct 13. PMID: 33047948
44. Kiiskinen J, Merivaara A, Hakkarainen T, Kääriäinen M, Miettinen S, Yliperttula M, Koivuniemi R (2019) Nanofibrillar cellulose wound dressing supports the growth and characteristics of human mesenchymal stem/stromal cells without cell adhesion coatings. Stem Cell Res Ther 10(1):292. https://doi.org/10.1186/s13287-019-1394-7. PMID: 31547864; PMCID: PMC6757411
45. Kirschbaum M, Karimian G, Adelmeijer J, Giepmans BN, Porte RJ, Lisman T (2015) Horizontal RNA transfer mediates platelet-induced hepatocyte proliferation. Blood 126(6):798–806. https://doi.org/10.1182/blood-2014-09-600312. Epub 2015 Jun 8. PMID: 26056167
46. Konop M, Sulejczak D, Czuwara J, Kosson P, Misicka A, Lipkowski AW, Rudnicka L (2017) The role of allogenic keratin-derived dressing in wound healing in a mouse model. Wound Repair Regener 25(1):62–74. https://doi.org/10.1111/wrr.12500
47. Konop M, Rybka M, Drapała A (2021) Keratin biomaterials in skin wound healing, an old player in modern medicine: a mini review. Pharmaceutics 13(12). https://doi.org/10.3390/pharmaceutics13122029
48. Lala R, Shinde A, Nandvikar N (2018) Solid lipid nanoparticles: A promising approach for combinational drug therapy in cancer. Int J Appl Pharmaceut 10:17. https://doi.org/10.22159/ijap.2018v10i5.27894
49. Lan D, Zhang Y, Zhang H, Zhou J, Chen X, Li Z, Dai F (2021) Silk fibroin/polycaprolactone nanofibrous membranes loaded with natural Manuka honey for potential wound healing. J Appl Polym Sci 139:51686. https://doi.org/10.1002/app.51686
50. Levin AA, Karalkin PA, Koudan EV, Senatov FS, Parfenov VA, Lvov VA, Petrov SV, Pereira FDAS, Kovalev AV, Osidak EO, Domogatsky SP, Manturova NE, Kasyanov VA, Sergeeva NS, Zorin VL, Khesuani YD, Mironov VA (2023) Commercial articulated collaborative in situ 3D bioprinter for skin wound healing. Int J Bioprint 9(2):675. https://doi.org/10.18063/ijb.v9i2.675. PMID: 37065657; PMCID: PMC10090815
51. Li Z, Man J, Wang X, Zhang Y (2023b) Biodegradable sodium alginate/carrageenan/cellulose composite hydrogel wound dressings containing herbal extracts for promoting blood coagulation and wound healing. Macromol Biosci 23(1):2200235. https://doi.org/10.1002/mabi.202200235
52. Li Z, Song P, Li G, Han Y, Ren X, Bai L, Su J (2024b) AI energized hydrogel design, optimization and application in biomedicine. Materials Today Bio 25:101014. https://doi.org/10.1016/j.mtbio.2024.101014
53. Li X, Xu M, Geng Z, Liu Y (2023a) Functional hydrogels for the repair and regeneration of tissue defects. Front Bioeng Biotechnol 11:1190171. https://doi.org/10.3389/fbioe.2023.1190171
54. Li Y, Zheng L, Xu X, Song L, Li Y, Li W, Zhang S, Zhang F, Jin H (2013) Mesenchymal stem cells modified with angiopoietin-1 gene promote wound healing. Stem Cell Res Ther 4:113. https://doi.org/10.1186/scrt324

55. Li D, Fan L, Liang D, Chen Y, Ding Y, Liu G, Huang G, Zeng Y, Feng J, Li A, Wu X, Mei J, Liu X, Zhang C (2024a) Evaluation of the effect of a new fully biodegradable external dressing for promoting diabetic ulcerative wound healing. BIO web of conferences, vol 111. https://doi.org/10.1051/bioconf/202411102001
56. Li Y, Fu R, Duan Z, Zhu C, Fan D (2022) Artificial nonenzymatic antioxidant mxene nanosheet-anchored injectable hydrogel as a mild photothermal-controlled oxygen release platform for diabetic wound healing. ACS Nano 16:7486. https://doi.org/10.1021/acsnano.1c10575
57. Liang Y, Li M, Yang Y, Qiao L, Xu H, Guo B (2022) PH/glucose dual responsive metformin release hydrogel dressings with adhesion and self-healing via dual-dynamic bonding for athletic diabetic foot wound healing. ACS Nano 16:3194. https://doi.org/10.1021/acsnano.1c11040
58. Liu Z, Tong H, Li J, Wang L, Fan X, Song H, Yang M, Wang H, Jiang X, Zhou X, Yuan H, Wang Y (2022) Low-stiffness hydrogels promote peripheral nerve regeneration through the rapid release of exosomes. Front Bioeng Biotechnol 10:922570. https://doi.org/10.3389/fbioe.2022.922570
59. Liu X, Zhou S, Cai B, Wang Y, Deng D, Wang X (2022) An injectable and self-healing hydrogel with antibacterial and angiogenic properties for diabetic wound healing. Biomater Sci 10(13):3480–3492. https://doi.org/10.1039/d2bm00224h. PMID: 35593179
60. Liu Z, Lu L (2024) New Advances in the Treatment of Diabetic Wounds. International Journal of Biology and Life Sciences (IJBLS) 6:1. https://doi.org/10.54097/1strve65
61. Liu S, Jia Z, Yang F, Ning T, Gu X, Niu X, Fan Y (2023) Flexible transient bioelectronic system enables multifunctional active-controlled drug delivery. Adv Funct Mater 33. https://doi.org/10.1002/adfm.202215034
62. Liu Y, Fan J, Lv M, She K, Sun J, Lu Q, Han C, Ding S, Zhao S, Wang G, Zhang Y, Zang G (2021) Photocrosslinking silver nanoparticles-aloe vera-silk fibroin composite hydrogel for treatment of full-thickness cutaneous wounds. Regener Biomater 8:rbab048. https://doi.org/10.1093/rb/rbab048
63. Malihi L, Hüsers J, Richter M, Moelleken M, Przysucha M, Busch D, Heggemann J, Hafer G, Wiemeyer S, Heidemann G, Dissemond J, Erfurt-Berge C, Hübner U (2022) Automatic wound type classification with convolutional neural networks. Stud Health Technol Inf 295:281–284. https://doi.org/10.3233/SHTI220717
64. Manuel R, Ratanavaraporn J, Mb F (2020) Comprehensive review of hybrid collagen and silk fibroin for cutaneous wound healing. Materials 13:3097. https://doi.org/10.3390/ma13143097
65. Mariello M, Eş I, Proctor CM (2024) Soft and flexible bioelectronic micro-systems for electronically controlled drug delivery. Adv Healthcare Mater 13(24):2302969. https://doi.org/10.1002/adhm.202302969
66. Mazurek Ł, Szudzik M, Rybka M, Konop M (2022) Silk fibroin biomaterials and their beneficial role in skin wound healing. Biomolecules 12:1852. https://doi.org/10.3390/biom12121852
67. Milne SD, Seoudi I, Al Hamad H, Talal TK, Anoop AA, Allahverdi N, Zakaria Z, Menzies R, Connolly P (2016) A wearable wound moisture sensor as an indicator for wound dressing change: an observational study of wound moisture and status. Int Wound J 13(6):1309–1314. https://doi.org/10.1111/iwj.12521
68. Min S, Lee J, Nah H, Moon H-J, Sang Jin L, Kang H, Hwang Y-S, Kwon IK, Heo DN (2023) Degradable and tunable keratin-fibrinogen hydrogel as controlled release system for skin tissue regeneration. J Bionic Eng 20:https://doi.org/10.1007/s42235-022-00317-7
69. Misbah Ul Haq M, Shahidulla D, Rahman MA, Razzak M, Uddin MA (2023) Revolutionizing drug delivery: targeted approaches and innovations for effective treatment. 5:1–8. https://doi.org/10.23880/pdraj-16000138
70. Mittal AK, Bhardwaj R, Arora R, Singh A, Mukherjee M, Rajput S (2020) Acceleration of wound healing in diabetic rats through poly dimethylaminoethyl acrylate–hyaluronic acid polymeric hydrogel impregnated with a didymocarpus pedicellatus plant extract. ACS Omega 5:24239–24246. https://doi.org/10.1021/acsomega.0c02040

71. Modrejewski J, Walter J-G, Kretschmer I, Kemal E, Green M, Belhadj H, Blume C, Scheper T (2016) Aptamer-modified polymer nanoparticles for targeted drug delivery. 17(1–2):43–51. https://doi.org/10.1515/bnm-2015-0027
72. Mostafalu P, Kiaee G, Giatsidis G, Khalilpour A, Nabavinia M, Dokmeci M, Sonkusale S, Orgill D, Tamayol A, Khademhosseini A (2017) A textile dressing for temporal and dosage controlled drug delivery. Adv Func Mater 27:1702399. https://doi.org/10.1002/adfm.201702399
73. Mostafalu P, Tamayol A, Rahimi R, Ochoa M, Khalilpour A, Kiaee G, Yazdi IK, Bagherifard S, Dokmeci MR, Ziaie B, Sonkusale SR, Khademhosseini A (2018) Smart bandage for monitoring and treatment of chronic wounds. Small 14(33):1703509. https://doi.org/10.1002/smll.201703509
74. Nandhini J, Karthikeyan E, Shanmugam R (2024) Eco-friendly bio-nanocomposites: pioneering sustainable biomedical advancements in engineering. Discover Nano 19:86. https://doi.org/10.1186/s11671-024-04007-7
75. Negut I, Bogdan B (2023) Exploring the potential of artificial intelligence for hydrogel development—A short review. Gels 9:845. https://doi.org/10.3390/gels9110845
76. Negut I, Dorcioman G, Grumezescu V (2020) Scaffolds for wound healing applications. Polymers 12:2010. https://doi.org/10.3390/polym12092010
77. Nussbaum S, Carter M, Fife C, DaVanzo J, Haught R, Nusgart M, Cartwright D (2017) An economic evaluation of the impact, cost, and medicare policy implications of chronic nonhealing wounds. Value Health 21:27. https://doi.org/10.1016/j.jval.2017.07.007
78. Nuutila K, Eriksson E (2022) Moist wound healing with commonly available dressings. Adv Wound Care 10(12):685–698. https://doi.org/10.1089/wound.2020.1232
79. Nuutila K, Samandari M, Endo Y, Zhang Y, Quint J, Schmidt T, Tamayol A, Sinha I (2021) In vivo printing of growth factor-eluting adhesive scaffolds improves wound healing. Bioactive Mater 8:296. https://doi.org/10.1016/j.bioactmat.2021.06.030
80. Paladini F, Pollini M (2019) Antimicrobial silver nanoparticles for wound healing application: progress and future trends. Materials 12:2540. https://doi.org/10.3390/ma12162540
81. Palani N, Vijaykumar P, Monisha P, Saravanakumar A, Rajadesingu S (2024) Electrospun nanofibers synthesized from polymers incorporated with bioactive compounds for wound healing. J Nanobiotechnol 22:211. https://doi.org/10.1186/s12951-024-02491-8
82. Papaioannou A, Liakopoulou A, Papoulis D, Gianni E, Gkolfi P, Zygouri E, Letsiou S, Hatziantoniou S (2023) Effect of peptides on the synthesis, properties and wound healing capacity of silver nanoparticles. Pharmaceutics 15:2471. https://doi.org/10.3390/pharmaceutics15102471
83. Pathi BK, Mishra S, Moharana N, Kanungo A, Mishra A, Sahu S, Dash RK, Dubey R, Das MK (2024) Effect of topical silver nanoparticle formulation on wound bacteria clearance and healing in patients with infected wounds compared to standard topical antibiotic application: a randomized open-label parallel clinical trial. Cureus 16(5):e60569. https://doi.org/10.7759/cureus.60569
84. Polera N, Badolato M, Perri F, Carullo G, Aiello F (2018) Quercetin and its natural sources in wound healing management. Curr Med Chem 25:5825. https://doi.org/10.2174/0929867325666180713150626
85. Qiao Z, Zhang K, Liu H, Roh Y, Kim MG, Lee HJ, Koo B, Lee EY, Lee M, Park CO, Shin Y (2024) CSMP: a self-assembled plant polysaccharide-based hydrofilm for enhanced wound healing. Adv Healthc Mater 13(6):2303244. https://doi.org/10.1002/adhm.202303244
86. Radhakumary C, Antonty M, Sreenivasan K (2011) Drug loaded thermoresponsive and cytocompatible chitosan based hydrogel as a potential wound dressing. Carbohyd Polym 83:705–713. https://doi.org/10.1016/j.carbpol.2010.08.042
87. Safaee MM, Gravely M, Roxbury D (2021) A wearable optical microfibrous biomaterial with encapsulated nanosensors enables wireless monitoring of oxidative stress. Adv Func Mater 31(13):2006254. https://doi.org/10.1002/adfm.202006254
88. Saghazadeh S, Schot M, Kashaf S, Sharifi F, Jalilian E, Nuutila K, Giatsidis G, Mostafalu P, Derakhshandeh H, Yue K, Swieszkowski W, Memic A, Tamayol A, Khademhosseini A

(2018) Drug delivery systems and materials for wound healing applications. Adv Drug Deliv Rev 127:138. https://doi.org/10.1016/j.addr.2018.04.008

89. Sanjay S, Shanmugarajan D (2024) Development and formulation of drug-loaded hydrogel for skin regenerative potential. J Adv Zool 45. https://doi.org/10.53555/jaz.v45i5.4692
90. Seaton P, Cant R, Trip H (2020) Quality indicators for a community-based wound care centre: an integrative review. Int Wound J 17:587. https://doi.org/10.1111/iwj.13308
91. Shafei S, Khanmohammadi M, Heidari R, Ghanbari H, Taghdiri Nooshabadi V, Farzamfar S, Akbariqomi M, Sanikhani N, Absalan M, Tavoosidana G (2019) Exosome loaded alginate hydrogel promotes tissue regeneration in full-thickness skin wounds: an in vivo study. J Biomed Mater Res Part A 108:545. https://doi.org/10.1002/jbm.a.36835
92. Sharma R, Gupta N, Kumar V, Pal S, Kaundal V, Sharma V (2017) Silver colloid dressings score over conventional dressings in diabetic foot ulcer: a randomized clinical trial. Int Surgery J 4:2627. https://doi.org/10.18203/2349-2902.isj20173401
93. Singh H, Li W, Kazemian MR, Yang R, Yang C, Logsetty S, Liu S (2019) Lipase-responsive electrospun theranostic wound dressing for simultaneous recognition and treatment of wound infection. ACS Appl Bio Mater 2(5):2028–2036. https://doi.org/10.1021/acsabm.9b00076
94. Soni RS, Padwad YS (2022) Nanotechnological interventions and mechanistic insights into wound-healing events. In: Maurício AC, Alvites R, Gönül M (eds) Wound healing. IntechOpen, p. Ch. 10. https://doi.org/10.5772/intechopen.106481
95. Stan D, Tanase C, Avram M, Apetrei R, Mincu N-B, Mateescu AL, Stan D (2021) Wound healing applications of creams and "smart" hydrogels. Exp Dermatol 30(9):1218–1232. https://doi.org/10.1111/exd.14396
96. Tabatabaei Hosseini BS, Meadows K, Gabriel V, Hu J, Kim K (2024) Biofabrication of cellulose-based hydrogels for advanced wound healing: a special emphasis on 3D bioprinting. Macromol Biosci 24(5):2300376. https://doi.org/10.1002/mabi.202300376
97. Tan M, Chin JS, Madden L, Becker D (2022) Challenges faced in developing an ideal chronic wound model. Expert Opin Drug Discov 18:1–16. https://doi.org/10.1080/17460441.2023.2158809
98. Tang L, Guo Z, Zhao Q, Fan X, Pu Y, He B, Chen J (2024) A biodegradable Janus sponge for negative pressure wound therapy. Biomacromol 25:2542. https://doi.org/10.1021/acs.biomac.4c00046
99. Thomas J, Sharma A, Panwar V, Chopra V (2019) Polysaccharide-based hybrid self-healing hydrogel supports paracrine response of mesenchymal stem cells. ACS Appl Bio Mater 2. https://doi.org/10.1021/acsabm.9b00074
100. Tiwari R, Pathak K (2023) Local drug delivery strategies towards wound healing. Pharmaceutics 15(2):634. https://doi.org/10.3390/pharmaceutics15020634
101. Tu Z, Chen M, Wang M, Shao Z, Jiang X, Wang K, Yao Z, Yang S, Zhang X, Gao W, Lin C, Lei B, Mao C (2021) Engineering bioactive M2 macrophage-polarized anti-inflammatory, antioxidant, and antibacterial scaffolds for rapid angiogenesis and diabetic wound repair. Adv Func Mater 31:2100924. https://doi.org/10.1002/adfm.202100924
102. Vasile C, Pamfil D, Stoleru E, Baican M (2020) New developments in medical applications of hybrid hydrogels containing natural polymers. Molecules:25(7). https://doi.org/10.3390/molecules25071539
103. Vázquez-González M, Willner I (2020) Stimuli-responsive biomolecule-based hydrogels and their applications. Angew Chem Int Ed 59(36):15342–15377. https://doi.org/10.1002/anie.201907670
104. Wang H, Wan J, Zhang Z, Hou R (2024) Recent advances on 3D-bioprinted gelatin methacrylate hydrogels for tissue engineering in wound healing: A review of current applications and future prospects. Int Wound J 21(4):e14533. https://doi.org/10.1111/iwj.14533
105. Wang S, Lee JM, Yeong WY (2015) Smart hydrogels for 3D bioprinting. Int J Bioprint 1:https://doi.org/10.18063/IJB.2015.01.005
106. Wang X, Wang Y, Teng Y, Shi J, Yang X, Ding Z, Guo X, Hou S, Lv Q (2023) 3D bioprinting: opportunities for wound dressing development. Biomed Mater 18:052001. https://doi.org/10.1088/1748-605X/ace228

107. Wen H, Zeng X, Xu X, Li W, Xie F, Xiong Z, Song S, Li X, Cao Y (2021) Reversible data encryption-decryption using ph stimuli-responsive hydrogel. J Mater Chem C 9:2455. https://doi.org/10.1039/D0TC05578F
108. Wu S-Y, Tsai W-B (2023) Development of an in situ photo-crosslinking antimicrobial collagen hydrogel for the treatment of infected wounds. Polymers 15:4701. https://doi.org/10.3390/polym15244701
109. Xu G, Lu Y, Cheng C, Li X, Xu J, Liu Z, Liu J, Liu G, Shi Z, Chen Z, Zhang F, Jia Y, Xu D, Yuan W, Cui Z, Low SS, Liu Q (2021) Battery-free and wireless smart wound dressing for wound infection monitoring and electrically controlled on-demand drug delivery. Adv Func Mater 31(26):2100852. https://doi.org/10.1002/adfm.202100852
110. Yamane T, Nakagami G, Yoshino S, Shimura M, Kitamura A, Kobayashi-Hattori K, Oishi Y, Nishijima Y, Minematsu T, Sanada H (2014) Hydrocellular foam dressings promote wound healing associated with decrease in inflammation in rat periwound skin and granulation tissue, compared with hydrocolloid dressings. Biosci Biotechnol Biochem 79:1–5. https://doi.org/10.1080/09168451.2014.968088
111. Yang X, Zhang C, Deng D, Gu Y, Wang H, Zhong Q (2022) Multiple stimuli-responsive MXene-based hydrogel as intelligent drug delivery carriers for deep chronic wound healing. Small 18(5):2104368. https://doi.org/10.1002/smll.202104368
112. Zengjie F, Liu B, Wang J, Zhang S, Lin Q, Gong P, Ma L, Yang S (2014) A Novel wound dressing based on Ag/Graphene polymer hydrogel: effectively kill bacteria and accelerate wound healing. Adv Func Mater 24:3933. https://doi.org/10.1002/adfm.201304202
113. Zhang J, Guan J, Niu X, Hu G, Guo S-C, Li Q, Xie Z, Zhang C, Wang Y (2015) Exosomes released from human induced pluripotent stem cells-derived MSCs facilitate cutaneous wound healing by promoting collagen synthesis and angiogenesis. J Transl Med 13:49. https://doi.org/10.1186/s12967-015-0417-0
114. Zhao X, Pei D, Yang Y, Xu K, Yu J, Zhang Y, Zhang Q, He G, Zhang Y, Li A, Cheng Y, Chen X (2021) Green tea derivative driven smart hydrogels with desired functions for chronic diabetic wound treatment. Adv Func Mater 31(18):2009442. https://doi.org/10.1002/adfm.202009442
115. Zhou Y, Layani M, Wang S, Hu P, Ke Y, Magdassi S, Long Y (2018) Fully printed flexible smart hybrid hydrogels. Adv Func Mater 28(9):1705365. https://doi.org/10.1002/adfm.201705365
116. Zhu Y-S, Tang K, Lv J (2021) Peptide–drug conjugate-based novel molecular drug delivery system in cancer. Trends Pharmacol Sci 42:857. https://doi.org/10.1016/j.tips.2021.07.001
117. Zuniga K, Isaac A, Christy S, Wrice N, Mangum L, Natesan S, Burnett L, Christy R, Kowalczewski C (2022) Characterization of a human platelet lysate-loaded keratin hydrogel for wound healing applications in vitro. Int J Mol Sci 23:4100. https://doi.org/10.3390/ijms23084100

Recombinant Antimicrobial Peptides-Embedded Bacterial Cellulose Hydrogels for Advanced Wound Healing

Tarek Agha, Derya İlke Gungor, Zeynep Beyatli, Gamze Balci, and Ahmet Kati

Abstract Bacterial cellulose (BC), a highly pure and biocompatible biopolymer produced by microbial fermentation, offers exceptional properties such as high water retention, mechanical strength, and a nanofibrous structure resembling the extracellular matrix. These features make it an excellent candidate for wound dressing applications. In recent years, the incorporation of recombinant antimicrobial peptides (AMPs) into BC hydrogels has gained increasing attention as an innovative strategy for enhancing wound healing outcomes. AMPs are short, cationic peptides that possess potent broad-spectrum antimicrobial activity and the ability to modulate immune responses and promote tissue regeneration. This chapter presents an overview of the production of recombinant AMPs through microbial expression systems and their subsequent integration into BC matrices using various methods such as physical entrapment, adsorption, and covalent binding. The resulting AMP-loaded BC hydrogels not only provide a moist and protective wound environment but also ensure sustained antimicrobial activity, which is particularly advantageous in the treatment of chronic and infected wounds. The chapter also reviews the physicochemical and biological properties of these composite hydrogels, emphasizing their structural integrity, biocompatibility, and ability to support cell proliferation and tissue repair. Overall, the synergistic combination of bacterial cellulose and recombinant antimicrobial peptides represents a promising platform for the development of next-generation wound dressings with enhanced therapeutic efficacy.

T. Agha
Department of Molecular Biology and Genetics, Uskudar University, Istanbul 34664, Turkey

T. Agha · Z. Beyatli · G. Balci · A. Kati (✉)
Department of Biotechnology, Institution of Health Sciences, University of Health Sciences, Istanbul 34664, Turkey
e-mail: ahmet.kati@sbu.edu.tr

D. İ. Gungor · Z. Beyatli · G. Balci · A. Kati
Experimental Medicine Research and Application Center, Validebag Research Park, University of Health Sciences, Istanbul 34664, Turkey

A. Kati
Center for Targeted Therapy Technologies, Bogazici University, Istanbul 34684, Türkiye

A. K. Mandal et al. (eds.), *Functionalized Nanoparticles Hydrogels for Wound Healing*, Smart Nanomaterials Technology, https://doi.org/10.1007/978-981-95-0720-7_13

Keywords Bacterial cellulose · Recombinant antimicrobial peptides · Wound healing · Hydrogel · Infection Control · Tissue regeneration

1 Introduction

Bacterial cellulose hydrogels have garnered significant attention as advanced biomaterials for wound healing applications due to their unique physicochemical properties [36]. These hydrogels are characterized by high water retention, excellent biocompatibility, and mechanical stability, making them ideal for creating a moist wound environment that accelerates tissue regeneration [11, 86]. Additionally, bacterial cellulose is naturally non-toxic, non-immunogenic, and has a nanofibrillar structure that closely mimics the extracellular matrix (ECM) of human skin [32]. This ECM-like structure facilitates cellular adhesion, proliferation, and migration, thereby promoting wound healing [86]. The incorporation of antimicrobial peptides into these hydrogels can enhance their therapeutic efficacy by preventing microbial colonization and infection at the wound site [51, 75, 77, 78].

Antimicrobial peptides (AMPs) are a class of small, cationic molecules that exhibit broad-spectrum antimicrobial activity against bacteria, fungi, and viruses, including drug-resistant strains [28]. Unlike conventional antibiotics, AMPs target microbial membranes through electrostatic interactions, leading to rapid disruption and lysis of pathogenic cells [53]. This mechanism reduces the likelihood of developing resistance. Moreover, AMPs play a pivotal role in modulating the immune response by recruiting immune cells to the wound site, promoting angiogenesis, and accelerating re-epithelialization [75, 77, 78]. These multifunctional properties make AMPs a promising alternative to conventional antibiotics in wound management [70]

The synergistic integration of recombinant antimicrobial peptides into bacterial cellulose hydrogels presents a novel therapeutic approach for advanced wound healing applications [57]. This combination offers a dual-action mechanism: the hydrogel matrix maintains an optimal moist environment and provides structural support, while the AMPs deliver localized antimicrobial activity and immunomodulatory effects [79]. This chapter explores the design, synthesis, and evaluation of recombinant antimicrobial peptide-embedded bacterial cellulose hydrogels, focusing on their mechanisms of action, applications, and future potential in wound care.

This chapter will further discuss the various techniques employed for synthesizing and characterizing these functionalized hydrogels. It will also delve into the in vitro and in vivo models used to evaluate their efficacy. The challenges associated with scaling up production and maintaining peptide stability will be addressed, along with emerging trends in functionalizing bacterial cellulose with additional bioactive molecules to enhance wound healing. Ultimately, this chapter aims to provide a comprehensive overview of the potential of recombinant antimicrobial peptide-embedded bacterial cellulose hydrogels as next-generation wound dressings.

2 Design and Synthesis

The design and synthesis of recombinant antimicrobial peptide-embedded bacterial cellulose hydrogels involve a multi-step process that integrates biotechnology, materials science, and biomedical engineering principles. The recombinant production of antimicrobial peptides (AMPs) is typically achieved using engineered microbial systems, such as *Escherichia coli* or *Pichia pastoris* [16, 37]. These systems offer high yield and cost-effective synthesis, enabling the large-scale production of specific peptides with desired antimicrobial properties [54]. The recombinant expression of AMPs involves constructing a suitable expression vector containing the gene encoding the target peptide. The vector is then introduced into the host microorganism, where the peptide is expressed under the control of an inducible promoter. Fusion tags are often included to facilitate purification and prevent proteolytic degradation, enhancing peptide yield and stability. Once expressed, the peptides are purified using chromatography techniques, such as affinity or ion-exchange chromatography. These purified peptides are then concentrated and characterized to ensure structural integrity and antimicrobial activity. Incorporating AMPs into bacterial cellulose hydrogels can be achieved using several methods, including physical entrapment, covalent immobilization, and electrospinning [62, 64]. Physical entrapment involves diffusing the peptides into the hydrogel matrix, where they are retained through electrostatic interactions and hydrogen bonding [58]. This method is straightforward but may result in a burst release profile (Fig. 1).

Alternatively, covalent immobilization provides a more controlled release by chemically linking the peptides to the cellulose fibers. This approach maintains peptide activity while minimizing rapid diffusion [69]. Conversely, electrospinning involves generating nanofibrous hydrogels with embedded AMPs, enhancing the

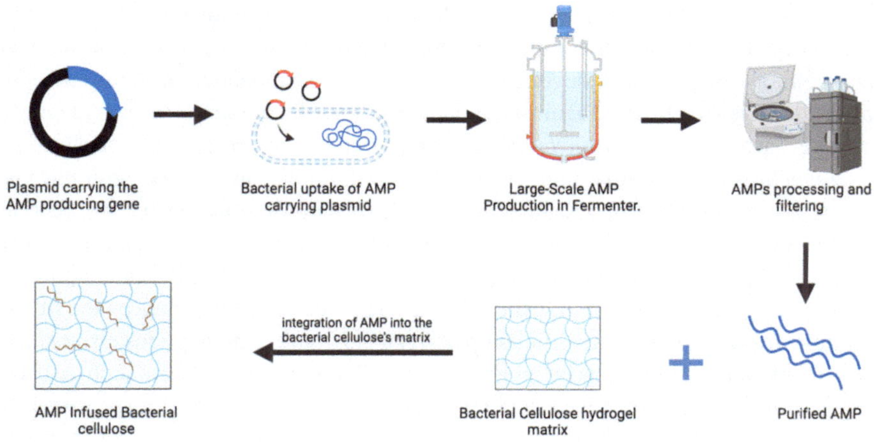

Fig. 1 Recombinant production and integration of antimicrobial peptides (AMPs) into bacterial cellulose hydrogels

surface area for microbial interaction and providing a scaffold for cell attachment and proliferation [4, 33]. Comprehensive characterization using various analytical techniques ensures the hydrogels' functionality. Scanning electron microscopy (SEM) and atomic force microscopy (AFM) were used to observe the nanofibrous structure and surface morphology. Fourier-transform infrared spectroscopy (FTIR) and X-ray diffraction (XRD) confirm the chemical interactions and crystallinity of the embedded peptides. Rheological studies evaluate the mechanical properties, while swelling and degradation assays assess the hydrogel's stability and water retention capacity. Optimizing the synthesis parameters, including peptide concentration, crosslinking density, and hydrogel porosity, is crucial for achieving the desired therapeutic effects. Additionally, scaling up the production process requires careful consideration of bioreactor design and process economics. Genetically engineered strains with high productivity and the development of cost-effective downstream processing techniques can significantly enhance scalability. Future research will optimize these processes to enable large-scale manufacturing and clinical translation.

3 Mechanism of Action

The mechanism of action of recombinant antimicrobial peptide-embedded bacterial cellulose hydrogels is multifaceted, involving direct antimicrobial activity, immune modulation, and enhancement of tissue regeneration. Understanding these mechanisms is crucial for designing effective wound dressings that prevent infection and promote rapid healing. The primary mechanism of action involves the potent antimicrobial activity of the embedded peptides. These recombinant peptides are typically cationic, allowing them to interact electrostatically with negatively charged microbial membranes [9]. This interaction disrupts the lipid bilayer, leading to increased membrane permeability, leakage of cellular contents, and eventual cell lysis. Unlike conventional antibiotics, this rapid and non-specific membrane disruption reduces the likelihood of resistance development. The peptides embedded in bacterial cellulose hydrogels are effective against a broad spectrum of pathogens, including Gram-positive bacteria (e.g., *Staphylococcus aureus*) and Gram-negative bacteria (e.g., *Pseudomonas aeruginosa*), both of which are commonly associated with chronic wound infections [28, 48, 53]. In addition to their antimicrobial properties, recombinant AMPs play a significant role in modulating the host's immune response (Fig. 2). They act as immunomodulators by recruiting neutrophils, macrophages, and dendritic cells to the wound site [19, 57]. This recruitment enhances the innate immune response, leading to efficient pathogen clearance. The peptides also exhibit anti-inflammatory properties by inhibiting pro-inflammatory cytokines such as TNF-α and IL-6, reducing excessive inflammation that can impede the healing process [20, 75, 77, 78]. By balancing the inflammatory response, these peptides create a favorable environment for tissue repair and regeneration.

Recombinant AMPs have been shown to stimulate angiogenesis by promoting the proliferation and migration of endothelial cells, leading to the formation of new blood

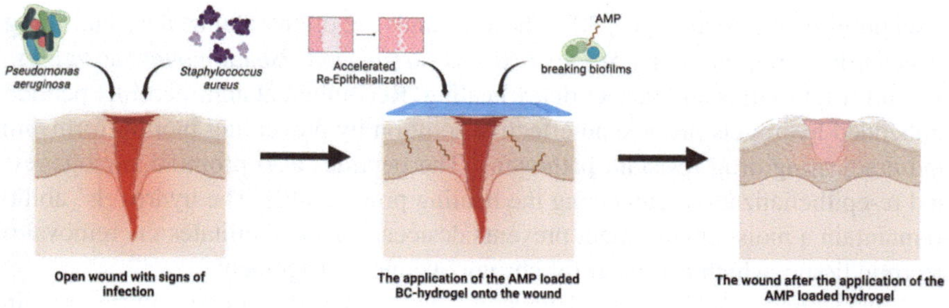

Fig. 2 Mechanism of action of AMP-embedded bacterial cellulose hydrogels in wound healing

vessels [85] This neovascularization enhances oxygen and nutrient supply to the wound bed, accelerating granulation tissue formation and wound closure. Additionally, AMPs facilitate epithelialization by inducing keratinocyte migration and proliferation, which restores the skin barrier function [7, 41]. The nanofibrillar structure of bacterial cellulose further supports this process by providing a scaffold for cellular attachment and growth. A critical advantage of using bacterial cellulose hydrogels as delivery vehicles is their ability to provide controlled and sustained release of AMPs. The hydrogels maintain a moist wound environment, promoting cellular activities and gradually releasing the peptides over time, ensuring a continuous antimicrobial effect [34, 49] This controlled release profile minimizes the risk of peptide degradation and reduces the frequency of dressing changes, thereby enhancing patient compliance and comfort.

These combined mechanisms of action make recombinant antimicrobial peptide-embedded bacterial cellulose hydrogels highly effective for advanced wound healing applications. Future research will explore using multifunctional peptides that combine antimicrobial, immunomodulatory, and regenerative properties, paving the way for next-generation wound dressings.

4 Applications and Advantages

Recombinant antimicrobial peptide-embedded bacterial cellulose hydrogels have demonstrated significant potential in managing various wound types, including acute wounds, chronic wounds, and burn injuries. Their multifunctional properties make them highly suitable for addressing complex wound healing requirements, including infection control, moisture retention, and tissue regeneration [25, 72]. These hydrogels offer a novel approach by combining the biocompatibility and structural support of bacterial cellulose with the potent antimicrobial and immunomodulatory effects of recombinant peptides.

Chronic wounds, such as diabetic foot ulcers, pressure ulcers, and venous leg ulcers, are characterized by prolonged inflammation, impaired angiogenesis, and high susceptibility to infection [13, 30]. These wounds are often colonized by multidrug-resistant bacteria, including *Staphylococcus aureus* and *Pseudomonas aeruginosa*, which form biofilms and further delay healing. Recombinant antimicrobial peptide-embedded hydrogels provide an effective solution by preventing biofilm formation and eradicating drug-resistant pathogens. The peptides also promote angiogenesis and re-epithelialization, enhancing the healing process [40]. The hydrogels' ability to maintain a moist environment prevents desiccation and facilitates the removal of necrotic tissue, which is crucial for chronic wound management.

Acute wounds, such as surgical incisions, abrasions, and traumatic injuries, require rapid healing and protection against microbial contamination. Burn injuries, on the other hand, are highly susceptible to infections due to the loss of the skin barrier. The recombinant antimicrobial peptide-embedded hydrogels provide a protective barrier while delivering localized antimicrobial action to prevent infection [46, 67]. Their excellent water retention capability keeps the wound hydrated, essential for cellular activities, including keratinocyte migration and fibroblast proliferation. In burn injuries, these hydrogels prevent microbial colonization and reduce pain and scarring by promoting rapid re-epithelialization [71].

One of the significant challenges in wound management is preventing biofilm formation, which can lead to chronic infections and antibiotic resistance. Recombinant antimicrobial peptides disrupt microbial membranes and inhibit quorum sensing, preventing biofilm development at the wound site [3, 60]. This mechanism is particularly effective against polymicrobial biofilms involving multiple species of bacteria and fungi. Additionally, the controlled release of peptides from the hydrogel matrix ensures sustained antimicrobial action, minimizing the risk of infection recurrence.

Compared to traditional wound dressings, recombinant antimicrobial peptide-embedded bacterial cellulose hydrogels offer several advantages, including enhanced biocompatibility, targeted antimicrobial activity, and improved wound healing outcomes. Conventional dressings, such as gauze and hydrocolloids, lack antimicrobial properties and may require frequent changes, leading to patient discomfort and disrupting the healing process (Fig. 3). In contrast, these advanced hydrogels release sustained peptide, reducing dressing changes and enhancing patient compliance [24]. Additionally, the biodegradable nature of bacterial cellulose eliminates the need for painful removal, further improving patient comfort.

These hydrogels can be customized to meet the specific requirements of different wound types. For instance, the hydrogels' mechanical strength and degradation rate can be tailored by adjusting the peptide concentration and crosslinking density [75, 77, 78]. This versatility makes them suitable for various applications, from superficial to deep, exudative wounds. Furthermore, the modular design allows additional bioactive agents, such as growth factors and anti-inflammatory compounds, to enhance therapeutic efficacy [83].

The multifunctional properties and customization potential of recombinant antimicrobial peptide-embedded bacterial cellulose hydrogels make them a promising platform for advanced wound care. Their ability to address the challenges of infection control, biofilm prevention, and tissue regeneration positions them as a superior alternative to conventional wound dressings.

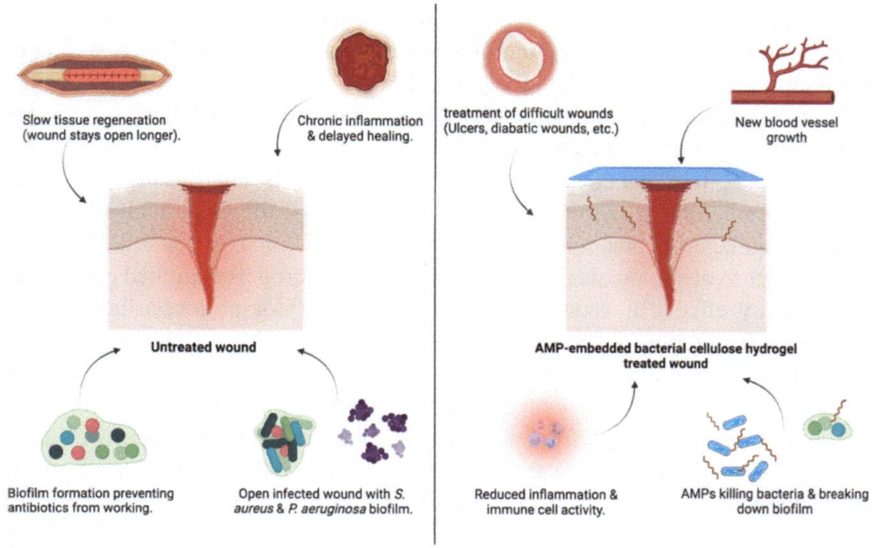

Fig. 3 Comparative effect of AMP-embedded hydrogels versus untreated wounds

5 Case Studies and Experimental Evidence

Numerous studies have demonstrated the efficacy of recombinant antimicrobial peptide-embedded bacterial cellulose hydrogels in promoting wound healing and preventing infections [1, 31, 50]. These studies include both in vitro and in vivo experiments, highlighting the potential of these hydrogels as next-generation wound dressings. The following sections discuss some of the most relevant case studies and experimental findings that support the clinical application of these advanced biomaterials.

In vitro studies have shown that recombinant antimicrobial peptide-embedded bacterial cellulose hydrogels exhibit potent antimicrobial activity against a broad spectrum of wound-related pathogens. For example, hydrogels containing recombinant LL-37 peptides demonstrated significant bactericidal activity against *Staphylococcus aureus*, *Pseudomonas aeruginosa*, and *Escherichia coli* [27, 80, 81]. The antimicrobial mechanism involved disruption of the bacterial membrane integrity, leading to rapid cell lysis. These findings were supported by electron microscopy images showing membrane blebbing and cytoplasmic leakage. Additionally, the hydrogels effectively inhibited biofilm formation, which is crucial for preventing chronic wound infections [38, 65]. To evaluate the safety and biocompatibility of these hydrogels, cytocompatibility studies were conducted using human keratinocytes and fibroblasts. The results indicated that the recombinant peptide-embedded hydrogels supported cell adhesion, proliferation, and migration, demonstrating excellent biocompatibility [42, 63]. Furthermore, the hydrogels showed no signs of cytotoxicity, confirming their safety for wound healing applications. Hemocompatibility tests also revealed minimal hemolysis, indicating that these hydrogels are suitable for use in contact with blood [23].

In vivo studies using animal models have provided strong evidence for the wound healing efficacy of recombinant antimicrobial peptide-embedded bacterial cellulose hydrogels [42]. In a diabetic mouse model, the hydrogels significantly accelerated wound closure compared to conventional dressings. Histological analysis showed enhanced collagen deposition, increased angiogenesis, and re-epithelialization in the hydrogel-treated wounds [61]. Additionally, the hydrogels exhibited potent anti-inflammatory effects by reducing pro-inflammatory cytokine levels, leading to a balanced inflammatory response conducive to healing. Comparative studies have demonstrated the superior performance of recombinant peptide-embedded bacterial cellulose hydrogels over conventional wound dressings [80]. In a burn wound model, the hydrogels outperformed silver sulfadiazine dressings in terms of infection control, pain relief, and scar reduction [12]. The controlled release of recombinant peptides ensured sustained antimicrobial activity, while the bacterial cellulose matrix provided optimal moisture balance and structural support [87]. These combined effects resulted in faster wound closure and improved cosmetic outcomes. Although preclinical studies have shown promising results, clinical trials are essential for validating the safety and efficacy of these hydrogels in human subjects [43, 73]. Ongoing clinical trials are investigating the use of recombinant antimicrobial peptide-embedded bacterial cellulose hydrogels in the treatment of diabetic foot ulcers and surgical wounds [31, 35]. These trials aim to evaluate healing rates, infection control, pain management, and patient satisfaction. The results of these clinical studies will pave the way for regulatory approval and commercial translation of these advanced wound dressings.

6 Challenges and Future Perspectives

Despite the remarkable potential of recombinant antimicrobial peptide-embedded bacterial cellulose hydrogels for advanced wound healing, several challenges must be addressed to fully realize their clinical and commercial applications. These challenges range from technical and manufacturing limitations to regulatory and economic considerations. However, they also present opportunities for innovation and research, paving the way for next-generation wound care solutions.

One of the primary challenges lies in the large-scale production and cost-effective synthesis of recombinant antimicrobial peptides. Although microbial systems such as *Escherichia coli* and *Pichia pastoris* provide high yields and cost-efficiency, maintaining peptide stability and bioactivity during the production process remains a significant hurdle [15]. The use of fusion tags and optimized expression systems can improve peptide stability, however, these modifications may impact the antimicrobial efficacy. Furthermore, the downstream purification process is complex and expensive, contributing to the overall cost of production [56].

To overcome these challenges, future research should focus on developing robust bioprocessing strategies, including continuous fermentation and cost-effective purification techniques [84]. Advances in synthetic biology and metabolic engineering could also enhance peptide yield and stability. Additionally, exploring alternative production platforms such as cell-free systems and plant-based expression systems may provide scalable and sustainable solutions for large-scale peptide manufacturing [15, 47].

A critical aspect of recombinant antimicrobial peptide-embedded hydrogels is the controlled and sustained release of peptides to ensure therapeutic efficacy at the wound site. However, achieving a precise and consistent release profile is challenging due to the complex interactions between the peptides and the hydrogel matrix. Factors such as peptide charge, hydrophobicity, and molecular weight significantly influence the release kinetics. Inadequate control over peptide release can lead to burst release, reduced therapeutic efficacy, and potential cytotoxicity [52, 55].

Advanced drug delivery technologies, including smart hydrogels and stimuli-responsive systems, are being explored to address this. These systems utilize external triggers such as pH, temperature, or enzyme activity to control the release of peptides, providing on-demand delivery at the wound site [8, 21]. Developing hybrid hydrogels incorporating nanoparticles or liposomes as peptide carriers can also enhance the release profile and bioavailability [44, 66]. Future research should focus on tailoring the physicochemical properties of bacterial cellulose to achieve tunable and programmable release mechanisms.

Another major challenge is maintaining the structural stability and antimicrobial activity of recombinant peptides during storage and after incorporation into hydrogels [10, 14]. Peptides are prone to degradation due to hydrolysis, oxidation, and proteolytic cleavage, leading to reduced efficacy and shelf life [2]. Moreover, incorporating peptides into hydrogels can alter their secondary structure, potentially compromising their antimicrobial function.

Innovative approaches, such as peptide engineering and stabilization techniques, are required to enhance peptide stability. These include non-natural amino acids, cyclization, and peptide fusion with stabilizing proteins [26, 59, 68]. Encapsulation strategies using nanocarriers, such as liposomes or polymeric nanoparticles, can protect peptides from environmental degradation and improve their bioavailability [18, 82]. Developing lyophilized hydrogel formulations can also extend the shelf life by minimizing moisture-induced degradation [45].

Although recombinant antimicrobial peptide-embedded bacterial cellulose hydrogels have demonstrated excellent biocompatibility in vitro and in vivo, the potential for cytotoxicity and immunogenicity remains a concern [29]. High concentrations of cationic peptides may cause membrane disruption in mammalian cells, leading to cytotoxic effects [6]. Also, prolonged peptide exposure can trigger immune responses, potentially leading to hypersensitivity or allergic reactions [17].

Systematic biocompatibility and immunogenicity studies are essential to ensure safety and regulatory compliance. Future research should focus on optimizing peptide concentrations and release rates to minimize cytotoxicity while maintaining antimicrobial efficacy. Advanced in silico modeling and computational chemistry approaches can predict peptide-membrane interactions, guiding the design of safer peptide analogs [39]. Moreover, using bioinspired peptides with enhanced selectivity and reduced toxicity offers a promising direction for safer wound healing applications [5].

Translating recombinant antimicrobial peptide-embedded bacterial cellulose hydrogels from the laboratory to clinical practice requires rigorous regulatory approval. These hydrogels must comply with stringent safety, efficacy, and quality standards set by regulatory agencies such as the FDA and EMA [43]. However, the complex nature of peptide-based therapeutics poses unique regulatory challenges, including stability testing, bioavailability, and manufacturing consistency.

Early engagement with regulatory agencies and adopting standardized testing protocols are crucial to streamlining the regulatory pathway. Collaborative efforts between academia, industry, and regulatory bodies can facilitate the development of regulatory guidelines specific to peptide-based wound dressings. Additionally, cost-effective manufacturing strategies and strategic partnerships with healthcare providers and wound care companies are essential for successful commercialization.

The future of recombinant antimicrobial peptide-embedded bacterial cellulose hydrogels lies in their multifunctionality and customization potential. Emerging trends include the development of hybrid hydrogels that combine peptides with other bioactive agents, such as growth factors, anti-inflammatory compounds, and extracellular vesicles [76]. These multifunctional hydrogels can synergistically enhance wound healing through immunomodulation, angiogenesis, and tissue regeneration [22]

Another exciting direction is the application of 3D bioprinting and electrospinning technologies to create patient-specific wound dressings with tailored structures and properties [74]. These technologies enable the precise deposition of peptides and other biomolecules, allowing for personalized wound care solutions. Furthermore, integrating smart biosensors into hydrogels can provide real-time monitoring of wound healing progress, offering a holistic approach to wound management.

Artificial intelligence (AI) and machine learning (ML) in the design and optimization of peptide-based hydrogels are also gaining traction. AI-driven algorithms can predict peptide structure–function relationships, optimize formulation parameters, and accelerate the discovery of novel peptide candidates with enhanced antimicrobial and regenerative properties.

Finally, interdisciplinary research combining biotechnology, materials science, and computational modeling will be crucial in advancing this field. Collaborations between academic institutions, biotechnology companies, and healthcare providers will accelerate the clinical translation and commercialization of these innovative wound-healing platforms.

7 Conclusion

Recombinant antimicrobial peptide-embedded bacterial cellulose hydrogels represent a paradigm shift in wound care by combining the biocompatibility of bacterial cellulose with the multifunctional properties of recombinant peptides. While significant challenges exist, the rapid advancements in biotechnology, materials science, and smart delivery systems offer promising solutions. By addressing these challenges and leveraging emerging trends, these advanced hydrogels have the potential to revolutionize wound healing and establish a new standard of care in chronic wound management.

References

1. Ahovan ZA, Esmaeili Z, Eftekhari BS, Khosravimelal S, Alehosseini M, Orive G, Dolatshahi-Pirouz A et al (2022) Antibacterial smart hydrogels: new hope for infectious wound management. Mater Today Bio 17(November):100499. https://doi.org/10.1016/j.mtbio.2022.100499
2. Akbarian M, Chen S-H (2022) Instability challenges and stabilization strategies of pharmaceutical proteins. Pharmaceutics 14(11):2533. https://doi.org/10.3390/pharmaceutics14112533
3. Alisigwe CV, Ikpa CS, Otuonye UJ (2025) Examining alternative approaches to antibiotic utilisation: a critical evaluation of phage therapy and antimicrobial peptides combination as potential alternatives. Microbe January:100254. https://doi.org/10.1016/j.microb.2025.100254
4. Amariei G, Kokol V, Karina Boltes P, Letón F, Rosal R (2018) Incorporation of antimicrobial peptides on electrospun nanofibres for biomedical applications. RSC Adv 8(49):28013. https://doi.org/10.1039/c8ra03861a
5. Azadi S, Yazdanpanah M, Afshari A, Alahdad N, Chegeni S, Angaji SA, Rezayat SM, Tavakol S (2024) Bioinspired synthetic peptide-based biomaterials regenerate bone through biomimicking of extracellular matrix. J Tissue Eng 15(January). https://doi.org/10.1177/20417314241303818
6. Bacalum M, Radu M (2014) Cationic antimicrobial peptides cytotoxicity on mammalian cells: an analysis using therapeutic index integrative concept. Int J Pept Res Ther 21(1):47. https://doi.org/10.1007/s10989-014-9430-z
7. Barman P, Joshi S, Sharma S, Preet S, Sharma S, Saini A (2023) Strategic approaches to improvise peptide drugs as next generation therapeutics. Int J Peptide Res Therapeutics (Springer). https://doi.org/10.1007/s10989-023-10524-3
8. Bordbar-Khiabani A, Ueda M (2022) Smart hydrogels for advanced drug delivery systems. Int J Mol Sci 23(7):3665. https://doi.org/10.3390/ijms23073665
9. Bradshaw JP (2003) Cationic antimicrobial peptides. BioDrugs 17(4):233. https://doi.org/10.2165/00063030-200317040-00002
10. Bucataru C, Ciobănaşu C (2024) Antimicrobial peptides: opportunities and challenges in overcoming resistance. Microbiol Res 286(June):127822. https://doi.org/10.1016/j.micres.2024.127822
11. Carvalho T Guedes G, Sousa FL, Freire CSR, Santos HA (2019) Latest advances on bacterial cellulose-based materials for wound healing, delivery systems, and tissue engineering. Biotechnol J 14(12). https://doi.org/10.1002/biot.201900059
12. Çelik E, Akelma H (2023) Hydrogel burn dressing effectiveness in burn pain. Burns 50(1):190. https://doi.org/10.1016/j.burns.2023.08.012

13. Cherng J-H, Chou S-C, Chen C-L, Wang Y-W, Chang S-J, Fan G-Y, Leung F-S, Meng E (2021) Bacterial cellulose as a potential bio-scaffold for effective re-epithelialization therapy. Pharmaceutics 13(10):1592. https://doi.org/10.3390/pharmaceutics13101592
14. Cross ER, Coulter S, Pentlavalli S, Laverty G (2021) Unravelling the antimicrobial activity of peptide hydrogel systems: current and future perspectives. Soft Matter 17(35):8001. https://doi.org/10.1039/d1sm00839k
15. Cruz JC, Reyes LH (2025) Biological platforms for the production of antimicrobial peptides: bacterial, yeast, and cell-free systems Elsevier eBooks January:131. https://doi.org/10.1016/b978-0-443-15393-8.00007-5
16. Deo S, Turton KL, Kainth T, Kumar A, Wieden H-J (2022) Strategies for improving antimicrobial peptide production. Biotechnol Adv 59(April):107968. https://doi.org/10.1016/j.biotechadv.2022.107968
17. Deptuła M, Wardowska A, Dzierżyńska M, Rodziewicz-Motowidło S, Pikuła M (2018) Antibacterial peptides in dermatology-strategies for evaluation of allergic potential. Molecules 23(2):414. https://doi.org/10.3390/molecules23020414
18. Dumont C, Bourgeois S, Fessi H, Jannin V (2018) Lipid-based nanosuspensions for oral delivery of peptides, a critical review. Int J Pharm 541(February):117. https://doi.org/10.1016/j.ijpharm.2018.02.038
19. Duong L, Gross SP, Siryaporn A (2021) Developing antimicrobial synergy with AMPs. Front Med Technol (Frontiers Media). https://doi.org/10.3389/fmedt.2021.640981
20. Dutta P, Das S (2015) Mammalian antimicrobial peptides: promising therapeutic targets against infection and chronic inflammation. Curr Top Med Chem 16(1):99. https://doi.org/10.2174/1568026615666150703121819
21. El-Husseiny HM, Mady EA, Hamabe L, Abugomaa A, Shimada K, Yoshida T, Tanaka T, Yokoi A, Elbadawy M, Tanaka R (2021) Smart/Stimuli-responsive hydrogels: cutting-edge platforms for tissue engineering and other biomedical applications. Mater Today Bio 13(December):100186. https://doi.org/10.1016/j.mtbio.2021.100186
22. Fadilah NIMd, Shahabudin NA, Razif RAM, Sanyal A, Ghosh A, Baharin KI, Ahmad H, Maarof M, Motta A, Fauzi MhB (2024) Discovery of bioactive peptides as therapeutic agents for skin wound repair. J Tissue Eng 15(January). https://doi.org/10.1177/20417314241280359
23. Fei R, Kailang M, Liu G, Liu Y, Pang Y, Guo F, Zhou L, Peng L (2024) Preparation, characterization, and wound healing promotion of hydrogels containing glucosyloxybenzyl 2-isobutylmalates extract from bletilla striata (thunb.) reichb.f. Int J Mol Sci 25(19):10563. https://doi.org/10.3390/ijms251910563
24. Ferraz MP (2025) Wound dressing materials: bridging material science and clinical practice. Appl Sci 15(4):1725. https://doi.org/10.3390/app15041725
25. Fjell CD, Hiss JA, Hancock REW, Schneider G (2011) Designing antimicrobial peptides: form follows function. Nat Rev Drug Discovery 11(1):37. https://doi.org/10.1038/nrd3591
26. Furman JL, Chiu ML, Hunter MJ (2014) Early Engineering approaches to improve peptide developability and manufacturability. AAPS J 17(1):111. https://doi.org/10.1208/s12248-014-9681-9
27. Gan W, Cui Y, Liu H, Tian Y, Li S, Fan Y, Sun S, Dankai W, Peng C (2023) Antibacterial peptides-loaded bioactive materials for the treatment of bone infection. Colloids Surf B 225(March):113255. https://doi.org/10.1016/j.colsurfb.2023.113255
28. Garvey M (2023) Antimicrobial peptides demonstrate activity against resistant bacterial pathogens. Infectious Disease Rep. https://doi.org/10.3390/idr15040046
29. Girard V-D, Chaussé J, Borduas M, Dubuc E, Iorio-Morin C, Brisebois S, Vermette P (2024) In vitro and in vivo biocompatibility of bacterial cellulose. J Biomed Mater Res Part B Appl Biomater 112(10). https://doi.org/10.1002/jbm.b.35488
30. Goldberg SR, Diegelmann RF (2020) What makes wounds chronic. Surg Clin North Am (Elsevier BV). https://doi.org/10.1016/j.suc.2020.05.001
31. Gomes A, Teixeira C, Ferraz R, Prudêncio C, Gomes P (2017) Wound-healing peptides for treatment of chronic diabetic foot ulcers and other infected skin injuries. Molecules 22(10):1743. https://doi.org/10.3390/molecules22101743

32. Gorgieva S (2020) Bacterial cellulose as a versatile platform for research and development of biomedical materials. Processes 8(5):624. https://doi.org/10.3390/pr8050624
33. Goudarzi ZM, Zaszczyńska A, Kowalczyk T, Sajkiewicz P (2024) Electrospun antimicrobial drug delivery systems and hydrogels used for wound dressings. Pharmaceutics 16(1):93. https://doi.org/10.3390/pharmaceutics16010093
34. Guan T, Li J, Chen C, Liu Y (2022) Self-assembling peptide-based hydrogels for wound tissue repair. Adv Sci 9(10). https://doi.org/10.1002/advs.202104165
35. Guiza-Arguello V, Solarte-David VA, Pinzón-Mora AV, Ávila-Quiroga JE, Becerra-Bayona S (2022) Current advances in the development of hydrogel-based wound dressings for diabetic foot ulcer treatment. Polymers 14(14):2764. https://doi.org/10.3390/polym14142764
36. Gupte Y, Kulkarni A, Raut B, Sarkar P, Choudhury RP, Chawande A, Raja Krishna Kumar G, et al (2021) Characterization of nanocellulose production by strains of komagataeibacter sp. isolated from organic waste and Kombucha. Carbohydrate Polym 266(May):118176. https://doi.org/10.1016/j.carbpol.2021.118176.
37. Han Y, Zhang M, Lai R, Zhang Z (2021) Chemical modifications to increase the therapeutic potential of antimicrobial peptides. Peptides (Elsevier BV). https://doi.org/10.1016/j.peptides.2021.170666
38. Hartmann M, Marina Berditsch J, Hawecker MF, Ardakani DG, Ulrich AS (2010) Damage of the bacterial cell envelope by antimicrobial peptides gramicidin S and PGLa as revealed by transmission and scanning electron microscopy. Antimicrob Agents Chemother 54(8):3132. https://doi.org/10.1128/aac.00124-10
39. Hashemi ZS, Zarei M, Fath MK, Ganji M, Farahani MS, Afsharnouri F, Pourzardosht N, et al (2021). In silico approaches for the design and optimization of interfering peptides against protein–protein interactions. Front Mol Biosci 8(April). https://doi.org/10.3389/fmolb.2021.669431
40. Huang L-C, Wang H-C, Chen L-H, Ho C-Y, Hsieh P-H, Huang M-Y, Hsi-Chin W, Wang T-W (2019) Bioinspired self-assembling peptide hydrogel with proteoglycan-assisted growth factor delivery for therapeutic angiogenesis. Theranostics 9(23):7072. https://doi.org/10.7150/thno.35803
41. Huang Y, Yang N, Teng D, Mao R, Hao Y, Ma X, Wei L, Wang J (2022) Antibacterial peptide NZ2114-loaded hydrogel accelerates staphylococcus aureus-infected wound healing. Appl Microbiol Biotechnol (Springer). https://doi.org/10.1007/s00253-022-11943-w.
42. Jia Q, Zhe F, Li Y, Kang Z, Yutong W, Zeqiong R, Peng Y et al (2024) Hydrogel loaded with peptide-containing nanocomplexes: symphonic cooperation of photothermal antimicrobial nanoparticles and prohealing peptides for the treatment of infected wounds. ACS Appl Mater Interfaces 16(11):13422. https://doi.org/10.1021/acsami.3c16061
43. Karami P, Stampoultzis T, Guo Y, Pioletti DP (2023) A guide to preclinical evaluation of hydrogel-based devices for treatment of cartilage lesions. Acta Biomater 158(January):12. https://doi.org/10.1016/j.actbio.2023.01.015
44. Kass L, Nguyen J (2021) Nanocarrier-hydrogel composite delivery systems for precision drug release. Wiley Interdisc Rev Nanomed Nanobiotechnol 14(2). https://doi.org/10.1002/wnan.1756
45. Koivunotko E, Merivaara A, Niemelä A, Valkonen S, Manninen K, Mäkinen H, Viljanen M et al (2021) Molecular insights on successful reconstitution of freeze-dried nanofibrillated cellulose hydrogel. ACS Appl Bio Mater 4(9):7157. https://doi.org/10.1021/acsabm.1c00739
46. Kopecki Z (2021) Development of next-generation antimicrobial hydrogel dressing to combat burn wound infection Biosci Rep 41(2). https://doi.org/10.1042/bsr20203404
47. Kulshreshtha A, Sharma S, Padilla CS, Mandadi KK (2022) Plant-based expression platforms to produce high-value metabolites and proteins. Front Plant Sci 13(November). https://doi.org/10.3389/fpls.2022.1043478
48. Li L, Zhou Y, Li P, Xu Q, Li K, Hu H, Wei B, Zhang Z (2022) Peptide hydrogel based sponge patch for wound infection treatment. Front Bioeng Biotechnol 10(December). https://doi.org/10.3389/fbioe.2022.1066306

49. Liang Y, He J, Guo B (2021) Functional hydrogels as wound dressing to enhance wound healing. ACS Nano 15(8):12687. https://doi.org/10.1021/acsnano.1c04206
50. Liang Y, Liang Y, Zhang H, Guo B (2022) Antibacterial biomaterials for skin wound dressing. Asian J Pharm Sci 17(3):353. https://doi.org/10.1016/j.ajps.2022.01.001
51. Liu K, Wang Y, Liu W, Zheng C, Ting X, Haishun D, Yuan Z, Si C (2024) Bacterial cellulose/chitosan composite materials for biomedical applications. Chem Eng J 494(June):153014. https://doi.org/10.1016/j.cej.2024.153014
52. Lombardi L, Li J, Williams DR (2024) Peptide-based biomaterials for combatting infections and improving drug delivery. Pharmaceutics 16(11):1468. https://doi.org/10.3390/pharmaceutics16111468
53. Lopes BS, Hanafiah A, Nachimuthu R, Saravanan M, Md Nesran ZN, Patil S (2022) The role of antimicrobial peptides as antimicrobial and antibiofilm agents in tackling the silent pandemic of antimicrobial resistance. Molecules (Multidisciplinary Digital Publishing Institute). https://doi.org/10.3390/molecules27092995
54. Lou M, Ji S, Wu R, Zhu Y, Wu J, Zhang J (2025) Microbial production systems and optimization strategies of antimicrobial peptides: a review. World Microbiol Biotechnol 41(2). https://doi.org/10.1007/s11274-025-04278-x
55. Lu P, Ruan D, Huang M, Tian M, Zhu K, Gan Z, Xiao Z (2024) Harnessing the potential of hydrogels for advanced therapeutic applications: current achievements and future directions. Signal Transduct Targeted Therapy 9(1). https://doi.org/10.1038/s41392-024-01852-x
56. Mazurkiewicz-Pisarek A, Baran J, Ciach T (2023) Antimicrobial peptides: challenging journey to the pharmaceutical, biomedical, and cosmeceutical use. Int J Mol Sci 24(10):9031. https://doi.org/10.3390/ijms24109031
57. Miao F, Liu Y, Tai Z, Zhang Y, Gao Y, Hu M, Zhu Q (2021) Antimicrobial peptides: the promising therapeutics for cutaneous wound healing. Macromol Biosci (Wiley). https://doi.org/10.1002/mabi.202100103
58. Mondal S, Das SK, Nandi AK (2020) A review on recent advances in polymer and peptide hydrogels. Soft Matter 16(6):1404. https://doi.org/10.1039/c9sm02127b
59. Musaimi OA, Lombardi L, Williams DR, Albericio F (2022) Strategies for improving peptide stability and delivery. Pharmaceuticals 15(10):1283. https://doi.org/10.3390/ph15101283
60. Pfalzgraff A, Brandenburg K, Weindl G (2018) Antimicrobial peptides and their therapeutic potential for bacterial skin infections and wounds. Front Pharmacol 9(March). https://doi.org/10.3389/fphar.2018.00281.
61. Ponrasu T, Ramachandran B, Chakraborty S, Kannan R, Suguna L, Muthuvijayan V (2017) Accelerated healing of diabetic wounds treated with l-glutamic acid loaded hydrogels through enhanced collagen deposition and angiogenesis: an in vivo study. Sci Rep 7(1). https://doi.org/10.1038/s41598-017-10882-1
62. Rai A, Ferrão R, Palma P, Patrício T, Parreira P, Anes E, Chiara Tonda-Turo M, Martins CL, Alves N, Ferreira L (2022) Antimicrobial peptide-based materials: opportunities and challenges. J Mater Chem B 10(14):2384. https://doi.org/10.1039/d1tb02617h
63. Ribeiro M, Simões M, Vitorino C, Mascarenhas-Melo F (2024) Hydrogels in cutaneous wound healing: insights into characterization, properties, formulation and therapeutic potential. Gels 10(3):188. https://doi.org/10.3390/gels10030188
64. Rodríguez AA, Otero-González AJ, Ghattas M, Ständker L (2021) Discovery, optimization, and clinical application of natural antimicrobial peptides. Biomedicines (Multidisciplinary Digital Publishing Institute). https://doi.org/10.3390/biomedicines9101381
65. Sacco F, Bitossi C, Casciaro B, Loffredo MR, Fabiano G, Torrini L, Raponi F, Raponi G, Mangoni ML (2022) The antimicrobial peptide Esc(1–21) synergizes with colistin in inhibiting the growth and in killing multidrug resistant acinetobacter baumannii strains. Antibiotics 11(2):234. https://doi.org/10.3390/antibiotics11020234
66. Sanati M, Yavari SA (2024) Liposome-integrated hydrogel hybrids: promising platforms for cancer therapy and tissue regeneration. J Control Release 368(March):703. https://doi.org/10.1016/j.jconrel.2024.03.008

67. Shahriari-Khalaji M, Sattar M, Cao R, Zhu M (2023) Angiogenesis, hemocompatibility and bactericidal effect of bioactive natural polymer-based bilayer adhesive skin substitute for infected burned wound healing. Bioactive Mater 29(July):177. https://doi.org/10.1016/j.bioactmat.2023.07.008
68. Shi M, McHugh KJ (2023) Strategies for overcoming protein and peptide instability in biodegradable drug delivery systems. Adv Drug Deliv Rev 199(May):114904. https://doi.org/10.1016/j.addr.2023.114904
69. Shishparenok AN, Furman VV, Dobryakova NV, Zhdanov DD (2024) Protein immobilization on bacterial cellulose for biomedical application. Polymers 16(17):2468. https://doi.org/10.3390/polym16172468
70. Skowron K, Bauza-Kaszewska J, Kraszewska Z, Wiktorczyk-Kapischke N, Grudlewska-Buda K, Kwiecińska-Piróg J, Wałecka-Zacharska E, Radtke L, Gospodarek-Komkowska E (2021) Human skin microbiome: impact of intrinsic and extrinsic factors on skin microbiota. Microorganisms 9(3):543. https://doi.org/10.3390/microorganisms9030543
71. Smith R, Russo J, Fiegel J, Brogden NK (2020) Antibiotic delivery strategies to treat skin infections when innate antimicrobial defense fails. Antibiotics 9(2):56. https://doi.org/10.3390/antibiotics9020056
72. Țiplea RE, Popa G-ML, Trușcă R, Holban AM, Georgiana M, Kaya MGA, et al (2020) Antimicrobial films based on chitosan, collagen, and ZnO for skin tissue regeneration. Biointerface Res Appl Chem 11(4):11985. https://doi.org/10.33263/briac114.1198511995
73. Truong D, Wu KY, Nguyen L, Tran SD (2024) Advancements in hydrogel technology for ocular drug delivery 1(5):331. https://doi.org/10.37349/ebmx.2024.00023
74. Umur E, Bayrak E, Arslan F, Bulut SB, Baysoy E, Kaleli-Can G, Ayan B (2023) Advances in three dimensional bioprinting for wound healing: a comprehensive review. Appl Sci 13(18):10269. https://doi.org/10.3390/app131810269
75. Wang M, Li T, Tian J, Zhang L, Wang Y, Li S, Lei B, Peng X (2023) Engineering single-component antibacterial anti-inflammatory polyitaconate-based hydrogel for promoting methicillin-resistant staphylococcus aureus-infected wound healing and skin regeneration. ACS Nano 18(1):395. https://doi.org/10.1021/acsnano.3c07638
76. Wang P, Cai F, Li Y, Yang X, Feng R, H L, Bai X, Han J (2024) Emerging trends in the application of hydrogel-based biomaterials for enhanced wound healing: a literature review. Int J Biol Macromol 261(January):129300. https://doi.org/10.1016/j.ijbiomac.2024.129300
77. Wang Q, Zhang Y, Ma Y, Wang M, Pan G (2023) Nano-crosslinked dynamic hydrogels for biomedical applications. Materials Today Bio 20(April):100640. https://doi.org/10.1016/j.mtbio.2023.100640
78. Wang XM, Duan H, Li M, Xu W, Wei L (2023) Characterization and mechanism of action of amphibian-derived wound-healing-promoting peptides. Front Cell Dev Biol (Frontiers Media). https://doi.org/10.3389/fcell.2023.1219427
79. Yao X, Miner H, Zhai X, Lin J, Xia X, Wei W, Xia S (2024) A bioinspired regenerative antimicrobial hydrogel coating for cardiovascular implantable electronic devices. Chem Eng J 494(June):153020. https://doi.org/10.1016/j.cej.2024.153020
80. Zhang W, Liu L, Cheng H, Zhu J, Li X, Ye S, Li X (2023) Hydrogel-based dressings designed to facilitate wound healing. Mater Adv 5(4):1364. https://doi.org/10.1039/d3ma00682d
81. Zhang X, Li W, Wei G, Yan Y, He R, Wang Y, Chen D, Qin X (2023) A rapid-crosslinking antimicrobial hydrogel with enhanced antibacterial capabilities for improving wound healing. Front Physiol 14 (May). https://doi.org/10.3389/fphys.2023.1206211
82. Zhang X, Li X, Zhao Y, Zheng Q, Wu Q, Yu Y (2022) Nanocarrier system: an emerging strategy for bioactive peptide delivery. Front in Nutr 9(December). https://doi.org/10.3389/fnut.2022.1050647
83. Zhao S, Zhao W, Wang N, Ling J, Ouyang X-K (2025). A sustained H2S-releasing nanocellulose-based hydrogel with anti-inflammatory and antibacterial properties for promoting infected wound healing. Carbohydrate Polym 355(February):123424. https://doi.org/10.1016/j.carbpol.2025.123424

84. Zheng L, Yang F, Wang C, Zafir M, Gao Z, Liu P, El-Gohary FA, Zhao X, Xue H (2025) High-level biosynthesis and purification of the antimicrobial peptide Kiadin based on non-chromatographic purification and acid cleavage methods. Biotechnol Biofuels Bioprod 18(1). https://doi.org/10.1186/s13068-025-02607-8
85. Zheng S, Li W, Xu M-J, Bai X, Zhou Z, Han J-Y, Shyy JY-J, Wang X (2010) Calcitonin gene-related peptide promotes angiogenesis via AMP-activated protein kinase. AJP Cell Physiol 299(6). https://doi.org/10.1152/ajpcell.00173.2010
86. ZhengYue L, Chen X, Bao C, Liu C, Liu C, Li D, Yan H, Lin Q (2021) Fabrication and evaluation of alginate/bacterial cellulose nanocrystals–chitosan–gelatin composite scaffolds. Molecules 26(16):5003. https://doi.org/10.3390/molecules26165003
87. Zyl van EM, Coburn JM (2024) Functionalization of bacterial cellulose with the antimicrobial peptide KR-12 via chimerical cellulose-binding peptides. Int J Mol Sci 25(3):1462. https://doi.org/10.3390/ijms25031462

MIX
Papier aus verantwortungsvollen Quellen
Paper from responsible sources
FSC® C105338

If you have any concerns about our products,
you can contact us on
ProductSafety@springernature.com

In case Publisher is established outside the EU,
the EU authorized representative is:
Springer Nature Customer Service Center GmbH
Europaplatz 3, 69115 Heidelberg, Germany

Printed by Libri Plureos GmbH
in Hamburg, Germany